Handbook of International Electrical Safety Practices

Scrivener Publishing
3 Winter Street, Suite 3
Salem, MA 01970

Publishers at Scrivener
Martin Scrivener (martin@scrivenerpublishing.com)
Phillip Carmical (pcarmical@scrivenerpublishing.com)

Handbook of International Electrical Safety Practices

Princeton Energy Resources International

Scrivener

WILEY

Co-published by John Wiley & Sons, Inc. Hoboken, New Jersey, and Scrivener Publishing LLC, Salem, Massachusetts.
Published simultaneously in Canada.

For general information on our other products and services or for technical support, please contact our Customer Care Department within the United States at (800) 762-2974, outside the United States at (317) 572-3993 or fax (317) 572-4002.

Wiley also publishes its books in a variety of electronic formats. Some content that appears in print may not be available in electronic formats. For more information about Wiley products, visit our web site at www.wiley.com.

For more information about Scrivener products please visit www.scrivenerpublishing.com.

Cover design by Russell Richardson.

Library of Congress Cataloging-in-Publication Data:

ISBN 978-0-470-62589-7

Printed in the United States of America

10 9 8 7 6 5 4 3 2 1

Contents

Acknowledgements

A large number of organizations and individuals have contributed time and information that went into the creation of the standards. The following organizations are expressly thanked for their contribution of materials, reviews, critiques and guidance:

The U.S. Trade and Development Agency
The World Bank Organization
The U.S. Department of Energy
The U.S. Department of Defense
The U.S. Department of Homeland Security
The World Health Organization
The U.S. Occupational Safety and Health Organization
The National Institute of Occupational Safety and Health
The American Conference of Governmental Industrial Hygienists
SciTech Technical Services
Tennessee Valley Authority
The U.S. Department of Agriculture, Rural Utilities Service

List of Contributors

David Phillippi

Adolfo Menéndez

Andrea Duffy

José Menéndez

Nicholas P. Cheremisinoff, Ph.D.

Daniel Ancona

Stacy Herrick

Viola New

Abbreviations

AC	Alternating Current (electricity; physics)
ACGIH	American Conference of Governmental Industrial Hygienists
AED	Automated Eternal Defibrillator
AFFF	Aqueous Film Forming Foam
AFS	American Foundry Society
AIHI	American Industrial Hygiene Association
ANSI	American National Standards Institute
ASTDR	Agency for Toxic Substances and Disease Registry
ASTM	American Society for Testing and Materials
ATB	Anti-Two-Block
AU	Absorption Units
C	Celsius
CaF	Calcium Fluoride
CDC	Center for Disease Control
CERCLA	Comprehensive Environmental Response Compensation and Liability Act
CGA	Compressed Gas Association
CGI	Combustible Gas Indicators
CISD	Critical Incident Stress Debriefing
Cm	Centimeters
CNC	Condensation Nucleus Counter
CO	Carbon monoxide
CO_2	Carbon Dioxide
CPR	Cardiopulmonary Resuscitation
CSA	Construction Safety Association
CSHO	Compliance Safety and Health Officer
dB	Decibels

DC	Direct Current (electricity)
DUTs	Devices Under Test
EAR	Expired Air Resuscitation
EHSS	Environmental Health and Safety Services
EHV	Extremely High Voltage
EMS	Emergency Medical Services
EPA	Environmental Protection Agency
EPS	Electric Power Systems
ESCBA	Escape Self-Contained Breathing Apparatus
ESLI	End of Service Life Indicator
FID	Flame Ionization Detector
FMIS	Facilities Management Information System
GFCI	Ground Fault Circuit Interrupter
GHz	Gigahertz (thousands of MHz)
GM	Geiger-Muller
H2S	Hydrogen Sulfide
HASPs	Health and Safety Plans
HAZWOPER	Hazardous Waste Operations and Emergency Response Standard
HEPA	High Efficiency Particulate Air
Hg	Mercury
HR	Human Resource
HRT	Health Response Team
HSC	Health and Safety Coordinator
HSO	Health and Safety Officer
HV	High Voltage
Hz	Hertz
IDLH	Immediately Dangerous to Life and Health
Kg	Kilograms
kHz	Kilohertz (1000 Hertz)
kPa	Kilo Pascal
LEL	Lower Explosive Limit
LFL	Lower Flammable Limit
LiF	Lithium Fluoride
LMI	Load moment Indicators
m	Meters
MeV	Mega Electron Volt
MHz	Megahertz (million Hertz)

mR/hr	Milliroentgen Per Hour
MRLS	Minimal Risk Levels
MSDS	Material Safety Data Sheets
MUC	Maximum Use Concentration
MW	Molecular Weight
NFPA	National Fire Protection Agency
NAELC	North American Electric Liability Corporation
NHCA	National Hearing Conservation Association
NIOSH	National Institute of Occupational Safety and Health
NO	Nitric oxide
NRR	Noise Reduction Rating
NRTL	Nationally Recognized Testing Laboratories
NTOF	National Traumatic Occupational Facilities
O_3	Ozone
OH&S	Occupational Health and Safety
OSHA	Office of Safety and Health Administration
Pa	Pascal
PAPR	Powered Air Purifying Respirator
PEL	Permissible Exposure Limits
PID	Photo Ionization Detectors
PM	Particulate Matter
PPE	Personal Protective Equipment
ppm	Parts Per Million
PVC	polyvinyl chloride
QLFT	Qualitative Fit Test
QNFT	Quantitative Fit Test
RCRA	Resource Conservation and Recovery Acts
REL	Recommended Exposure Limits
RF	Radiated Frequency
RFC	Reference Concentration
RFD	Reference Dose
R/hr	Roentgens per Hour
RH	Relative Humidity
SA	Spectrum Analyzer
SAR	Supplied Air Respirator
SARA	Superfund Amendments and Reauthorization Act
SCBA	Self-Contained Breathing Apparatus
SHELS	Significant Human Exposure Levels

SLTC Salt Lake Technical Center
SO_2 Sulfur dioxide
SPL Sound Pressure Level
STEL Short-term Exposure Limits
SVOCs Semi-Volatile Organic Compounds
TD Thermal Desorption
TICs Toxicity Identified Compounds
TLD Thermoluminescent dosimeters
TLV Threshold Limit Value
TWA Time Weighted Average
UEL Upper Explosive Limits
U.K. United Kingdom
U.S. United States
UV Ultraviolet
VOCs Volatile Organic Compound
VOM Colt-Ohm-Milliameter
VSA Vector Signal Analyzer
WHO World Health Organization

Preface

The *Handbook of International Electrical Safety Practices* is intended for the electricity generating, power transmission and distribution sectors. Electricity is an essential ingredient of our lives and has vast social implications to all nations. It is perhaps among the greatest and most important developments of the 20th century that continues to touch our lives daily. The economic and social developments of nations have been intrinsically coupled to electricity use. The lack of electricity, or insufficient electricity, is a concern of virtually all governments. Electricity is in fact one of the prime measures of a nation's wealth and it defines the quality of life. Electricity generation does come with a cost to the environment and there are long term impacts to the health and well-being of communities. These however, are topics for examining in other publications.

This volume has been prepared by the staff of Princeton Energy Resources International (PERI). PERI has a long history of working in the power sector with staff and principals' experiences stretching back decades. PERI has assembled its own decades of knowledge and surveyed the international literature to assemble this reference for practitioners in the electricity sector. The handbook is an assemblage of best international management practices for the safe handling operations of personnel working in generation, transmission and dealing with power distribution issues. It is intended as a practical desk reference to which technicians, engineers, supervisors, line personnel and business managers with worker safety responsibility can refer. The handbook covers a broad spectrum of subjects related to physical, chemical, thermal, energizing, and other hazards that may be encountered while working with energizing sources and transmission systems. The handbook may also serve as a basis for developing site specific health and safety programs, industrial hygiene programs, worker safety training and recertification programs, in establishing safety and worker training recordkeeping and monitoring programs, and much more.

There are eight chapters to the handbook. Chapter one addresses procedures and provides guidance on how companies can effectively evaluate their safety programs. Good safety records can only be achieved when there is a cultural basis for working safely and applying best management practices within the corporate environment. This chapter covers the topics of creation of a safety culture within the organization, adoption of good housekeeping practices and employee orientation.

Chapter two covers instrumentation, specifically, meters and various types of monitors and sensors that are relied upon by industry for air quality testing, noise level testing, electronics testing, and others.

Chapter three covers general best safety practices in the areas of chemicals, performing job hazards assessments, the types and use of personal protective equipment, first aid and resuscitation, fire prevention and protection, and safely working in excavations and other confined spaces and other areas.

Chapter four addresses the safe use of various field equipment, including hand tools and workshop tools and machinery, ladder and scaffolding safety, forklift safety, crane operation, work platforms, safely working with compressed gas cylinders, drum handling safety, and safe welding practices.

Chapter five covers electrical safety. This chapter includes detailed discussions and best practices for electrical shock and lockout and tagout procedures along with sample forms to use in the field, lineman general safe work practices, electrical safe work practices planning, safe use of electrical equipment, working near power lines, and functional safety practices for electrical power transmission.

Chapter six covers the subject of worker safety rules. Topics covered include critical incident stress, working with toxic chemicals, working with electrical protective devices, hand protection, performing hazard assessments and additional guidance on personal protective equipment selection, sanitation and industrial hygiene. This section covers additional guidance and topical discussions on marking and labeling physical hazards, the use of accident prevention signs and posters, fire protection and other topics.

Chapter seven covers recommended protocols and practices for recordkeeping, worker training, performing safety inspections and accident investigation and reporting.

Chapter eight covers risk and vulnerability assessments. Vulnerability assessment practices have become much more critical since the days of 9/11 and should be an integral part of any safety program in order to ensure the protection of workers, communities and assets.

This volume has been the labor of the entire Princeton Energy Resources team under the overall guidance of its principal, Adolfo Menéndez. His years of experience have contributed greatly to the technical content of this work and his dedication of resources and staff time in the preparation of this industry reference is a testament to the company's commitment to safe work practices. Special thanks and appreciation is extended to Scrivener Publishing for their tireless efforts in the publication of this work.

Staff of Princeton Energy Resources International, LLC
Rockville, MD

1

How to Evaluate Safety Programs

1.1 Introduction

This chapter provides a roadmap for both employers and employees to creating a safe working environment. While the standards in chapters 2 through 6 are organized and presented in a highly prescriptive format, this first part provides workers within the industry an overall orientation to the philosophy, tools and corporate culture that the international community has adopted as a part of best management practices.

At the same time, this chapter of the standards explicitly defines the rights of workers to be kept informed of the hazards associated with their job assignments, to be provided with knowledge, engineering and management controls that eliminate unsafe working conditions, along with the actions that workers may take in order to ensure that they are never placed in situations that pose either immediate or long-term risks to their health and well-being.

Employees generally have little or no control over their working environment and must accept whatever environment employers offer. Employers and their designated corporate representatives have both a moral and legal obligation to ensure that both workers and the public at large are kept insulated from the hazards associated with the industry sector.

1.2 Creating a Culture of Safety

The design of a safe plant layout is beyond the responsibility of individual employees, but it nevertheless is essential for good power production practices and safe working conditions. Narrow aisles, blind intersections, insufficient overhead space and limited access for equipment repair and maintenance all are detrimental to a safe operating environment.

The National Safety Council in the United States has estimated that work-related accidents in the private sector in 1988 cost industry an average of $15,100 per disabling injury. Based on this figure and the U.S. Bureau of Labor Statistics – which reported that in 1988 private U.S. Industry, employing 90 million workers, had 6.2 million job-related accidents and injuries was in excess of $93 billion. Approximately, half of this total ($46 billion) was for such visible costs as damaged equipment and materials, production delays, time losses of other workers not involved in the accidents and accident reporting.

Similar statistics have been reported in the United Kingdom (UK) and in the European Community. The statistics support the premise that it is the responsibility of every employer to take a strong, proactive stance to ensure their employees' safety.

Designing for safe work environments also means proper scheduling of work activities. It should not be the operator's or worker's responsibility to determine the proper routing of work in process. To make this type of decision a worker's responsibility unfairly shifts to what is truly management's responsibility directly to the worker. It is management's responsibility to ensure that tight work standards are not only defined for each operating facility, but to ensure that procedures and policies are adopted and enforced. Establishing fair work standards through work measurement or some similar technique is, without question, a prerogative and a right of management. Establishing and enforcing tight work standards has resulted and will continue to result in operators taking dangerous short cuts while completing tasks. These short cuts often result in industrial accidents and injuries. By the same token, managers should use standards to ensure a fair day's pay for a fair day's work, but they should not use them as a whip to achieve maximum productivity through coercion.

Pressure placed on employees to meet tight production schedules results in the same type of problems as with tight work standards. Reasonable schedules based on reasonable capacity determinations and work standards eliminate the pressure and work-related stress placed on employees to overproduce because of unsafe short cuts.

Having a corporate culture that promotes and makes safety and environment a priority should be the goal of the industry. Creating a culture of safety first requires site-specific work practices and working environments to be carefully assessed with a focus on identifying high-risk areas, and then developing concrete plans for improved occupational and process safety performance. Management must focus on using employee insights to prevent costly and potentially deadly accidents before they occur, creating a safer workplace by taking into account both the environment in which employees work and the culture that drives their daily work experience.

As an employer, it is your responsibility to provide a safe work environment for all employees, free from any hazards, and complying with legal and recommended best practices defined in the standards. Health and safety in the workplace is about preventing work-related injury and disease, and designing an environment that promotes well-being for everyone at work.

Knowledge is the key ingredient in providing a safe work environment. If everyone knows the correct procedures, accidents and injuries will be kept to a minimum.

Both employers and employees should:

- Ensure that the way work is done is safe and does not affect employees' health.
- Ensure that tools, equipment and machinery are safe and are kept safe.
- Ensure that ways of storing, transporting or working with dangerous substances is safe and does not damage employees' health.

Employers must:

- Provide employees with the information, instruction and training they need to do their job safely and without damaging their health.

- Consult with employees about health and safety in the workplace.
- Monitor the work place regularly and keep a record of what is found during these checks.

Policies should be developed in consultation with employees, both with and without disability. It may be necessary to organize support persons or interpreters so that all employees may participate in the consultation.

Occupational Health and Safety (OH&S) procedures must be implemented wherever the work is being conducted, be that in an office, factory, construction site, substation, along transmission line work or home. As an employer, it is your responsibility to ensure all employees have access to information about safety procedures, and for any reasonable adjustments to be made.

It is crucial that **new** employees be:

- Briefed of all new staff on OH&S policy at induction.
- Be provided training on all safety procedures, including evacuation and other emergency procedures.
- Provided access to information about safety procedures, in appropriate formats.

It is crucial the **existing** employees:

- Have access to information in appropriate formats.
- Be provided with regular information updates and re-training sessions.
- Be provided access to information about safety procedures.
- Conduct relevant training on any new equipment or machinery.

The following are some anecdotal facts that can serve to raise a culture of safety among workers and management:

Factors That Make New Workers More Accident Prone:

- Lack of safety training;
- Don't ask questions;
- Unaware of hazards;

- Assume employer is responsible for safety;
- Don't understand rights/responsibilities;
- "It can't happen to me" attitude; and
- Fatigue.

Major Causes of Injury:

- Slips, trips and falls;
- Improper use of equipment;
- Faulty use of equipment;
- Improper lighting;
- Not turning off power while repairing equipment (lock out procedures);
- Entering unsafe confined spaces that do not normally accommodate people; and
- Improper use or storage of chemicals and other hazardous materials.

The Employer's Responsibilities:

- Provide a safe workplace.
- Ensure adequate training of workers.
- Keep written records of training: who, when and what type.
- Establish and maintain a comprehensive occupational safety program, including a written safety policy and an accident investigation program.
- Support supervisors, safety coordinators, and workers in their safety activities.
- Take action immediately when the worker or supervisor tells you about a potentially hazardous situation.
- Initiate an immediate investigation into accidents.
- Report serious accidents to the HSO (Health and Safety Officer) or to the Corporate Prevention Division.
- Provide adequate first aid facilities and services.
- Provide personal protective equipment where required.
- Make available to all workers copies of Industrial Hygiene Policy.
- Post Health and Safety Regulations and all other regulations.

The Supervisor's Responsibilities:

- Instruct new workers on safe work procedures.
- Train workers for all tasks assigned to them and check their progress.
- Ensure that only authorized, adequately trained workers operate tools & equipment and use hazardous chemicals.
- Ensure that equipment and materials are properly handled, stored and maintained.
- Enforce safety regulations.
- Correct unsafe acts.
- Identify workers with problems such as drugs or alcohol that could affect their performance, and follow up with interviews and referrals where necessary.
- Formulate safety rules and inspect for hazards in your own area.

Worker's Responsibilities:

- You have an obligation to make your workplace safe.
- Know and follow safety and health regulations affecting your job.
- If you don't know how, ask for training before you begin work.
- Work safely, and encourage your co-workers to do the same.
- Correct or immediately report any unsafe conditions to your supervisor.
- Report any injury immediately to a First Aid attendant or supervisor.
- Take the initiative; make suggestions for improved safety conditions.

Site Orientation:

- All new employees should have a site orientation from their supervisor and be given a brief on safety procedures including: the layout of the section, a safety orientation, fire exit locations and procedures, emergency equipment and location of first aid facilities or services.

- Always be aware of your surroundings;
- Look up for falling objects;
- Be mindful of up/down traffic;
- Be cautious of structural inadequacies;
- Do not race up or down flights of stairs;
- In a noisy environment, use visual cues to look for potential hazards;
- Only use designated walkways, they are there for a reason; and
- Often some sites require workers to work in pairs or teams in case of an accident.

Hazards:

- Hazards in a work place can never be eliminated, but they can be controlled with proper equipment, training and supervision.
- Hazards are divided into two general categories:
  - Physical hazards – those that can cause immediate damage to the body. Examples: Moving equipment, machinery, confined space; heat, cold, chemicals, electrical, visibility.
  - Delayed Injury hazards – those due to repeated exposure or, a delayed reaction of the human body causing injury. Examples: Noise, ergonomics, repetitive strain injury, back injury, inhalation, radiation.

Computer Workstations:

- Problems with lighting can cause eyestrain and may also contribute to muscle soreness and fatigue.
- Avoid glare, position your computer workstation so that when you face the monitor, you are sitting beside or parallel to the window.
- If you face the window, light from the outside shines directly into your eyes and makes it difficult for you to focus on the screen.
- If you sit with your back to a window, you may have problems with reflections, or high light levels. In either situation, adjust the blinds to reduce the amount of light on your screen.

- Overhead lights can shine directly in your eyes or reflect images onto your computer screen, making it difficult to view your screen. To avoid this adjust the angle of your screen so that it's at right angles to the work surface.
- The top line of text on your screen should be at eye level. Most of the screen will be slightly below eye level.
- The distance between your eyes and the screen should be about arm's length.
- Your chair should provide good lumbar support.

Lighting:

- Problems with lighting can cause eyestrain.
- Poor lighting can also contribute to muscle soreness and fatigue.

Back Injuries:

- Approximately 1/3 of workers will suffer back problems during their working life. Back injuries are the most common work related injury.
- The low back is subject to greater mechanical stress than any other part of the body and, because of this, is commonly injured.
- A majority of low-back problems are caused by simple strains that can heal within six to eight weeks with appropriate treatment. Fear and misunderstanding are the two main factors that can delay recovery.
- For engineers and computer scientists, computer use can cause lower back pain and repetitive strain injuries over time.

Everyone in a work force should wisely take a careful look at their work environment and their work habits. Both workers and management need to note whether their workplace is really safe. Both need to consider whether the job requires working with any toxic substances; and if so, are people adequately protected? Both managers and workers need to ask themselves whether they are constantly under stress. Does your environment accept work schedules that breach legally established limits or hours? Answers to questions such as these might reveal much about how safe people are at work.

Creating a corporate culture of safety means being conscious of the dangers at work. Trying to maintain an unreasonable work schedule can be dangerous. Indeed, tired workers are less efficient and make more mistakes. Companies that foster excessive work and actively seek out and reward workaholics result into consequences that are potentially devastating. Poor work habits, which may include lack of tidiness and cleanliness, pose another hazard. Leaving tools thrown on the floor or live electric wires exposed often lead to accidents, even fatalities. The same can be said of ignoring safety precautions when using power tools and machinery. Another cause of injury and death is failing to clean up spilled fluids, especially toxic ones. Many injuries have occurred when workers have slipped on oily or wet floors. So it might be said that the first law of good work is to be clean and orderly.

Yet, many are tempted to ignore safety procedures. Work pressure may lead to perceptions that short cuts are necessary to meet demands. Therefore, some may reason regarding a safety regulation, 'It has never caused any problem when I ignored it.' One of the worst things that you can do at work is ignore safety procedures and get away with it! Why? Because this fosters overconfidence and carelessness, leading to more accidents.

The explosion of the Chernobyl plant in Ukraine in 1986 is often described as "the world's worst nuclear accident." What went wrong? A report on the disaster speaks of a "catalogue of reckless operating procedures" and "the repeated flouting of safety precautions."

Both employer and employee can cooperate in foreseeing potential safety hazards. Yes, the wise one observes what could prove to be a dangerous situation and looks for ways to protect himself and others.

When employers do this, they benefit, and so do their employees. For example, a company that redesigned their office to avoid "sick building syndrome" found that before long, productivity was up and staff satisfaction levels had improved dramatically. It was also found that fewer people were out on sick leave. Such consideration for the health of others not only makes for a more pleasant atmosphere for employer and employee but, as seen in this case, can also make good sense economically.

The proper approach to creating the safety culture is to make sure every employee applies a common-sense approach to safety.

This includes being diligent in following local safety regulations. Doing this can go a long way towards making the workplace safer.

1.3 Good Housekeeping

1.3.1 General Information

Housekeeping can reduce and eventually eliminate accident causes. The National Safety Council in the United States reports that falls due to improper housekeeping result in between 200,000 and 300,000 disabling industrial accidents each year. Management must take the initiative to properly train employees in the essentials of good housekeeping and in the necessity of maintaining a hazard-free workplace.

Good housekeeping is one of the surest ways to identify a safe workplace. You can tell how workers' feel about safety just by looking at their housekeeping practices. Good housekeeping isn't the result of cleaning up once a week or even once a day. It's the result of keeping cleaned-up all the time. It's an essential factor in a good safety program, promoting safety, health, production, and morale.

Whose responsibility is housekeeping? It's everyone's. Clean work areas and aisles help eliminate tripping hazards. Respecting "wet floor" signs and immediately cleaning up spills prevents slipping injuries. Keeping storage areas uncluttered reduces the chances of disease and fire as well as slips, trips, and falls. Accumulated debris can cause fires, and clutter slows movement of personnel and equipment during fires.

Other housekeeping practices include keeping tools and equipment clean and in good shape or keeping hoses and cables or wires bundled when not in use. Broken glass should be picked up immediately with a broom and dustpan, never with bare hands. Be aware of open cabinet drawers, electric wires, sharp corners or protruding nails. Either correct the unsafe condition if you are able and it is safe to do so, or notify the person responsible for overall maintenance that something should be done.

How a workplace looks makes an impression on employees and visitors alike. A visitor's first impression of a business is important because that image affects the amount of business it does. Good housekeeping goes hand-in-hand with good public relations. It projects order, care, and pride.

Besides preventing accidents and injuries, good housekeeping saves space, time, and materials. When a workplace is clean, orderly, and free of obstruction; work can get done safely and properly. Workers feel better, think better, do better work, and increase the quantity and quality of their work.

1.3.1.1 *Establishing a Good Housekeeping Program*

Effective housekeeping can eliminate some workplace hazards and help get a job done safely and properly. Poor housekeeping can frequently contribute to accidents by hiding hazards that cause injuries. If the sight of paper, debris, used and spent parts, fuses, clutter and spills is accepted as normal, then other more serious health and safety hazards may be taken for granted.

Housekeeping is not just cleanliness. It includes keeping work areas neat and orderly; maintaining halls and floors free of slip and trip hazards; and removing of waste materials (e.g., paper, cardboard) and other fire hazards from work areas. It also requires paying attention to important details such as the layout of the whole workplace, aisle marking, the adequacy of storage facilities, and maintenance. Good housekeeping is also a basic part of accident and fire prevention.

Effective housekeeping is an ongoing operation: it is not a hit-and-miss cleanup done occasionally. Periodic "panic" cleanups are costly and ineffective in reducing accidents.

Poor housekeeping can be a cause of accidents, such as:

- tripping over loose objects on floors, stairs and platforms;
- being hit by falling objects;
- slipping on greasy, wet or dirty surfaces;
- striking against projecting, poorly stacked items or misplaced material; and
- cutting, puncturing, or tearing the skin of hands or other parts of the body on projecting nails, wire or steel strapping.

To avoid these hazards, a workplace must "maintain" order throughout a workday. Although this effort requires a great deal of management and planning, the benefits are many.

Effective housekeeping results in:

- reduced handling to ease the flow of materials;
- fewer tripping and slipping accidents in clutter-free and spill-free work areas;
- decreased fire hazards;
- lower worker exposures to hazardous substances;
- better control of tools and materials;
- more efficient equipment cleanup and maintenance;
- better hygienic conditions leading to improved health;
- more effective use of space;
- reduced property damage by improving preventive maintenance;
- less janitorial work; and
- improved morale.

A good housekeeping program plans and manages the orderly storage and movement of materials from point of entry to exit. It includes a material flow plan to ensure minimal handling. The plan also ensures that work areas are not used as storage areas by having workers move materials to and from work areas as needed. Part of the plan could include investing in extra bins and more frequent disposal.

The costs of this investment could be offset by the elimination of repeated handling of the same material and more effective use of the workers' time. Often, ineffective or insufficient storage planning results in materials being handled and stored in hazardous ways. Knowing the plant layout and the movement of materials throughout the workplace can help plan work procedures.

Worker training is an essential part of any good housekeeping program. Workers need to know how to work safely with the products they use. They also need to know how to protect other workers such as by posting signs (e.g., "Wet - Slippery Floor") and reporting any unusual conditions.

Housekeeping order is "maintained" not "achieved." This means removing the inevitable messes that occur from time to time and not waiting until the end of the shift to reorganize and clean up. Integrating housekeeping into jobs can help ensure this is done. A good housekeeping program identifies and assigns responsibilities for the following:

- clean up during the shift;
- day-to-day cleanup;

- waste disposal;
- removal of unused materials; and
- inspection to ensure cleanup is complete.

Do not forget out-of-the-way places such as shelves, basements, sheds, and boiler rooms that would otherwise be overlooked. The orderly arrangement of operations, tools, equipment and supplies is an important part of a good housekeeping program.

The final addition to any housekeeping program is inspection. It is the only way to check for deficiencies in the program so that changes can be made. The documents on workplace inspection checklists provide a general guide and examples of checklists for inspecting offices and manufacturing facilities.

The following elements constitute the basis for establishing a good housekeeping program.

1.3.1.2 Dust and Dirt Removal

In some jobs, enclosures and exhaust ventilation systems may fail to collect dust, dirt and chips adequately. Vacuum cleaners are suitable for removing light dust and dirt. Industrial models have special fittings for cleaning walls, ceilings, ledges, machinery, and other hard-to-reach places where dust and dirt may accumulate.

Dampening floors or using sweeping compounds before sweeping reduces the amount of airborne dust. The dust and grime that collect in places like shelves, piping, conduits, light fixtures, reflectors, windows, cupboards and lockers may require manual cleaning. Special-purpose vacuums are useful for removing hazardous substances. For example, vacuum cleaners fitted with HEPA (High Efficiency Particulate Air) filters may be used to capture fine particles of asbestos or fiberglass.

Compressed air should not be used for removing dust, dirt or chips from equipment or work surfaces. First, compressed air is extremely forceful. Depending on its pressure, compressed air can dislodge particles. These particles are a danger since they can enter a worker's eyes or abrade skin. The possible damage would depend on the size, weight, shape, composition, and speed of the particles. There have also been reports of hearing damage caused by the pressure of compressed air and by its sound.

Second, compressed air itself is also a serious hazard. On rare occasions, some of the compressed air can enter the blood stream through a break in the skin or through a body opening. An air bubble in the blood stream is known medically as an embolism, a dangerous medical condition in which a blood vessel is blocked, in this case, by an air bubble. An embolism of an artery can cause coma, paralysis or death depending upon its size, duration and location. While air embolisms are usually associated with incorrect diving procedures, they are possible with compressed air due to high pressures. While this seems improbable, the consequences of even a small quantity of air or other gas in the blood can quickly be fatal.

Horseplay has also been a cause of some serious workplace accidents caused by individuals not aware of the hazards of compressed air, or proper work procedures.

1.3.1.3 Employee Facilities

Employee facilities need to be adequate, clean and well maintained. Lockers are necessary for storing employees' personal belongings. Washroom facilities require cleaning once or more each shift. They also need to have a good supply of soap, towels, plus disinfectants, if needed.

If workers are using hazardous materials, employee facilities should provide special precautions such as showers, washing facilities and changing rooms. Some facilities may require two locker rooms with showers between. Using such double locker rooms allows workers to shower off workplace contaminants and prevents them from contaminating their "street clothes" by keeping their work clothes separated from the clothing that they wear home.

Smoking, eating or drinking in the work area should be prohibited where toxic materials are handled. The eating area should be separate from the work area and should be cleaned properly each shift.

1.3.1.4 Surfaces

Floors: Poor floor conditions are a leading cause of accidents so cleaning up spilled oil and other liquids at once is important. Allowing chips, shavings and dust to accumulate can also cause accidents. Trapping chips, shavings and dust before they reach the floor or cleaning them up regularly can prevent their accumulation.

Areas that cannot be cleaned continuously, such as entrance ways, should have anti-slip flooring. Keeping floors in good order also means replacing any worn, ripped, or damaged flooring that poses a tripping hazard.

Walls: Light-colored walls reflect light while dirty or dark-colored walls absorb light. Contrasting colors warn of physical hazards and mark obstructions such as pillars. Paint can highlight railings, guards and other safety equipment, but should never be used as a substitute for guarding. The program should outline the regulations and standards for colors.

1.3.1.5 *Light Fixtures*

Dirty light fixtures reduce essential light levels. Clean light fixtures can improve lighting efficiency significantly.

1.3.1.6 *Aisles and Stairways*

Aisles should be wide enough to accommodate people and vehicles comfortably and safely. Aisle space allows for the movement of people, products and materials. Warning signs and mirrors can improve sight-lines in blind corners. Arranging aisles properly encourages people to use them so that they do not take shortcuts through hazardous areas.

Keeping aisles and stairways clear is important. They should not be used for temporary "overflow" or "bottleneck" storage. Stairways and aisles also require adequate lighting.

1.3.1.7 *Spill Control*

The best way to control spills is to stop them before they happen. Regularly cleaning and maintaining machines and equipment is one way. Another is to use drip pans and guards where possible spills might occur. When spills do occur, it is important to clean them up immediately. Absorbent materials are useful for wiping up greasy, oily or other liquid spills. Used absorbents must be disposed of properly and safely.

1.3.1.8 *Tools and Equipment*

Tool housekeeping is very important, whether in the tool room, on the rack, in the yard, or on the bench. Tools require suitable fixtures with marked locations to provide orderly arrangement, both in the tool room and near the work bench. Returning them promptly after use reduces the chance of the tools being misplaced or lost. Workers should regularly inspect, clean and repair all tools and take any damaged or worn tools out of service.

1.3.1.9 Maintenance

The maintenance of buildings and equipment may be one of the most important elements of good housekeeping. Maintenance involves keeping buildings, equipment and machinery in safe, efficient working order and in good repair. This includes maintaining sanitary facilities and regularly painting and cleaning walls. Broken windows, damaged doors, defective plumbing and broken floor surfaces can make a workplace look neglected; these conditions can cause accidents and affect work practices. So, it is important to replace or fix broken or damaged items as quickly as possible. A good maintenance program provides for the inspection, maintenance, upkeep and repair of tools, equipment, machines and processes.

1.3.1.10 Waste Disposal

The regular collection, grading and sorting of scrap contribute to good housekeeping practices. It also makes it possible to separate materials that can be recycled from those going to waste disposal facilities.

Allowing material to build up on the floor wastes time and energy since additional time is required for cleaning it up. Placing scrap containers near where the waste is produced encourages orderly waste disposal and makes collection easier. All waste receptacles should be clearly labeled (e.g., recyclable glass, plastic, scrap metal, etc.).

1.3.1.11 Storage

Good organization of stored materials is essential for overcoming material storage problems whether on a temporary or permanent basis. There will also be fewer strain injuries if the amount of handling is reduced, especially if less manual materials handling is required. The location of the stockpiles should not interfere with work but they should still be readily available when required. Stored materials should allow at least one meter (or about three feet) of clear space under sprinkler heads.

Stacking cartons and drums on a firm foundation and cross tying them, where necessary, reduce the chance of their movement. Stored materials should not obstruct aisles, stairs, exits, fire equipment, emergency eyewash fountains, emergency showers, or first aid stations. All storage areas should be clearly marked.

Flammable, combustible, toxic and other hazardous materials should be stored in approved containers in designated areas that are appropriate for the different hazards that they pose. Storage of materials should meet all requirements specified in the fire codes and the regulations of environmental and occupational health and safety agencies in your jurisdiction.

1.4 New Employee Orientation

For employers with a safety manager, the manager can conduct the classroom part of orientation/training, prepare all the training materials (handouts, forms, checklists, lesson plan, etc.), conduct the employee evaluation, and maintain all documentation. The facility supervisor(s) can conduct the on-the-job training and observation, and determine when the employee is ready for the evaluation.

For employers or departments without a safety manager, the company safety committee can share responsibilities for conducting the job hazard analyses and the training program. The safety committee can put together the orientation/training materials, conduct the "classroom" training, and keep records. The department where employees will work can conduct the hands-on training.

During the orientation period, introduce new workers to all the basic safety information that applies to their work areas, such as:

- General hazards in the work area;
- Specific hazards involved in each task the employee performs;
- Hazards associated with other areas of the facility;
- Company safety policies and work rules;
- Proper safety practices and procedures to prevent accidents;
- The location of emergency equipment such as fire extinguishers, eyewash stations, first-aid supplies, etc.;
- Smoking regulations and designated smoking areas;
- Emergency evacuation procedures and routes;

- Who to talk to about safety questions, problems, etc.;
- What to do if there is an accident or injury;
- How to report emergencies, accidents, and near misses;
- How to select, use, and care for personal protective equipment;
- Safe housekeeping rules;
- Facility security procedures and systems;
- How to use tools and equipment safely;
- Safe lifting techniques and materials-handling procedures; and
- Safe methods for handling, using, or storing hazardous materials and the location of material safety data sheets.

Orientation programs can be updated and refined by reviewing accident near-miss reports. Near-miss reports offered early warning signs of new or recurrent hazards in the workplace that must be corrected before someone gets hurt or equipment is damaged. An evaluation of illness and injury reports are also a catalyst for changes in safety orientation and training programs.

Orientation can involve several levels of new employee involvement, from awareness information to formal training programs. Awareness orientation/training informs employees about a potential hazard in the workplace and their role in responding to the hazard, even though they are not directly exposed to the hazard. For example, "affected" employees can be told about locks and tags for electrical systems without being trained how to implement the lockout/tagout program.

It is useful to rely on a checklist to ensure that appropriate safety orientation is provided to new workers. Examples of a new employee safety orientation checklist can be found in Appendix B. These checklists should be modified to fit the needs of the organization or site.

1.5 Worker Rights

The following worker rights are adopted from the U.S. Occupational Safety and Health Act of 1970.

You have the right to a safe workplace. Employers are required to provide a workplace that is free of serious recognized hazards.

Specifically, you have the right to:

1. Get training from your employer.
   a. Get training from your employer on chemicals you are exposed to during your work and information on how to protect yourself from harm. Employers must establish a comprehensive, written hazard communication program (chemical hazard communication). Your employer must label chemical containers, make material safety data sheets with detailed hazard information available to employees, and train you about the health effects of the chemicals you work with and what the employer is doing and what you can do to protect yourself from these hazards.
   b. The program must list the hazardous chemicals in each work area, how the employer will inform employees of the hazards of non-routine tasks (for example, the cleaning of reactor vessels), and hazards associated with chemicals in unlabeled pipes and how the employer will inform other employers at a multi-employer worksite of the hazards to which their employees may be exposed.
   c. Get training from your employer on a variety of other health and safety hazards and standards that your employer must follow. These include electrical safety, lockout-tagout, blood-borne pathogens, confined spaces, construction hazards and a variety of other subjects.
2. Request information from your employer about standards, worker injuries and illnesses, job hazards and workers' rights.
   a. Request information from your employer on safety and health hazards in your workplace, chemicals used in your workplace, tests your employer has done to measure chemical, noise and radiation levels, precautions you should take and procedures to be followed if you or other employees are involved in an incident or are exposed to hazardous chemicals or other toxic substances.

b. Request copies of appropriate standards, rules, regulations and requirements that your employer should have available at the workplace.
c. Review the log and summary of occupational injuries and illnesses at a reasonable time and in a reasonable manner or have an authorized representative do so for you.
d. Access relevant exposure and medical records.
e. Employers must inform you of the existence, location and availability of your medical and exposure records when you first begin employment and at least annually thereafter. Employers also must provide these records to you or your designated representatives within 15 working days of your request. When an employer plans to stop doing business and there is no successor employer to receive and maintain these records, the employer must notify you of your right of access to records at least 3 months before the employer ceases to do business.
f. Observe any monitoring or measuring of toxic materials or chemicals, as well as harmful physical agents, such as noise, and see the resulting records. If the exposure levels are above the limit, the employer must tell you what will be done to reduce the exposure – the right to observe monitoring exists only where monitoring is performed pursuant to a standard that provides employees with the right to observe.

3. Request action from your employer to correct hazards or violations.
   a. You may ask your employer to correct hazards even if they are not violations of specific standards. Be sure to keep copies of any requests you make to your employer to correct hazards.
   b. File a complaint if you believe that there are either violations of standards or serious workplace hazards.
   c. File a complaint and request an inspection take place if you believe serious workplace hazards or violations of standards exist in your workplace. You can file a complaint online, in writing, by telephone or fax. If you want an inspector to come inspect your workplace, put your complaint in writing and send it to the proper authorities.

   d. Request in your written complaint to keep your name confidential if you do not want your employer to know who filed the complaint.
4. Be involved in the inspection of your workplace.
   a. Have an authorized employee representative (such as a union representative) accompany the compliance safety and health officer during the inspection tour.
   b. The authorized employee representative has a right to accompany a compliance safety and health officer during an inspection. Under no circumstances may the employer choose the workers' representative.
   c. Where there is no union or employee representative, the inspector must talk confidentially with a reasonable number of workers during the course of the investigation.
   d. Respond to questions from the compliance officer and tell the compliance officer about workplace hazards, particularly if there is no authorized employee representative accompanying the compliance officer on the inspection "walkaround."
   e. You and your co-workers have a right to talk privately and confidentially to the compliance officer whether or not a workers' representative has been chosen.
   f. You may point out hazards, describe injuries or illnesses or near misses that resulted from those hazards and describe past complaints about hazards. Inform the inspector if working conditions are not normal during the inspection. Make sure that the inspector is aware if equipment has been shut down, windows opened or other conditions changed from normal.
5. Find out results of inspections.
   a. Find out the results of inspections and request a review if a citation is not issued.
   b. If health hazards are present in your workplace, a special health inspection may be conducted by an industrial hygienist. This inspector may take samples to measure levels of chemicals or other hazardous materials.
   c. The inspecting agency will let the employee representative know whether your employer is in compliance. The inspector also will gather detailed

information about your employer's efforts to control health hazards, including results of tests your employer may have conducted.

6. Get involved in any meetings or hearings to discuss any objections your employer has with citations or changes in abatement deadlines.
   a. File a formal appeal of deadlines for correction of hazards.
   b. File an appeal of the deadlines that is set for your employer to correct any violation in the citation issued to the employer.
7. File a discrimination complaint.
   a. File a discrimination complaint within 30 days if you are punished or discriminated against for exercising your safety and health rights or for refusing to work when faced with an imminent danger of death or serious injury and there is insufficient time for an inspection.
8. Request a research investigation on possible workplace health hazards.
   a. Contact local government safety office to request a health hazard evaluation if you are concerned about toxic effects of a substance in the workplace.

2

Meters and Monitors

2.1 Air Quality Testing and Monitoring

2.1.1 Introduction

Indoor air quality testing may be necessary to ensure employee safety. Testing and monitoring may be applied to those conditions where employees may be exposed to:

- nitrogen dioxide and sulfur dioxide
- landfill gases
- noxious odors
- radon gas
- factory emissions
- odor complaints
- rainwater
- metals
- smoke levels
- dust
- volatile organic compounds
- indoor air quality (including Carbon Monoxide)

The results of air quality testing may be used to:

- Assign levels of worker respiratory protection
- For emergency planning

2.1.2 Methods of Sampling and Testing

Electric Power producers shall provide adequate means of carrying air monitoring in generator houses, transmitting stations, injection and switching substations, etc.

Three main methods are available to measure air pollution:

Passive Sampling: This refers to absorption or diffusion tubes or badges that provide a simple and inexpensive indication of average pollution levels over a period of weeks or months. Plastic tubes or discs open at one end to the atmosphere and with a chemical absorbent at the other, collect a sample for subsequent analysis in the laboratory. The low cost per tube allows sampling at a number of points and is useful in highlighting "hotspots" where more detailed study may be needed. The quality and accuracy of the data from passive sampling tubes does not make them suitable for precise measurements but they can give useful long term trend data.

Active Sampling: This involves the collection of samples, by physical or chemical means, for subsequent laboratory analysis. Typically, a known volume of air is pumped through a filter or chemical collector for a known period of time - the collection medium is then subjected to laboratory analysis. This method is not suitable for continuous or near-real time air quality monitoring.

Automatic Sampling: This is the most sophisticated method of air quality analysis, producing high-resolution measurement data of a range of pollutants. The pollutants that can be measured include, but are not limited to, NO_x, SO_2 CO, O_3, VOC's, PM10, PM2.5, Carbon Black, Hg, Benzene etc. The air quality is continuously sampled and measured on-line and in real-time. The real time data is stored, typically as one hourly averages, with data being collected remotely from individual monitoring stations by telemetry. Remote control of the monitoring and data system is also possible as is remote diagnostics for most of the analyzers.

2.1.3 Samplers and Monitors

Toxic gas monitors are among the most reliable in the industry. Air quality testing is crucial to ensure a safe work environment.

2.1.3.1 Diffusive Samplers for Air Monitoring

This is a reliable, simple and inexpensive measurement technique. The sampling technique is based on molecular diffusion of gases, hence the term "diffusive sampling". The gas molecules diffuse into the sampler where they are quantitatively collected on an impregnated filter or an adsorbent material, giving a concentration value integrated over time.

Since diffusive samplers are small, silent, have light weight and do not require electricity, their exposure site location can be relatively flexible. They can be used outdoors in tropical conditions, in rural to urban environments, or indoors e.g. for personal exposure. No technically skilled personnel is required at the sampling site since the samplers are easy to handle and no field calibration is needed. See Figure 2.1.

After sampling, the analytes are chemically desorbed by solvent extraction or thermally desorbed and analyzed. Passive sampling does not involve the use of heavy and encumbering pumping systems, is not impacted by power disruptions, does not require extensive supervision, is quiet, non-flammable and does not represent an explosion

Figure 2.1 Diffusive sampler. Photo downloaded from http://www.ivl.se/en/business/monitoring/diffusive_samplers.asp

hazard. It can be performed by anyone, anywhere and at a very low cost. Moreover, it is not susceptible to sample breakthrough, a common problem associated with active sampling performed with an air pump.

2.1.3.2 Color Diffusion Tubes

Using the principles of gas diffusion and colorimetric reaction, Dräger color Diffusion cubes allow the user to reliably measure time-weighted average (TWA) concentrations for up to eight hours. An easy-to-use tube holder allows safe tube end removal and secure sampling. Color diffusion tubes can be evaluated at any point during a workshift. The calibrated scale printed directly on each tube indicates parts-per-million-hours (ppm-hours). For a TWA reading, simply determine the point where the stain length ends and divide this reading by the total length of sampling time. No pumps, charts, or analyses are needed. To use a Diffusion Tube, place the tube in the tube holder with the red dot on the tube visible above the hinged portion of the holder (Figure 2.2a).

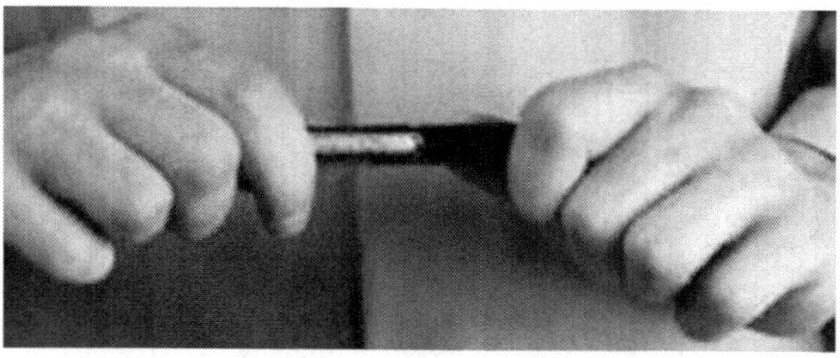

Figure 2.2a

Open the tube by grasping the ends of the tube holder, with the red dot on the tube away from the body, and bending the holder until the tube breaks (Figure 2.2b)

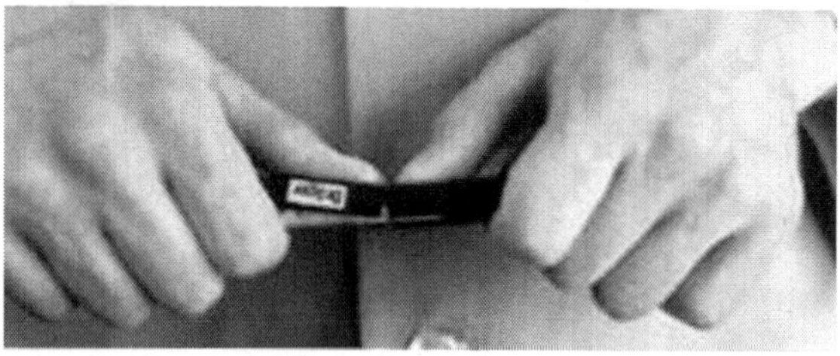

Figure 2.2b

Remove the end of the tube from the holder and discard. The diffusion tube is ready for use. Snap the tube into the tube holder with the open end positioned at the open end of the holder (Figure 2.2c).

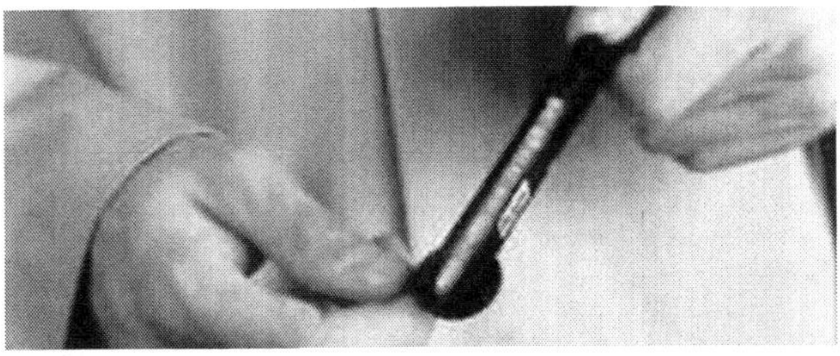

Figure 2.2c

Attach the diffusion tube holder to a worker's pocket or lapel and note the starting time of the test. At the end of the workshift, or any time during the workshift, the user can evaluate the tube.

When using diffusion tubes, the time-weighted average (TWA) can be calculated at any time during the measurement by simply dividing the length of the tube reading (in ppm-hours) by the elapsed sampling time (in hours).

Table 2.1 provides guidance on the commercially available measurement ranges for Dräger color diffusion tubes.

2.1.3.3 Active Samplers

Active samplers are normally used to measure the average concentration of pollutants over periods of 30 minutes to 24 hours. For indoor air, the measurement can be undertaken with normal working conditions in the building, but if the purpose is to identify the presence of chemicals that may be a cause of odor or a health problem then 'worst case' conditions may be appropriate. This is best achieved by keeping all windows and doors closed in the room/s of interest for 12 hours before and during sampling. It may be appropriate to heat the room as normally if the problem is influenced by temperature.

There are a range of active devices widely used to measure air pollutants. The Health and Safety Officer should first identify the nature of air pollutants that workers may be exposed to.

For volatile organic compounds (VOCs) – VOCs is a term used to cover a wide range of individual chemicals containing carbon that form a vapor in air. Most people are familiar with fumes from

Table 2.1 Ranges of practical measurement for Dräger color diffusion tubes.[1]

Chemical Hazard	Measuring Range (ppm-hrs)	Chemical Hazard	Measuring Range (ppm-hrs)
Acetic acid	1.3–200	Hydrocyanic acid	2.5–200
Ammonia	2.5–1500	Hydrogen sulfide	1.3–300
Butadiene	1.3–300	Nitrogen dioxide	1.3–200
Carbon dioxide	65–20000	Perchloroethylene	25–1500
Carbon dioxide	0.13–30 vol%	Sulfur dioxide	0.7–150
Carbon monoxide	6–600	Toluene	13–3000
Ethanol	125–25000	Trichloroethylene	25–1000
Hydrochloric acid	1.3–200		

Data downloaded from http://www.skcinc.com/prod/800-01071.asp

drying paint which fit into this category but many other products used in the workplace as well as consumer products (e.g. toiletries and polishes) are a source of VOCs. Outdoors pollution from traffic contains a wide range of VOCs including benzene.

A number of VOCs are irritants and several can have specific health effects e.g. benzene is a carcinogen. As with all chemicals the health effect, if any, depends upon the concentration of the pollutant and the period that people are exposed. There is evidence that people's sensitivity to pollutants varies considerably.

Commercially available sampling tubes are used to measure VOCs in the air. This is a stainless steel tube that contains a powdered adsorbent. When the end caps are removed and air pulled through the tube, VOCs are removed from the air by the sorbent. When returned to the laboratory, the tubes are analyzed by thermal desorption (TD) followed by gas chromatography with a flame ionization detector to measure amounts of VOCs collected and/or a mass spectrometer (MS) to confirm the identity of the individual VOCs collected.

The adsorbent used determines the range of VOCs collected. A porous polymer (Tenax TA) has been widely used and this is optimum for compounds that have a boiling point between 75°C and 280°C. Sampling and analysis using this sorbent are in accordance with the international standard ISO16000-6.

Other sorbents are more suitable for compounds of different volatility, for example, some detectors use Carboxen 569 (a carbon molecular sieve) for the analysis of carbon disulphide.

Polyurethane foam samplers are used for semi-volatile compounds (SVOCs), such as biocides and chemicals associated with oil mists.

Formaldehyde and other volatile carbonyl compounds – Formaldehyde is an irritating gas released by many products and through combustion of fuels. It is useful as a preservative and is a component of glues used to bind particles of wood in particleboard and similar products. In outdoor air it occurs as a product of combustion and photochemical reaction. Other carbonyl compounds in air can include acetaldehyde, acetone, acrolein and glutaraldehyde.

Special cartridge units can be used to measure carbonyls in the air. This is a circular plastic sampler that contains silica gel coated with a chemical that absorbs carbonyls from the air pulled through it by a pump. The sampling and analysis is undertaken in accordance with the International Standard ISO16000-3.

Other pollutants – There are a range of continuous monitoring equipment to determine gases including nitrogen dioxide, carbon monoxide, methane, carbon dioxide and ozone in air. In addition, measurement of particulates (PM10) and ultrafine particles can be undertaken.

The standards do not recommend or endorse any specific supplier or manufacturer. HSOs can identify specific suppliers and evaluate the most appropriate instruments based on cost, availability, dependability, reliability, and intended application.

Active sampling normally requires a specialist to visit your site with a range of equipment to undertake the air sampling. Measurements of temperature and humidity are routinely made at the same time. Alternatively, the HSO can undertake the sampling alone; however close consultation with instrument suppliers is recommended.

2.1.3.4 Continuous Emissions Monitors

Continuous emissions monitors have been built using most types of sensor technology. There are systems in use using electrochemical, infrared, ultra-violet, chemiluminescent and other measurements for the toxic gases, whilst oxygen measurements are mainly carried out with electrochemical, zirconium or paramagnetic sensors. All these methods have their own advantages and drawbacks, be it technological or financial.

Among the portable instruments most suitable for workplace environments are combustible gas indicators (CGI) which detect flammables, four gas meters which detect flammables, oxygen, carbon monoxide and hydrogen sulfide, and photoionization (PID) detectors and the colorimetric tubes discussed above which detect low level toxics. None of these technologies is capable of identifying unknown gases and vapors.

2.1.3.4.1 Photoionization Meters

Description and Applications. Ionization is based upon making a gas conductive by the creation of electrically charged atoms, molecules, or electrons and the collection of these charged particles under the influence of an applied electric field. The photoionization analyzer is a screening instrument used to measure a wide variety of organic and some inorganic compounds. It is also useful as a leak detector.

The limit of detection for most contaminants is approximately 0.1 ppm.

Calibration. The procedure for calibration involves applying the calibration gas (typically 100 ppm isobutylene) to the instrument and checking the reading.

Special Considerations. The specificity of the instrument depends on the sensitivity of the detector to the substance being measured, the number of interfering compounds present, and the concentration of the substance being measured relative to any interference. Many models now have built-in correction or correlation factors. After calibrating the unit on isobutylene, select the gas to be measured. The instrument will automatically correct for the relative sensitivity of the gas selected. Some instruments are listed by an NRTL for hazardous locations. Check the operating manual for specific conditions.

Maintenance. Keeping these instruments in top operating shape means charging the battery, cleaning the ultraviolet lamp window, light source and replacing the dust filter. The exterior of the instrument can be wiped clean with a damp cloth and mild detergent if necessary. Keep the cloth away from the sample inlet, however, and do not attempt to clean while the instrument is connected to line power.

2.1.3.4.2 Infrared Analyzers

Description and Applications. The infrared analyzer is used as a screening tool for a number of gases and vapors and is presently the recommended screening method for substances with no feasible sampling and analytical method.

These analyzers are often factory-programmed to measure many gases and are also user-programmable to measure other gases.

A microprocessor automatically controls the spectrometer, averages the measurement signal, and calculates absorbance values. Analysis results can be displayed either in parts per million (ppm) or absorbance units (AU). The variable path-length gas cell gives the analyzer the capability of measuring concentration levels from below 1 ppm up to percent levels.

Some typical screening applications are:

- Carbon monoxide and carbon dioxide, especially useful for indoor air assessments;
- Anesthetic gases including, e.g., nitrous oxide, halothane, enflurane, penthrane, and isoflurane;
- Ethylene oxide; and
- Fumigants including e.g. ethylene dibromide, chloropicrin, and methyl bromide.

Calibration. The analyzer and any strip-chart recorder should be calibrated before and after each use in accordance with the manufacturer's instructions.

Special Considerations. The infrared analyzer may be only semi-specific for sampling some gases and vapors because of interference by other chemicals with similar absorption wavelengths.

Maintenance. No field maintenance of this device should be attempted except items specifically detailed in the instruction book such as filter replacements and battery charging.

2.1.3.4.3 Toxic Gas Meters

Description and Application. This analyzer uses an electrochemical voltametric sensor or polarographic cell to provide continuous analyses and electronic recording. In operation, sample gas is drawn through the sensor and absorbed on an electrocatalytic sensing electrode, after passing through a diffusion medium. An electrochemical reaction generates an electric current directly proportional to the gas concentration. The sample concentration is displayed directly in parts per million. Since the method of analysis is not absolute, prior calibration against a known standard is required. Exhaustive tests have shown the method to be linear; thus, calibration at a single concentration, along with checking the zero point, is sufficient.

Types: Sulfur dioxide, hydrogen cyanide, hydrogen chloride, hydrazine, carbon monoxide, hydrogen sulfide, nitrogen oxides, chlorine, and ethylene oxide. These can be combined with combustible gas and oxygen meters.

Calibration. Calibrate the direct-reading gas monitor before and after each use in accordance with the manufacturer's instructions and with the appropriate calibration gases.

Special Considerations.

- Interference from other gases can be a problem. See manufacturer's literature.
- When calibrating under external pressure, the pump must be disconnected from the sensor to avoid sensor damage. If the span gas is directly fed into the instrument from a regulated pressurized cylinder, the flow rate should be set to match the normal sampling rate.
- Due to the high reaction rate of the gas in the sensor, substantially lower flow rates result in lower readings. This high reaction rate makes rapid fall time possible simply by shutting off the pump. Calibration from a sample bag connected to the instrument is the preferred method.

2.1.3.4.4 Ozone Meter

Description and Application. The detector uses a thin-film semiconductor sensor. A thin-film platinum heater is formed on one side of an alumina substrate. A thin-film platinum electrode is formed on the other side, and a thin-film semiconductor is formed over the platinum electrode by vapor deposition. The semiconductor film, when kept at a high temperature by the heater, will vary in resistance due to the absorption and decomposition of ozone. The change in resistance is converted to a change of voltage by the constant-current circuit.

The measuring range of the instrument is 0.01 ppm to 9.5 ppm ozone in air. The readings are displayed on a liquid crystal display that reads ozone concentrations directly. The temperature range is 0°-40° C, and the relative humidity range is 10%-80% RH.

Calibration. Calibrate instrument before and after each use. Be sure to use a well-ventilated area since ozone levels may exceed the PEL for short periods. Calibration requires a source of ozone. Controlled ozone concentrations are difficult to generate in the field, and this calibration is normally performed at SLTC. Gas that is either specially desiccated or humidified must not be used for preparing calibration standards, as readings will be inaccurate.

Special Considerations.

- The instrument is not intrinsically safe.
- The instrument must not be exposed to water, rain, high humidity, high temperature, or extreme temperature fluctuation.
- The instrument must not be used or stored in an atmosphere containing silicon compounds, or the sensor will be poisoned.
- The instrument is not to be used for detecting gases other than ozone. Measurements must not be performed when the presence of organic solvents, reducing gases (such as nitrogen monoxide, etc.), or smoke is suspected; readings may be low.

Maintenance. The intake-filter unit-Teflon sampling tube should be clean and connected firmly. These should be checked before each operation. Check the pump aspiration and sensitivity before each operation.

2.1.3.4.5 Mercury Analyzer-Gold Film Analyzer

Description and Application. This instrument relies on a gold-film analyzer that draws a precise volume of air over a gold-film sensor. A microprocessor computes the concentration of mercury in milligrams per cubic meter and displays the results on the digital meter. The meter is selective for mercury and eliminates interference from water vapor, sulfur dioxide, aromatic hydrocarbons, and particulates. However, hydrogen sulfide is an interferant.

Calibration. Calibration should be performed by the manufacturer or a laboratory with the special facilities to generate known concentrations of mercury vapor. Instruments should be returned to the manufacturer or a calibration laboratory on a scheduled basis.

Special Considerations. In high concentrations of mercury vapor the gold film saturates quickly. Check operating manual for more specific information.

Maintenance. Mercury vapor instruments generally contain rechargeable battery packs, filter medium, pumps, and valves which require periodic maintenance. Except for routine charging of the battery pack, most periodic maintenance will be performed during the scheduled annual calibrations. However, depending on usage, routine maintenance should also include burning mercury deposits off of the gold-film and changing the zero filter when necessary. See operating manual for specific instructions.

2.1.3.4.6 Direct-Reading Particle Monitors

Description and Applications. Condensation-nuclei counters are based upon a miniature, continuous-flow condensation nucleus counter (CNC) that takes particles too small to be easily detected, enlarges them to a detectable size, and counts them. Submicrometer particles are grown to supermicrometer alcohol droplets by first saturating the particles with alcohol vapor as they pass through a heated saturator lined with alcohol soaked felt, and then condensing the alcohol on the particles in a cooled condenser. Optics focus laser light into a sensing volume.

As the droplets pass through the sensing volume, the particles scatter the light. The light is directed onto a photodiode which generates an electrical pulse from each droplet. The concentration of particles is counted by determining the number of pulses generated. Applications include the testing of respirators and dust monitors.

The counter counts individual airborne particles from sources such as smoke, dust, and exhaust fumes. Models typically operate in one of three possible modes, each with a particular application. In the "count" mode, the counter measures the concentration of these airborne particles. In the "test" (or fit test) mode, measurements are taken inside and outside a respirator and a fit factor is calculated. In the "sequential" mode, the instrument measures the concentration on either side of a filter and calculates filter penetration.

This instrument is sensitive to particles as small as 0.02 micrometers. However, it is insensitive to variations in size, shape, composition, and refractive index.

Calibration. Check the counter before and after each use in accordance with the manufacturer's instructions. This usually involves checking the zero of the instrument.

Special Considerations. Reagent-grade isopropyl alcohol for use in these types of instruments is available from HSO.

- Dry the saturator felt by installing a freshly charged battery pack without adding alcohol. Allow the instrument to run until the LO message (low battery) or the E-E message (low particle count) appears. Some instruments allow you to remove the alcohol cartridge for storage purposes.
- Remove the battery pack.
- Install the tube plugs into the ends of the twin-tube assembly.

Maintenance. Isopropyl alcohol must be added to the unit every 5-6 hours of operation, per the manufacturer's instructions. Take care not to overfill the unit. Under normal conditions, a fully charged battery pack will last for about 5 hours of operation. Low battery packs should be charged for at least 6 hours, and battery packs should not be stored in a discharged condition.

2.1.3.4.7 Combustible Gas Meters

Description and Applications. These meters use elements which are made of various materials such as platinum or palladium as an oxidizing catalyst. The element is one leg of a Wheatstone bridge circuit. These meters measure gas concentration as a percentage of the lower explosive limit of the calibrated gas.

The oxygen meter displays the concentration of oxygen in percent by volume measured with a galvanic cell. Other electrochemical sensors are available to measure carbon monoxide, hydrogen sulfide, and other toxic gases. Some units have an audible alarm that warns of low oxygen levels or malfunction.

Calibration. Before using the monitor each day, calibrate the instrument to a known concentration of combustible gas (usually methane) equivalent to 25%-50% LEL full-scale concentration.

The monitor must be calibrated to the altitude at which it will be used. Changes in total atmospheric pressure from changes in altitude will influence the instrument's measurement of the air's oxygen content. The unit's instruction manual provides additional details on calibration of sensors.

Special Considerations.

- Silicone compound vapors, leaded gasoline, and sulfur compounds will cause desensitization of the combustible sensor and produce erroneous (low) readings.

- High relative humidity (90%-100%) causes hydroxylation, which reduces sensitivity and causes erratic behavior including inability to calibrate.
- Oxygen deficiency or enrichment such as in steam or inert atmospheres will cause erroneous readings for combustible gases.
- In drying ovens or unusually hot locations, solvent vapors with high boiling points may condense in the sampling lines and produce erroneous (low) readings.
- High concentrations of chlorinated hydrocarbons such as trichloroethylene or acid gases such as sulfur dioxide will depress the meter reading in the presence of a high concentration of combustible gas.
- High-molecular-weight alcohols can burn out the meters filaments.
- If the flash point is greater than the ambient temperature, an erroneous (low) concentration will be indicated. If the closed vessel is then heated by welding or cutting, the vapors will increase and the atmosphere may become explosive.
- For gases and vapors other than those for which a device was calibrated, users should consult the manufacturer's instructions and correction curves.

Maintenance. The instrument requires no short-term maintenance other than regular calibration and recharging of batteries. Use a soft cloth to wipe dirt, oil, moisture, or foreign material from the instrument. Check the bridge sensors periodically, at least every six months, for proper functioning. A thermal combustion-oxygen sensor uses electrochemical cells to measure combustible gases and oxygen. It is not widely used in the area offices.

2.1.3.4.8 Oxygen Meters

These oxygen-measuring devices can include coulometric and fluorescence measurement, paramagnetic analysis, and polarographic methods. The output of most electrochemical oxygen sensors is dependent on the partial pressure of oxygen in the atmosphere. They do not actually measure concentration directly. An instrument calibrated at sea level and used at higher elevations, such as mountains, will indicate a value lower than the actual concentration.

21.3.4.9 Bioaerosol Monitors

Description and Applications: A bioaerosol meter, usually a two-stage sampler, is also a multiorifice cascade impactor. This unit is used when size distribution is not required and only respirable-nonrespirable segregation or total counts are needed.

Ninety-five to 100 percent of viable particles above 0.8 microns in an aerosol can be collected on a variety of bacteriological agar. Trypticase soy agar is normally used to collect bacteria, and malt extract agar is normally used to collect fungi. They can be used in assessing sick- (or tight-) building syndrome and mass psychogenic illness.

These samplers are also capable of collecting virus particles. However, there is no convenient, practical method for cultivation and enumeration of these particles.

Calibration: Bioaerosol meters must be calibrated before use. This can be done using an electronic calibration system with a high-flow cell, available through the HRT.

Special Considerations: Prior to sampling, determine the type of collection media required and an analytical laboratory. The HRT can provide this information. This specialized equipment is available from the HRT with accompanying instructions.

Maintenance: The sampler should be decontaminated prior to use by sterilizer or chemical decontamination with isopropanol.

2.1.4 Batteries

Many of the instruments described are powered by batteries. Battery care is important in assuring uninterrupted sampling. A pump battery pack, for example, should be discharged to the recommended level before charging, at least after every use. If the pump is allowed to run down until the battery reaches the low battery Fault condition, the pump should be turned OFF soon after the Fault condition stops the pump. Leaving some pumps ON for a long time after this Fault condition can damage the battery pack. Also, avoid overcharging the battery pack.

Alkaline Batteries. Replace frequently before they become depleted, or carry fresh replacements. When replacing a battery, never mix types (alkaline, carbon zinc, etc.) or capacity and age. Doing so can have negative affects on all the batteries. Remove batteries if equipment will not be used for an extended period of time.

Rechargeable Ni-Cad Batteries.

- Check the batteries under load (e.g., turn pump on and check voltage at charging jack, if one is available and this can be done safely) before use. See manufacturer's instructions for locations to check voltage. Use 1.2 volts per Ni-Cad cell for an estimate of the fully charged voltage of a rechargeable battery pack.
- It is undesirable to discharge a multicell Ni-Cad battery pack to voltage levels that are below 1.0 volts per cell; doing this will drive a reverse current through some of the cells and can permanently damage them.
- Rechargeable Ni-Cad batteries should be charged only in accordance with manufacturer's instructions. Chargers are generally designed to charge batteries in approximately 8 to 16 hours at a high charge rate. A battery can be overcharged and ruined when a high charge rate is applied for too long a time. However, Ni-Cad batteries may be left on a proper trickle charge indefinitely to maintain them at peak capacity. In this case, discharging for a period equal to the longest effective field service time may be necessary, because of short-term memory imprinting. However, do not let the battery run down overnight or longer. Turn the instrument OFF when the battery reaches the proper discharge level.

Other Rechargeable Batteries: Other types of rechargeable batteries are being used in equipment such as lead-acid, nickel-metal hydride, etc. Make sure the manufacturer's instructions are followed concerning the handling and recharging of these types of batteries.

2.1.5 Adverse Conditions

Adverse Temperature Effects

High ambient temperature, above 37.8°C and/or radiant heat (e.g., from nearby molten metal) can cause flow faults in air sampling pumps. If these conditions are likely, use the pump with a higher operating temperature range, as opposed to a pump with a lower operating temperature range. Temperature can also affect the accuracy of instrument readings or operation. Check the operating manual for the proper operating temperature range.

Explosive Atmospheres

- Instruments shall not be used in atmospheres where the potential for explosion exists unless the instrument is listed by a nationally recognized testing laboratory for use in the type of atmosphere present. Check the class and division ratings.
- When batteries are being replaced, use only the type of battery specified on the safety approval label.
- Do not assume that an instrument is intrinsically safe. Verify by contacting the instrument's maker if uncertain.

Atmospheres Containing Carcinogens
A plastic bag should be used to cover equipment when carcinogens are present. Decontamination procedures for special environments are available and should be followed after using equipment in carcinogenic environments. If at all possible, decontaminate the equipment after use on-site.

2.1.6 Annex–Instrument Chart

The information shown in Table 2.2 below is for reference only. Not every compliance officer or field office will have every type of instrument. Many of the instruments can be found in and are available through a Corporate Loan program.

Table 2.2 Gas and vapor meters.

Type of Instrument	Measured Substance	Application
Double-range meters	Combustible gas, O_2	Confined spaces
Triple range meters	Toxic, O_2, combustible gas	Confined spaces
Quad range meters	2 toxics, O_2, combustible gas	
CO dosimeter	CO	Ggarages, indoor air quality

Table 2.2 (cont.) Gas and vapor meters.

Type of Instrument	Measured Substance	Application
Carbon dioxide meter	CO_2	Indoor air quality (IAQ)
Infrared analyzers	CO, CO_2, organic substances	Traces indoor air, leaks, spills
Hydrogen cyanide monitors	Hydrogen cyanide	Plants
Hydrogen sulfide meters	Hydrogen sulfide	Farms, sewers
Mercury vapor meters	Mercury	Mercury plants, spills
NO and NO_2 meters	NO and NO_2 combustion	
Ozone analyzers	O_3	Water or air purification, IAQ

2.1.7 Bibliography

A.M. Best Co. (AMB). 1990. Bests Safety Directory. AMB: Odwick, NJ.

Hering, S.V., Ed. 1989. Air Sampling Instruments for Evaluation of Atmospheric Contaminants, American Conference of Governmental Industrial Hygienists: Cincinnati, Ohio.

Cheremisinoff, N. P., 1999, Handbook of Industrial Toxicology and Hazardous Materials, Marcel Dekker Publishers, NY, NY.

2.2 Noise Testing and Monitoring

2.2.1 Introduction

Hearing protection should be issued to employees:

- where extra protection is needed above what can been achieved using noise control;
- as a short-term measure while other methods of controlling noise are being developed.

The health and safety officer (HSO) should not use hearing protection as an alternative to controlling noise by technical and organizational means.

The national electricity health and safety standards require employers to:

- provide employees with hearing protectors if they ask for it and their noise exposure is between the lower and upper exposure action values;
- provide employees with hearing protectors and make sure they use them properly when their noise exposure exceeds the upper exposure action values;
- identify hearing protection zones, i.e. areas where the use of hearing protection is compulsory, and mark them with signs if possible;
- provide your employees with training and information on how to use and care for the hearing protectors;
- ensure that the hearing protectors are properly used and maintained.

Hearing protection can be used effectively by applying the following guidelines:

- make sure the protectors give enough protection – aim at least to get below 85 dB at the ear;
- target the use of protectors to the noisy tasks and jobs in a working day;
- select protectors which are suitable for the working environment - consider how comfortable and hygienic they are;
- think about how they will be worn with other protective equipment (e.g. hard hats, dust masks and eye protection);
- provide a range of protectors so that employees can choose ones which suit them.

Hearing protection can be used effectively by avoiding the following:

- provide protectors which cut out too much noise – this can cause isolation, or lead to an unwillingness to wear them;

- make the use of hearing protectors compulsory where the law doesn't require it;
- have a 'blanket' approach to hearing protection – better to target its use and only encourage people to wear it when they need to.

2.2.2 Noise Monitors and Meters

2.2.2.1 Sound Level Meters

Octave Band Analyzer. Some sound level meters may have an octave or one-third octave band filter attached or integrated into the instrument. The filters are used to analyze the frequency content of noise. They are also valuable for the calibration of audiometers and to determine the suitability of various types of noise control.

Calibration. In normal operation, calibration of the instrument usually requires only checking. Prior to and immediately after taking measurements, it is a good practice to check, using a calibrator, the ability of the sound level instrument to correctly measure sound levels. As long as the sound level readout is within 0.2 dB of the known source, it is suggested that no adjustments to the calibration pot be made. If large fluctuations in the level occur (more than 1 dB) then either the calibrator or the instrument may have a problem.

Special Considerations

- Always check the batteries prior to use. Use the microphone windscreen to protect the microphone when the wearer will be outdoors or in dusty or dirty areas. (The windscreen will not protect the microphone from rain or extreme humidity.)
- Never use any other type of covering over the microphone (e.g., plastic bag or plastic wrap) to protect it from moisture. These materials will distort the noise pickup, and the readings will be invalid.
- Never try to clean a microphone, particularly with compressed air, since damage is likely to result. Although dirt and exposure will damage microphones, regular use of an acoustical calibrator will detect such damage so that the microphones can be replaced.
- Remove the batteries from any meter that will be stored for more than 5 days. Protect meters from extreme heat and humidity.

Maintenance. No field maintenance is required other than replacement of batteries.

2.2.2.2 *Personal Dosimeters*

Calibration. Field calibrate at the measurement site according to the manufacturer's instructions both before and after each use. Use an acoustical calibrator that was designed to be used with the particular model noise dosimeter being used.

Special Considerations

- Always check the batteries prior to use. Be very careful with the microphone cable. Never kink, stretch, pinch, or otherwise damage the cable.
- Use the microphone windscreen to protect the microphone when the wearer will be outdoors or in dusty or dirty areas. (The windscreen will not protect the microphone from rain or extreme humidity.)
- Never use any type of covering over the microphone (e.g., plastic bag or plastic wrap) to protect it from moisture. Such materials will distort the noise pickup, and the readings will be invalid.
- Never try to clean a microphone, particularly with compressed air, since damage is likely to result. Although dirt and exposure to industrial environments will damage the microphones, regular use of an acoustical calibrator will detect such damage so that microphones can be replaced.
- Remove the batteries when the dosimeter will be stored for more than 5 days. Protect dosimeters from extreme heat and humidity.

Maintenance. No field maintenance is required other than replacement of batteries.

2.2.3 Occupational Noise Exposure Standard

2.2.3.1 *Allowable Levels of Exposure*

Protection against the effects of noise exposure shall be provided when the sound levels exceed those shown in Table 2.3 when measured on the A scale of a standard sound level meter at slow response.

Table 2.3 Permissible noise exposures.[1]

Duration Per Day, Hours	Sound Level, dBA Slow Response
8	90
6	92
4	95
3	97
2	100
1.5	102
1	105
0.5	110
0.25 or less	115

1. When the daily noise exposure is composed of two or more periods of noise exposure of different levels, their combined effect should be considered, rather than the individual effect of each. If the sum of the following fractions: C(1)/T(1) + C(2)/T(2) C(n)/T(n) exceeds unity, then, the mixed exposure should be considered to exceed the limit value. Cn indicates the total time of exposure at a specified noise level, and Tn indicates the total time of exposure permitted at that level. Exposure to impulsive or impact noise should not exceed 140 dB peak sound pressure level.

When noise levels are determined by octave band analysis, the equivalent A-weighted sound level may be determined as follows.

When employees are subjected to sound exceeding those listed in Table 2.3, feasible administrative or engineering controls shall be utilized. If such controls fail to reduce sound levels within the levels of Table 2.3, personal protective equipment shall be provided and used to reduce sound levels within the levels of the table.

If the variations in noise level involve maxima at intervals of 1 second or less, it is to be considered continuous.

2.2.3.2 *Hearing Conservation Program*

The employer shall administer a continuing, effective hearing conservation program, as described below, whenever employee noise exposures equal or exceed an 8-hour time-weighted average sound level (TWA) of 85 decibels measured on the A scale (slow response) or, equivalently, a dose of fifty percent. For purposes of the hearing conservation program, employee noise exposures shall be established without regard to any attenuation provided by the use of personal protective equipment.

For purposes of establishing a program, an 8-hour time-weighted average of 85 decibels or a dose of fifty percent shall also be referred to as the action level. The following constitute the basic components of a hearing conservation program to be adopted and implemented:

Monitoring: When information indicates that any employee's exposure may equal or exceed an 8-hour time-weighted average of 85 decibels, the employer shall develop and implement a monitoring program.

Sampling: The sampling strategy shall be designed to identify employees for inclusion in the hearing conservation program and to enable the proper selection of hearing protectors.

Where circumstances such as high worker mobility, significant variations in sound level, or a significant component of impulse noise make area monitoring generally inappropriate, the employer shall use representative personal sampling to comply with the monitoring requirements of the program unless the employer can show that area sampling produces equivalent results.

All continuous, intermittent and impulsive sound levels from 80 decibels to 130 decibels shall be integrated into the noise measurements.

Instruments used to measure employee noise exposure shall be calibrated to ensure measurement accuracy.

Monitoring shall be repeated whenever a change in production, process, equipment or controls increases noise exposures to the extent that:

Employee Notification: The employer shall notify each employee exposed at or above an 8-hour time-weighted average of 85 decibels of the results of the monitoring.

Observation of Monitoring: The employer shall provide affected employees or their representatives with an opportunity to observe any noise measurements conducted pursuant to this section.

Audiometric Testing Program: The employer shall establish and maintain an audiometric testing program as provided in this paragraph by making audiometric testing available to all employees whose exposures equal or exceed an 8-hour time-weighted average of 85 decibels.

The program shall be provided at no cost to employees.

Audiometric tests should be performed by a licensed or certified audiologist, otolaryngologist, or other physician, or by a technician who has satisfactorily demonstrated competence in administering audiometric examinations, obtaining valid audiograms, and properly using, maintaining and checking calibration and proper functioning of the audiometers being used. A technician who performs audiometric tests must be responsible to an audiologist, otolaryngologist or physician.

Baseline Audiogram: Within 6 months of an employee's first exposure at or above the action level, the employer shall establish a valid baseline audiogram against which subsequent audiograms can be compared.

Testing to establish a baseline audiogram should be preceded by at least 14 hours without exposure to workplace noise. Hearing protectors may be used as a substitute for the requirement that baseline audiograms be preceded by 14 hours without exposure to workplace noise.

The employer shall notify employees of the need to avoid high levels of non-occupational noise exposure during the 14-hour period immediately preceding the audiometric examination.

Annual Audiogram: At least annually after obtaining the baseline audiogram, the employer shall obtain a new audiogram for each employee exposed at or above an 8-hour time-weighted average of 85 decibels.

Evaluation of Audiogram: Each employee's annual audiogram shall be compared to that employee's baseline audiogram to determine if the audiogram is valid and if a standard threshold shift has occurred. This comparison may be done by a technician.

If the annual audiogram shows that an employee has suffered a standard threshold shift, the employer may obtain a retest within 30 days and consider the results of the retest as the annual audiogram.

The audiologist, otolaryngologist, or physician shall review problem audiograms and shall determine whether there is a need for further evaluation. The employer shall provide to the person performing this evaluation the following information:

- A copy of the requirements for hearing conservation as set forth in this section;

- The baseline audiogram and most recent audiogram of the employee to be evaluated;
- Measurements of background sound pressure levels in the audiometric test room;
- Records of audiometer calibrations are required.

Follow-up Procedures: If a comparison of the annual audiogram to the baseline audiogram indicates a standard threshold shift the employee shall be informed of this fact in writing, within 21 days of the determination.

Unless a physician determines that the standard threshold shift is not work related or aggravated by occupational noise exposure, the employer shall ensure that the following steps are taken when a standard threshold shift occurs:

- Employees not using hearing protectors shall be fitted with hearing protectors, trained in their use and care, and required to use them.
- Employees already using hearing protectors shall be refitted and retrained in the use of hearing protectors and provided with hearing protectors offering greater attenuation if necessary.
- The employee shall be referred for a clinical audiological evaluation or an otological examination, as appropriate, if additional testing is necessary or if the employer suspects that a medical pathology of the ear is caused or aggravated by the wearing of hearing protectors.
- The employee is informed of the need for an otological examination if a medical pathology of the ear that is unrelated to the use of hearing protectors is suspected.

If subsequent audiometric testing of an employee whose exposure to noise is less than an 8-hour TWA of 90 decibels indicates that a standard threshold shift is not persistent, the employer:

- Shall inform the employee of the new audiometric interpretation; and
- May discontinue the required use of hearing protectors for that employee.

Standard Threshold Shift: A standard threshold shift is a change in hearing threshold relative to the baseline audiogram of an average of 10 dB or more at 2000, 3000, and 4000 Hz in either ear. In determining whether a standard threshold shift has occurred, allowance may be made for the contribution of aging (presbycusis) to the change in hearing level by correcting the annual audiogram.

Hearing Protectors: Employers shall make hearing protectors available to all employees exposed to an 8-hour time-weighted average of 85 decibels or greater at no cost to the employees. Hearing protectors shall be replaced as necessary.

Employers shall ensure that hearing protectors are worn:

- By an employee who is required to wear personal protective equipment; and
- By any employee who is exposed to an 8-hour time-weighted average of 85 decibels or greater, and who:
  - Has not yet had a baseline audiogram; or
  - Has experienced a standard threshold shift.

Employees shall be given the opportunity to select their hearing protectors from a variety of suitable hearing protectors provided by the employer.

The employer shall provide training in the use and care of all hearing protectors provided to employees.

The employer shall ensure proper initial fitting and supervise the correct use of all hearing protectors.

Training Program: The employer shall institute a training program for all employees who are exposed to noise at or above an 8-hour time-weighted average of 85 decibels, and shall ensure employee participation in such program.

The training program shall be repeated annually for each employee included in the hearing conservation program. Information provided in the training program shall be updated to be consistent with changes in protective equipment and work processes.

The employer shall ensure that each employee is informed of the following:

- The effects of noise on hearing;
- The purpose of hearing protectors, the advantages, disadvantages, and attenuation of various types, and instructions on selection, fitting, use, and care; and

- The purpose of audiometric testing, and an explanation of the test procedures.

The employer shall make available to affected employees or their representatives copies of this standard and shall also post a copy in the workplace.

2.2.3.3 *Recordkeeping*

2.2.3.3.1 Exposure Measurements

The employer shall maintain an accurate record of all employee exposure measurements described in this section.

2.2.3.3.2 Audiometric Tests

The employer shall retain all employee audiometric test records described in this section: This record shall include:

- Name and job classification of the employee;
- Date of the audiogram;
- The examiner's name;
- Date of the last acoustic or exhaustive calibration of the audiometer; and
- Employee's most recent noise exposure assessment.
- The employer shall maintain accurate records of the measurements of the background sound pressure levels in audiometric test rooms.

2.2.3.3.3 Record Retention

The employer shall retain records for at least the following periods:

- Noise exposure measurement records shall be retained for two years.
- Audiometric test records shall be retained for the duration of the affected employee's employment.

2.2.3.3.4 Access to Records

All records required by this section shall be provided upon request to employees, former employees, and representatives designated by the individual employee.

2.2.3.3.5 Transfer of Records

If the employer ceases to do business, the employer shall transfer to the successor employer all records required to be maintained by this section, and the successor employer shall retain them for the remainder of the period prescribed in this section.

2.2.4 Bibliography

The following references were consulted. Internet sources can be accessed by typing the identified reference on a well-known search engine such as "Google".

OSHA Standards

Recording and Reporting Occupational Injuries and Illnesses (29 CFR 1904)

- 1904.10, Recording Criteria for Cases Involving Occupational Hearing Loss (for Recordkeeping)

General Industry (29 CFR 1910)

- 1910 Subpart G, Occupational health and environmental control
  - 1910.95, Occupational noise exposure

Construction Industry (29 CFR 1926)

- 1926 Subpart D, Occupational health and environmental controls
  - 1926.52, Occupational noise exposure
- 1926 Subpart E, Personal protective and lifesaving equipment
  - 1926.101, Hearing protection

Additional References (Internet Resources)

- Acoustical Society of America
- American Industrial Hygiene Association (AIHA)
- American Speech-Language-Hearing Association
- Guide for Selection and Use of Personal Protective Equipment & Special Clothing for Foundry Operators.

OSHA and American Foundry Society (AFS) Safety and Health Committee (10Q), (2005), 914 KB PDF, 29 pages. This guide was developed by the American Foundry Society's (AFS) Safety and Health Committee (10Q) and is a product of the OSHA and AFS Alliance. It describes special considerations for the selection and use of personal protective equipment and special clothing for work situations in metal melting and pouring operations that present a risk of exposure to foundry hazards

- E-A-R Hearing Conservation (and Hearing Loss Prevention)
- Laborers' Health & Safety Fund of North America (Topics of Interest: Noise)
- National Hearing Conservation Association (NHCA) (see also OSHA's NHCA Alliance page)
- National Institute of Occupational Safety and Health (NIOSH) (Hearing Loss Prevention)

Printed Resources

- Alberti, P.W. Personal Hearing Protection in Industry. New York: Raven Press, 1982.
- Beranek, L.L. Noise and Vibration Control. New York: McGraw-Hill, 1971.
- Berger, E. H., et al. The Noise Manual. 5th ed. Fairfax, VA: American Industrial Hygiene Association, 2000.
- Cheremisinoff, N. P., Noise Control in Industry, Noyes Publications, Park Ridge, N.J. (1996).
- Code of Federal Regulations. Title 29, Parts 1900 to 1990. Washington, DC: US Government Printing Office, 1992.
- Harris, C.M. Handbook for Noise Control. 2nd ed. New York: McGraw-Hill, 1957.
- Harris, C.M. Shock and Vibration Handbook. 3rd ed. New York: McGraw-Hill, 1988.
- Kryter, K.D. The Effects of Noise on Man. New York: Academic Press, 1970.
- Kryter, K.D. The Effects of Noise on Man. 2nd ed. New York: Academic Press, 1985.
- Olishifski, P.E. and E.R. Harford. Industrial Noise and Hearing Conservation. Chicago: National Safety Council, 1975.

- Field Operations Manual. OSHA Directive CPL 02-00-045 [CPL 2.45B]. Washington, DC: US Government Printing Office, 1992.
- Royster, Julia and Larry. Guide 15: A Guide to Developing and Maintaining an Effective Hearing Conservation Program. Raleigh, NC: North Carolina Department of Labor, Division of Occupational Safety and Health, 1998.
- Sataloff, J., and P. Michael. Hearing Conservation. Springfield, IL: Charles C Thomas, 1973.
- Sataloff, R.T., and J.T. Sataloff. Occupational Hearing Loss. New York: Marcel Dekker, 1987.

2.3 Radiation Monitors and Meters

2.3.1 Introduction

Instruments used for radiation measurement fall into two broad categories:

- rate measuring instruments and
- personal dose measuring instruments.

Rate measuring instruments measure the rate at which exposure is received (more commonly called the radiation intensity). Survey meters, audible alarms and area monitors fall into this category. These instruments present a radiation intensity reading relative to time, such as R/hr or mR/hr. An analogy can be made between these instruments and the speedometer of a car because both are measuring units relative to time.

Dose measuring instruments are those that measure the total amount of exposure received during a measuring period. The dose measuring instruments, or dosimeters, that are commonly used in industrial radiography are small devices which are designed to be worn by an individual to measure the exposure received by the individual. An analogy can be made between these instruments and the odometer of a car because both are measuring accumulated units.

The following provides general information on the selection and application of field instruments for monitoring radiation sources in the work place.

2.3.2 Light

Description and Applications. The light meter is a portable unit designed to measure visible, UV, and near-UV light in the workplace.

The light meter is capable of reading any optical unit of energy or power level if the appropriate detector has been calibrated with the meter. The spectral range of the instrument is limited only by the choice of detector.

Steady-state measurements can be made from a steady-state source using the "normal operation" mode. Average measurements can be obtained from a flickering or modulated light source with the meter set in the "fast function" position. Flash measurements can be measured using the "integrate" function.

Calibration. No field calibration is available. These instruments are generally very stable and require only periodic calibration.

Special Considerations. Exposure of the photomultiplier to bright illumination when the power is applied can damage the sensitive cathode or conduct excessive current.

Maintenance. Little maintenance is required unless the unit is subjected to extreme conditions of corrosion or temperature. Clean the optical unit with lens paper to avoid scratching.

2.3.3 Ionizing Radiation

Description and Applications. The ionizing radiation survey meter is useful for measuring radon decay products from air samples collected on filters. Wipe samples collected on a filter can also be counted with this detector, and general area sampling can be done. Several types of ionizing radiation meters are commercially available.

The survey meter with the scintillation detector can be used to measure the presence of radon-decay products in a dust sample. The barometric pressure should be noted for ionizing radiation chambers.

Calibration. No field calibration is available. Periodic calibration by a laboratory is essential and should be handled by the manufacturer.

2.3.4 Nonionizing Radiation

Description and Applications. Various nonionizing radiation survey meters are available for measuring electromagnetic fields.

The frequency ranges covered by instruments are: 10 Hz to 300 kHz, 0.5 MHz to 6000 MHz, 6 GHz to 40 GHz, and the 2.45 GHz microwave oven frequency. These instruments are capable of measuring the electric field strength (E-field), magnetic field strength (H-field), or both depending on the instrument.

Depending on the instrument, electromagnetic field strengths from power lines, transformers, video display terminals, RF induction heaters, RF heat sealers, radio & television transmitters, microwave ovens and other sources can be measured.

Calibration. No field calibration is available. Periodic calibration by the manufacturer is essential.

Special Considerations.

- Some of the instruments have an automatic instrument zeroing. Other instruments may require "zeroing" the instrument in a "zero-field" condition. Check the operating manual for guidance.
- Some units have a peak memory-hold circuit that retains the highest reading in memory.
- Some units operate with either electric (E) or magnetic (H) field probes based on diode-dipole antenna design. Total field strength is measured at the meter regardless of the field orientation or probe receiving angle. The diode-dipole antenna design of the probe is much more resistant to burnout from overload than the thermocouple design of probes used with other meters.

Maintenance. No field maintenance is required other than replacing the alkaline batteries when needed.

2.3.5 Survey Meters for Radiation Detection

There are many different models of survey meters available to measure radiation in the field. They all basically consist of a detector and a readout display. Analog and digital displays are available. Most of the survey meters used for industrial radiography use a gas filled detector.

Gas filled detectors consist of a gas filled cylinder with two electrodes. Sometimes, the cylinder itself acts as one electrode, and a needle or thin taut wire along the axis of the cylinder acts as the other electrode (Figure 2.3). A voltage is applied to the device so that the central needle or wire become an anode (+ charge) and the other electrode or cylinder wall becomes the cathode (- charge). The gas becomes ionized whenever the counter is brought near radioactive substances. The electric field created by the potential difference between the anode and cathode causes the electrons of each ion pair to move to the anode while the positively charged gas atom is drawn to the cathode. This results in an electrical signal that is amplified, correlated to exposure and displayed as a value.

Depending on the voltage applied between the anode and the cathode, the detector may be considered an ion chamber, a proportional counter, or a Geiger-Müller (GM) detector. Each of these types of detectors have their advantages and disadvantages. A brief summary of each of these detectors is as follows.

2.3.5.1 *Ion Chamber Counter*

Ion chambers have a relatively low voltage between the anode and cathode, which results in a collection of only the charges produced in the initial ionization event. This type of detector produces a weak output signal that corresponds to the number of ionization events. Higher energies and intensities of radiation will produce more ionization, which will result in a stronger output voltage.

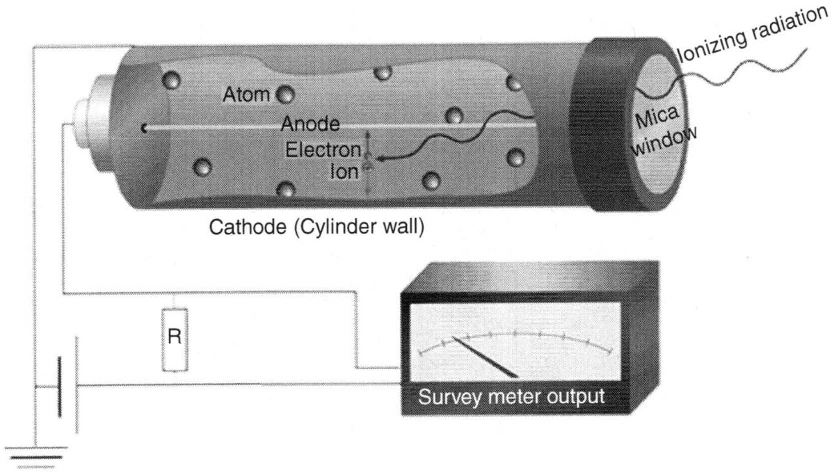

Figure 2.3 Gas filled detector.

Collection of only primary ions provides information on true radiation exposure (energy and intensity). However, the meters require sensitive electronics to amplify the signal, which makes them fairly expensive and delicate. The additional expense and required care is justified when it is necessary to make accurate radiation exposure measurements over a range of radiation energies. This might be necessary when measuring the Bremsstrahlung radiation produced by an X-ray generator. An ion chamber survey meter is sometimes used in the field when performing gamma radiography because it will provide accurate exposure measurements regardless of the radioactive isotope being used.

2.3.5.2 Proportional Counter

Proportional counter detectors use a slightly higher voltage between the anode and cathode. Due to the strong electrical field, the charges produced in the initial ionization are accelerated fast enough to ionize other electrons in the gas. The electrons produced in these secondary ion pairs, along with the primary electrons, continue to gain energy as they move towards the anode, and as they do, they produce more and more ionizations. The result is that each electron from a primary ion pair produces a cascade of ion pairs. This effect is known as gas multiplication or amplification. In this voltage regime, the number of particles liberated by secondary interactions is proportional to the number of ions produced by the passing ionizing particle. Hence, these gas ionization detectors are called proportional counters.

Like ion chamber detectors, proportional detectors discriminate between types of radiation. However, they require very stable electronics which are expensive and fragile. Proportional detectors are usually only used in a laboratory setting.

2.3.5.3 Geiger-Müller (GM) Counter

Geiger-Müller counters operate under even higher voltages between the anode and the cathode, usually in the 800 to 1200 volt range. Like the proportional counter, the high voltage accelerates the charges produced in the initial ionization to where they have enough energy to ionize other electrons in the gas. However, this cascading of ion pairs occurs to a much larger degree and continues until the counter is saturated with ions. This all happens in a fraction of a second and results in an electrical current pulse of constant voltage. The collection of the large number of secondary ions in the

GM region is known as an avalanche and produces a large voltage pulse. In other words, the size of the current pulse is independent of the size of the ionization event that produced it.

The electronic circuit of a GM counters counts and records the number of pulses and the information is often displayed in counts per minute. If the instrument has a speaker, the pulses can also produce an audible click. When the volume of gas in the chamber is completely ionized, ion collection stops until the electrical pulse discharges. Again, this only takes a fraction of a second, but this process slightly limits the rate at which individual events can be detected.Because they can display individual ionizing events, GM counters are generally more sensitive to low levels of radiation than ion chamber instruments. By means of calibration, the count rate can be displayed as the exposure rate over a specified energy range. When used for gamma radiography, GM meters are typically calibrated for the energy of the gamma radiation being used. Most often, gamma radiation from Cs-137 at 0.662 MeV provides the calibration. Only small errors occur when the radiographer uses Ir-192 (average energy about 0.34 MeV) or Co-60 (average energy about 1.25 MeV).

Since the Geiger-Müller counter produces many more electrons than an ion chamber counter or a proportional counter, it does not require the same level of electronic sophistication as other survey meters. This results in a meter that is relatively low cost and rugged. The disadvantages of GM survey meters are the lack of ability to account for different amounts of ionization caused by different energy photons and noncontinuous measurement (need to discharge).

2.3.5.4 *Comparison of Gas Filled Detectors*

In the ion chamber region, the voltage between the anode and cathode is relatively low and only primary ions are collected. In the proportional region, the voltage is higher, and primary ions and a number of secondary ions (proportional to the primary ions originally formed) are collected. In the GM region, a maximum number of secondary ions are collected when the gas around the anode is completely ionized. Note that discrimination between kinds of radiation (E1 and E2) is possible in the ion chamber and proportional regions. Radiation at different energy levels forms different numbers of primary ions in the detector. However in the GM region, the number of secondary ions collected per event remains the same no matter what the energy of the radiation that initiated the event. The GM counter gives up the ability to accurately measure the exposure due to different energies of radiation in exchange for a large signal pulse. This large signal pulse simplifies the electronics that are necessary for instruments such as survey meters.

2.3.6 Pocket Dosimeters

Pocket dosimeters are used to provide the wearer with an immediate reading of his or her exposure to x-rays and gamma rays. As the name implies, they are commonly worn in the pocket. The two types commonly used are the Direct Read Pocket Dosimeter and the Digital Electronic Dosimeter.

2.3.6.1 Direct Read Pocket Dosimeter

A direct reading pocket ionization dosimeter is generally of the size and shape of a fountain pen (Figure 2.4). The dosimeter contains a small ionization chamber with a volume of approximately two milliliters. Inside the ionization chamber is a central wire anode, and attached to this wire anode is a metal coated quartz fiber. When the anode is charged to a positive potential, the charge is distributed between the wire anode and quartz fiber. Electrostatic repulsion deflects the quartz fiber, and the greater the charge, the greater the deflection of the quartz fiber. Radiation incident on the chamber

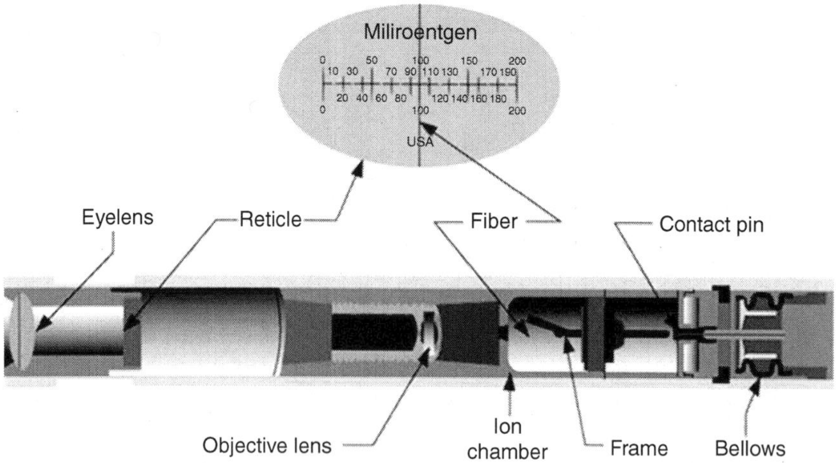

Figure 2.4 Direct read pocket dosimeter.

produces ionization inside the active volume of the chamber. The electrons produced by ionization are attracted to, and collected by, the positively charged central anode. This collection of electrons reduces the net positive charge and allows the quartz fiber to return in the direction of the original position. The amount of movement is directly proportional to the amount of ionization which occurs.

By pointing the instrument at a light source, the position of the fiber may be observed through a system of built-in lenses. The fiber is viewed on a translucent scale which is graduated in units of exposure. Typical industrial radiography pocket dosimeters have a full scale reading of 200 milliroentgens but there are designs that will record higher amounts. During the shift, the dosimeter reading should be checked frequently. The measured exposure should be recorded at the end of each shift.

The principal advantage of a pocket dosimeter is its ability to provide the wearer an immediate reading of his or her radiation exposure. It also has the advantage of being reusable. The limited range, inability to provide a permanent record, and the potential for discharging and reading loss due to dropping or bumping are a few of the main disadvantages of a pocket dosimeter. The dosimeters must be recharged and recorded at the start of each working shift. Charge leakage, or drift, can also affect the reading of a dosimeter. Leakage should be no greater than 2 percent of full scale in a 24 hour period.

2.3.6.2 *Digital Electronic Dosimeter*

Another type of pocket dosimeter is the Digital Electronic Dosimeter. These dosimeters record dose information and dose rate. These dosimeters most often use Geiger-Müller counters. The output of the radiation detector is collected and, when a predetermined exposure has been reached, the collected charge is discharged to trigger an electronic counter. The counter then displays the accumulated exposure and dose rate in digital form.

Some Digital Electronic Dosimeters include an audible alarm feature which emits an audible signal or chirp with each recorded increment of exposure. Some models can also be set to provide a continuous audible signal when a preset exposure has been reached. This format helps to minimize the reading errors associated with direct reading pocket ionization chamber dosimeters and allows the instrument to achieve a maximum readout before resetting is necessary.

2.3.7 Audible Alarm Rate Meters and Digital Electronic Dosimeters

Audible alarms are devices that emit a short "beep" or "chirp" when a predetermined exposure has been received. It is required that these electronic devices be worn by an individual working with gamma emitters. These devices reduce the likelihood of accidental exposures in industrial radiography by alerting the radiographer to dosages of radiation above a preset amount. Typical alarm rate meters will begin sounding in areas of 450-500 mR/h. It is important to note that audible alarms are not intended to be and should not be used as replacements for survey meters.

Most audible alarms use a Geiger-Müller detector. The output of the detector is collected, and when a predetermined exposure has been reached, this collected charge is discharged through a speaker. Hence, an audible "chirp" is emitted. Consequently, the frequency or chirp rate of the alarm is proportional to the radiation intensity. The chirp rate varies among different alarms from one chirp per milliroentgen to more than 100 chirps per milliroentgen.

2.3.8 Film Badges

Personnel dosimetry film badges (Figure 2.5) are commonly used to measure and record radiation exposure due to gamma rays, X-rays and beta particles. The detector is, as the name implies, a piece of radiation sensitive film. The film is packaged in a light proof, vapor proof envelope preventing light, moisture or chemical vapors from affecting the film.

A special film is used which is coated with two different emulsions. One side is coated with a large grain, fast emulsion that is sensitive to low levels of exposure. The other side of the film is coated with a fine grain, slow emulsion that is less sensitive to exposure. If the radiation exposure causes the fast emulsion in the processed film to be darkened to a degree that it cannot be interpreted, the fast emulsion is removed and the dose is computed using the slow emulsion.

The film is contained inside a film holder or badge. The badge incorporates a series of filters to determine the quality of the radiation. Radiation of a given energy is attenuated to a different extent by various types of absorbers. Therefore, the same quantity of radiation incident on the badge will produce a different degree of darkening under each filter. By comparing these results, the energy of the radiation can be determined and the dose can be calculated knowing the

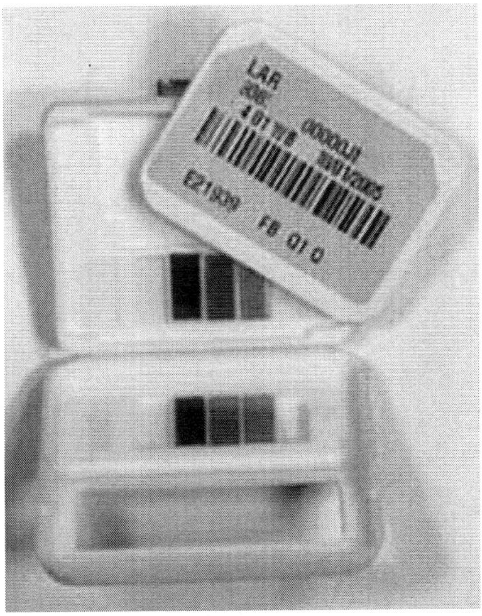

Figure 2.5 Personnel dosimerty film badge.

film response for that energy. The badge holder also contains an open window to determine radiation exposure due to beta particles. Beta particles are effectively shielded by a thin amount of material.

The major advantages of a film badge as a personnel monitoring device are that it provides a permanent record, it is able to distinguish between different energies of photons, and it can measure doses due to different types of radiation. It is quite accurate for exposures greater than 100 millirem. The major disadvantages are that it must be developed and read by a processor (which is time consuming), prolonged heat exposure can affect the film, and exposures of less than 20 millirem of gamma radiation cannot be accurately measured.

Film badges need to be worn correctly so that the dose they receive accurately represents the dose the wearer receives. Whole body badges are worn on the body between the neck and the waist, often on the belt or a shirt pocket. The clip-on badge is worn most often when performing X-ray or gamma radiography. The film badge may also be worn when working around a low curie source. Ring badges are worn on a finger of the hand most likely to be exposed to ionizing radiation. A LIXI system with its culminated and directional beam would be one example where monitoring the hands would be more important than the whole body.

2.3.9 Thermoluminescent Dosimeters

Thermoluminescent dosimeters (TLD) are often used instead of the film badge. Like a film badge, it is worn for a period of time (usually 3 months or less) and then must be processed to determine the dose received, if any. Thermoluminescent dosimeters can measure doses as low as 1 millirem, but under routine conditions their low-dose capability is approximately the same as for film badges. TLDs have a precision of approximately 15% for low doses. This precision improves to approximately 3% for high doses. The advantages of a TLD over other personnel monitors are its linearity of response to dose, its relative energy independence, and its sensitivity to low doses. It is also reusable, which is an advantage over film badges. However, no permanent record or re-readability is provided and an immediate, on the job readout is not possible.

A TLD is a phosphor, such as lithium fluoride (LiF) or calcium fluoride (CaF), in a solid crystal structure. When a TLD is exposed to ionizing radiation at ambient temperatures, the radiation interacts with the phosphor crystal and deposits all or part of the incident energy in that material. Some of the atoms in the material that absorb that energy become ionized, producing free electrons and areas lacking one or more electrons, called holes. Imperfections in the crystal lattice structure act as sites where free electrons can become trapped and locked into place.

Heating the crystal causes the crystal lattice to vibrate, releasing the trapped electrons in the process. Released electrons return to the original ground state, releasing the captured energy from ionization as light, hence the name thermoluminescent. Released light is counted using photomultiplier tubes and the number of photons counted is proportional to the quantity of radiation striking the phosphor.

Instead of reading the optical density (blackness) of a film, as is done with film badges, the amount of light released versus the heating of the individual pieces of thermoluminescent material is measured. The "glow curve" produced by this process is then related to the radiation exposure. The process can be repeated many times.

2.3.10 Annex – Guide to Meter Selection and Applications

Table 2.4 and Table 2.5 will help with the selection of appropriate meters through the comparison of their attributes and by their applications.

Table 2. 4 Radiation detection device selection chart.

Attribute	**Geiger-Mueller (GM) Tube**	**Ion Chamber**	**Scintillation**	**Proportional Counter**	**Semiconductor Detector**
Radiation Detected	alpha beta x ray gamma	alpha beta x ray gamma	alpha beta x ray gamma neutron	alpha beta x ray gamma neutron	alpha beta x ray gamma
Sensitive Medium	Gas	Gas	Solid Liquid	Gas	Solid
Ranges	0.04 mR/hr to 500 mR/hr	3 mR/hr to 10,000 R/hr	0.005 mR/hr to 200 mR/hr or to 800,000 c/m	to 500,000 c/m	to 100,000 c/m
Output Signal (V)	1	10^{-6}	1	10^{-2}	10^{-3}
Resolving Time (s)	10^{-4}	10^{-4}	10^{-7}	10^{-6}	10^{-9}
Energy Resolution (%)	N/A	N/A	10	15	1

Table 2. 4 (cont.) Radiation detection device selection chart.

Attribute	Geiger-Mueller (GM) Tube	Ion Chamber	Scintillation	Proportional Counter	Semiconductor Detector
Use	Low dose-rate surveys Area monitors Personnel radiation monitors Low-level contamination surveys	Medium and high dose-rate surveys Area monitors	Low-level contamination surveys LSC for tritium	Low-level contamination surveys Neutron survey	Laboratory Some field use
Advantages	Large output signal Moderate sensitivity	Low energy dependence Simple to operate	High sensitivity Rapid response Good energy resolution	Rapid response	Excellent energy response Short dead time
Disadvantages	Long dead time Energy dependent	Slow response Low sensitivity	Fragile Expensive	Requires stable high-voltage supply	Requires high amplification

Comments	Radiation detected depends on the type of GM tube may be energy dependent	Wide dose-rate range on a single instrument	High sensitivity Rapid response	Primary use is for alpha detection or neutron surveys	Primary use for alpha and gamma counting environmental samples
	Some models saturate—do not use in high radiation fields Sensitive to microwave fields Rate meter and audible pulse Rapid response Rugged, dependable	Low energy dependence Some models can be used in RF fields Some models slow to respond	Fragile Audible signal and rate meter Radiation detected depends on instrument and crystal Fast neutron detector where dose rate is not required	Alpha detector can discriminate between alpha and beta-gamma Neutron detector can discriminate against gamma radiation Maintenance may be a problem	May be used for in situ gamma fields

Table 2.5 Radiation meter application chart.

Type of Instrument	Measured Substance	Application
Heat stress meters	Ambient (environmental) heat	Foundries, furnaces, and ovens
Photoionization	Ionizable substances	Indoor air, leaks, spills
Light meters	Light	Indoor lighting, UV exposure
Nonionizing radiation meters	Nonionizing radiation	Communications, microwaves, heaters
Ionizing radiation meters	Ionizing radiation	Nuclear waste or plants
Electrostatic field tester	Static electric fields	Hazardous locations
RF instruments	Electromagnetic fields	RF heat sealers, VDT'S, induction motors

Source: http://hps.org/publicinformation/ate/faqs/devices.html

2.3.11 Bibliography

The following Internet references were consulted.

www.iop.org/EJ/abstract/0022-3735/20/11/029/
www.iop.org/EJ/article/0022-3735/20/11/029/jev20i11p1426.pdf
www.ki4u.com/nuclearsurvival/survival/detectors/index.htm
www.regtron.kfkipark.hu/history.htm
http://www.ndt-ed.org/EducationResources/CommunityCollege/RadiationSafety/radiation_safety_equipment/radiation_detectors.htm
Radiation safety videos available on the Internet can be found at: http://www.ndt-ed.org/EducationResources/CommunityCollege/RadiationSafety/Video/RTVideo.htm

2.4 Electrical/Electronics Testing Meters

2.4.1 Introduction

Electronic test equipment (called 'testgear') is used to create stimulus signals and capture responses from electronic devices under tests (DUTs). In this way, the proper operation of the DUT can be proven or faults in the device can be traced and repaired. Use of electronic test equipment is essential to any serious work on electronics systems.

Practical electronics engineering and assembly requires the use of many different kinds of electronic test equipment ranging from the very simple and inexpensive (such as a test light consisting of just a light bulb and a test lead) to extremely complex and sophisticated such as Automatic Test Equipment.

Generally, more advanced test gear is necessary when developing circuits and systems than is needed when doing production testing or when troubleshooting existing production units in the field.

The following is a general overview of the common test equipment that is available.

2.4.2 Electronic Test Equipment Types

2..4.2.1 *Basic Test Equipment*

The following items are used for basic measurement of voltages, currents, and components in the circuit under test.

- Voltmeter (Measures voltage)
- Ohmmeter (Measures resistance)
- Ammeter, e.g. Galvanometer or Milliameter (Measures current)
- Multimeter e.g., VOM (Volt-Ohm-Milliameter) or DVM (Digital "Volt" Meter) (Measures all of the above)

The following are used for stimulus of the circuit under test:

- Power supplies
- Signal generator
- Pulse generator

The following analyze the response of the circuit under test:

- Oscilloscope (Measures all of the above as they change over time)
- Frequency counter (Measures frequency)

Connecting it all together:

- Test probes

Other commonly used meters are:

- Solenoid voltmeter (Wiggy)
- Clamp meter (current transducer)
- Wheatstone bridge (Precisely measures resistance)
- Capacitance meter (Measures capacitance)
- EMF Meter (Measures Electric and Magnetic Fields)
- Electrometer (Measures charge)

2.4.2.2 *Probes*

These include:

- Multimeters
- RF probe
- Signal tracer

2.4.2.3 *Analyzers*

These include:

- Logic analyzer (Tests digital circuits)
- Spectrum analyzer (SA) (Measures spectral energy of signals)
- Vector signal analyzer (VSA) (Like the SA but it can also perform many more useful digital demodulation functions)
- Time-domain reflectometer for testing integrity of long cables

2.4.2.4 *Signal-generating devices*

These include:

- Signal generator
- Frequency synthesiser
- Function generator
- Pulse generator
- Signal injector

2.4.2.5 *Miscellaneous Devices*

These include:

- Continuity tester
- Cable tester
- Hipot tester
- Network analyzer (used to characterize components or complete computer networks)
- Test light
- Transistor tester

2.4.3 General Information on Safe Use

Description and Applications. Electrical testing meters include multimeters, clip-on current meters, megohmmeters, battery testers, ground-wire impedance testers, 120-V AC receptacle testers, ground fault interrupt testers, electrostatic meters, and AC voltage detectors.

Multimeters measure AC or DC voltage or current and resistance. They can check for AC leakage, proper line voltage, batteries, continuity, ground connection, integrity of shielded connections, fuses, etc.

Calibration. Few, if any, field calibrations are available. Check manufacturer's manual.

Maintenance. No field maintenance is required other than battery-pack servicing.

3

General Safety Practices

3.1 Safe Chemical Handling

3.1.1 Introduction

Health and safety standards have been established throughout the world to protect the safety, health and welfare of people engaged in work or employment. These standards also protect co-workers, family members, employers, customers, nearby communities, and other members of the public that may be impacted by the workplace environment. In the United States, several federal agencies, including the Occupational Safety and Health Administration (OSHA), the National Institute for Occupational Safety and Health (NIOSH), the American Conference of Governmental Industrial Hygienists (ACGIH), the Agency for Toxic Substances and Disease Registry (ATSDR), the Environmental Protection Agency (EPA), and the World Health Organization (WHO), have developed and recommended health and safety standards. These standards are internationally recognized as a basis to provide safe levels of protection for workers. This subsection provides a summary of the standards as recommended by the above agencies.

3.1.2 Reference Standards

3.1.2.1 Explanation of Material Safety Data Sheets (MSDS)

A material safety data sheet or MSDS is a multi-page document that contains information on the safe handling procedures and practices for chemical products. Employers should maintain a file of MSDS for all chemicals handled at a site, train and educate workers on understanding how to use the information in an MSDS, and make the file available and accessible to all employees. The MSDS for virtually any chemical product can be downloaded from multiple sources on the world wide web.

The following is a detailed explanation of the types of information found on a typical MSDS. The information is usually organized into distinct sections which may include:

I. Product Identification;
II. Component Data;
III. Precautions for Safe Handling and Storage;
IV. Physical Properties Data;
V. Personal Protective Equipment;
VI. Fire and Explosion Hazard Information;
VII. Reactivity Information;
VIII. First Aid;
IX. Toxicology and Health Information;
X. Transportation Information;
XI. Spill and Leak Procedures;
XII. Waste Disposal;
XIII. Additional Regulatory Status Information;
XIV. Additional Information; and
XV. Major References.

3.1.2.1.1 Product Identification

The product name and product code are used to identify the product. The file number and revision number identify the material safety data sheet (MSDS) itself. The chemical family or name and synonyms are given with formula when applicable. A brief use description of the product is presented below along with the UN hazard classifications.

3.1.2.1.2 Component Data

Most materials are evaluated to determine if they are hazardous. A hazardous chemical refers to any chemical that presents a physical hazard if it is combustible, flammable, pyrophoric, chemically unstable, water reactive or explosive, compressed gas, organic peroxide or other oxidizer.

A chemical may present a health hazard if exposure could result in acute or chronic adverse health effects. If it has been determined that the product is a health hazard then all components that present a health hazard and that comprise 1% or more of the material are listed in this section. Also, any component that is a carcinogen is listed if it comprises 0.1% or more of the product. If it has been determined that the product is a physical hazard, then any component that presents a physical hazard is listed. Components in a product that the manufacturer believes are not hazardous are often referred to as inert ingredients.

Normally, the chemical name and chemical abstracts service (CAS) numbers are used to identify a component. CAS numbers are assigned to chemicals and mixtures by the chemical abstracts service (published by the American Chemical Society) as a specific identification, but they are generally recognized internationally. Where the identity of a component is a trade secret, a descriptive name is used instead of the chemical name and a trade secret access number is given to that component. Disclosure of the identity of the trade secret component will be made to health professional upon request, subject to the conditions specified in the standard.

Exposure limits are given for each component where these have been established. Definitions of these exposure limits are as follow:

- **ACGIG TLV (Threshold Limit Value):** A term used by the American Conference of Governmental Industrial Hygienists to express the airborne concentration of a material to which nearly all persons can be exposed day after day without adverse effects. ACGIH expresses TLVs in three ways:
  - **TLV-TWA:** The allowable time-weighted average concentration for a normal 8-hour workday of a 40-hour workweek.
  - **TLV-STEL:** The Short-Term Exposure Limit, or maximum concentration for a continuous 15-minute exposure period. A maximum of four such periods per day, with at least 60 minutes between exposure periods are allowed, provided that the daily TLV is not exceeded.

- **TLV-C:** The Ceiling exposure limit; the concentration that should not be exceeded even instantaneously.
- **Skin:** A notation used to indicate that the stated substance may be absorbed by the skin, mucous membranes and eyes, either by air or direct contact, and that this additional exposure must be considered part of the total exposure to avoid exceeding the TLV for that substance.

The value quoted is the TWA unless another category is stated.

- **OSHA PEL (Permissible Exposure limits):** An exposure limit established by the Occupational Safety and Health Administration. May be a time-weighted average (TWA), short-term (STEL) or ceiling (C) exposure limit. A skin notation has the same meaning as for the TLV.

3.1.2.1.3 Precautions for Safe Handling and Storage
This section provides vital information for handling and storing a product. It is important that all recommendations be followed.

3.1.2.1.4 Physical Data
Knowledge of the physical properties of a substance is necessary for all safety and industrial hygiene decisions. Definitions of terms that apply to the physical data presented in this section are given below:

- **Freezing Point/ Melting Point:** The temperature at which a substance changes state from liquid to solid or solid to liquid. For mixtures, a range may be given.
- **Boiling Point:** The temperature at which a liquid changes to a vapor state at a given pressure. (Usually 760 mmHg, or one atmosphere). For mixtures, the initial boiling point or the boiling range may be given. Flammable materials with low boiling points generally present special fire hazards.
- **Decomposition Temp:** The temperature at which a substance will break down, or decompose, into smaller fragments.
- **Specific Gravity:** The weight of a material compared to the weight of an equal volume of water; an expression of the density (or heaviness) of the material. Example: if a volume of material weighs 3.6 kilograms, and an

equal volume of water weighs 4.5 kilograms, the material has a specific gravity of 0.8:

$$3.6 \text{ kg}/4.5 \text{ kg} = 0.8$$

Insoluble materials with a specific gravity of less than 1.0 may float in (or on) water. Insoluble materials with a specific gravity greater than 1.0 may sink in water. Most insoluble flammable liquids having a specific gravity of less than 1.0 will float on water, an important consideration for fire suppression.

- **Bulk Density:** Weight of material per unit volume.
- **pH:** A value presenting the acidity or alkalinity of an aqueous solution.

1 ----------------------- 7 -----------------------14

Acidic Neutral Alkaline

- **Vapor Pressure:** The pressure (usually expressed in millimeters of mercury) characteristic at any given temperature of a vapor in equilibrium with its liquid or solid form.
- **Solubility in Water:** The ability of a material to dissolve in water or another liquid. Solubility may be expressed as a ratio or may be described using words such as insoluble, very soluble or miscible.
- **Evaporation Rate:** The rate at which a particular material will vaporize (evaporate) when compared to the rate of vaporization of a known material. The evaporation rate can be useful in evaluating the health and fire hazards of a material. The known material is usually either normal butyl acetate or water, with a vaporization rate designated as 1.0.
- **Odor Threshold:** The odor threshold is the lowest concentration of a chemical in air that is detectable by smell. The ability to detect the odor of a chemical varies from person to person and depends on conditions such as the presence of other odorous materials. Odor cannot be used as a warning of unsafe conditions since workers may become used to the smell (adaptation), or the chemical may numb the sense of smell.

- **Vapor Density (Air = 1):** A relative comparison of the density of the vapor compared to the density of air (Air = 1). If the vapor density is greater than 1, then the vapor is heavier than air.
- **Molecular Weight:** The molecular weight of a chemical is the sum of the atomic weights of the atoms making up one molecule of the chemical.
- **Coefficient of Oil/Water Distribution:** If a substance which is soluble both in oil and in water is added to a two-phase oil/water system, then the ratio of the concentration of that substance in oil to its concentration in water is called the Coefficient of Oil/Water distribution.

3.1.2.1.5 Personal Protective Equipment Requirements

The proper use of personal protective equipment is of the utmost importance, and the guidelines presented in this section must be closely followed. Descriptions of specific equipment (goggles, gloves, respirators, etc.) required for routine use are given. Use of additional protective equipment, as required for fire-fighting and for spill and leak cleanup, is outlined in chapter 4.

Use of some products may require specific ventilation requirements. The following definitions apply to ventilation systems:

- **General Exhaust:** A system for exhausting air containing contaminants from a general work area. General exhaust may be referred to as dilution ventilation.
- **Local Exhaust:** A system for capturing and exhausting contaminants from the air at the point where the contaminants are produced (welding, grinding, sanding, other processes or operations). Typical local exhausts include the fume hood, canopy hood, slot bench, dust collector and other devices engineered to remove contaminants from workers' breathing zones.

3.1.2.1.6 Fire and Explosion Hazard Information

Most of the terms that follow are defined in 29 CFR1910.1200(c) which should be consulted for the complete text. Note that some of these same terms have different definitions for transportation information. Shortened forms of the definitions are as follow:

- **Explosive:** A chemical that causes an almost instantaneous release of gas and heat when subjected to certain conditions.

- **Incompatible Materials:** Materials that react with the product or with components of the product and may destroy the structure or function of a product; cause a fire, explosion or violent reaction; or cause the release of hazardous chemicals.
- **Extinguishing Media:** Agents which can put out fires involving the material. Common extinguishing agents are water, carbon dioxide, dry chemical, "alcohol" foam, and halogenated gases (Halons). It is important to know which extinguishers can be used so they can be made available at the worksite.
- **Oxidizer:** A material that gives up oxygen easily or can readily oxidize other materials. Examples of oxidizing agents are oxygen, chlorine and peroxide compounds. These chemicals will support a fire and are highly reactive
- **Pyrophoric:** A substance that burns spontaneously in air at a temperature of 54.4°C or below.
- **Flammable:** There are four classes of flammable chemicals as follows:
  - An aerosol flammable is an aerosol which yields a flame projection exceeding 45.7 centimeters or a flashback under certain test conditions.
  - A flammable gas is a gas which can ignite readily and burn rapidly or explosively.
  - A flammable liquid is any liquid having a flash point below 37.8°C, with the exception of mixtures in which 99% of the components have flash points of 37.8°C or higher.
  - A flammable solid is a solid (with certain exceptions) that is liable to cause fire through friction, absorption of water or other reasons, or which can be ignited readily, and when ignited burns in such a manner as to create a serious hazard.
- **Combustible:** A combustible liquid is any liquid having a flash point at or above 37.8°C, but below 93.3°C, with the exception of mixtures in which 99% of the components have flash points of 93.3°C or higher.
- **Flash Point:** The temperature at which a liquid will give off enough flammable vapors to ignite in the presence of an ignition source.

There are several flash point test methods. Because flash points may vary for the same material depending on the method used, the method is indicated when the flash point is given. The methods most frequently quoted are:

- **PMCC:** Pensky-Martens Closed Cup – ASTM D93;
- **SETA**: Setaflash Closed Cup – ASTM D3278; and
- **TCC:** Tag (Tagliabue) Closed Cup – ASTM D56.

Details of these methods can be found in Section B of the Annual Book of ASTM Standards.

- **Autoignition Temperature:** The lowest temperature at which a liquid will give off enough flammable vapors and heat energy to ignite spontaneously and maintain combustion.
- **UEL and LEL:** Upper Explosive Limit and Lower Explosive Limit (sometimes referred to as Upper and Lower Flammable Limits) are the highest concentration and lowest concentration respectively that will produce a flash of fire when an ignition source is present. At higher concentrations than the UEL, the mixture is too "rich" to burn. At concentrations lower than the LEL, the mixture is too "lean" to burn.
- **NFPA Rating:** The U.S. National Fire Protection Association Standard System for the Identification of the Fire Hazards of Materials (NFPA 704M). The NFPA ratings provide a general idea both of the hazards and of the degree of the hazards associated with a material relative to fire protection and control. The Standard addresses the hazards under the three categories of "Health", "Flammability" and "Reactivity" and assigns numeric ratings using a scale of 0 to 4 with 0 indicating no particular hazard, and 4 the most hazardous. It should be noted that health hazard ratings refer specifically to short-term exposure under fire conditions. The Standard also includes provisions for special hazard warnings, such as water reactivity. For further details see 'Fire Protection Guide on Hazardous Materials' – National Fire Protection Association, Quincy, MA.
- **HMIS Ratings:** The Hazard Materials Identification System of the U.S. National Paint and Coatings

Association. The system is similar to the NFPA Standard in utilizing a 0-4 scale, rating the degree of hazard under the same three categories of health, flammability and reactivity, with 0 being the least hazardous and 4 the most. It should be noted that unlike NFPA ratings, HMIS ratings are not intended for emergency situations. The flammability and reactivity ratings will, however, usually be the same as the NFPA ratings. The health hazard rating is based on the acute toxicity of the chemical. For further information on these ratings, see 'HMIS Rating Manual' – National Paint and Coatings Association, Washington, DC.

3.1.2.1.7 Reactivity Information

A substance is said to be reactive if it readily enters into chemical reactions and undergoes chemical change. For MSDS purposes, the reactions can be grouped into three broad categories:

- **Decomposition** - Stable/Unstable. A substance is stable if it is resistant to decomposition or possesses the ability to remain unchanged. For MSDS purposes, a material is stable if it remains in the same form under expected and reasonable conditions of use. A substance is considered unstable if it tends to suffer decomposition under these conditions. Some materials may become unstable at higher temperatures. Whenever relevant, the temperature at which a material can be said to be unstable is stated. Other conditions that may cause instability, such as shock from dropping or static electricity, are noted when applicable.
- **Polymerization** - Hazardous Polymerization. A polymerization reaction is hazardous when it takes place at a rate that releases large amounts of energy. If hazardous polymerization can occur with a given material, the MSDS usually will list conditions that could start the reaction. In addition, since the material usually contains a polymerization inhibitor, the expected time period before the inhibitor is used up is also given.
- **Reactions with Other Chemicals** - Incompatible Materials. Materials that could cause dangerous reactions from direct contact with one another are described as incompatible. Common chemicals that react with

the product are usually listed in the MSDS. Hazardous products of decomposition, including combustion products, are listed.

3.1.2.1.8 First Aid

First aid procedures are described for each of the normal routes of exposure. It is important that first aid be administered as soon as possible after exposure has occurred. If in any doubt regarding the victim's condition, a physician should be called.

3.1.2.1.9 Toxicology and Health Information

The consequences of exposure, if any, by inhalation, skin or eye contact, or ingestion are outlined in this section. The signs, symptoms and effects that the exposure could produce are described so that any exposure would be recognized as quickly as possible and the appropriate action taken. The organs that are more susceptible to attack are referred to as target organs. The effects and damage that exposure could produce on these organs are given together with the symptoms. Some of the terms used that may be less familiar or which may have a specific inference in MSDS are defined below:

- **Acute Effect:** An adverse effect on a human or animal resulting from a single exposure with symptoms developing almost immediately after exposure. The effect is often of short duration.
- **Chronic Effect:** An adverse effect on a human or animal body resulting from repeated low level exposure, with symptoms that develop slowly over a long period of time or that reoccur frequently.
- **Corrosive:** A liquid or solid that causes visible destruction or irreversible alterations in human or animal tissue.
- **Irritation:** An inflammatory response or reaction of the eye, skin or respiratory system.
- **Allergic Sensitization:** A process whereby on first exposure a substance causes little or no reaction in humans or test animals, but which on repeated exposure may cause a marked response not necessarily limited to the contact site. Skin sensitization is the most common form of sensitization in the industrial setting, although respiratory sensitization is also known to occur.

- **Teratogen:** A substance or agent to which exposure of a pregnant female can result in malformations (birth defects) to the skeleton and or soft tissue of the fetus.
- **Mutagen:** A substance or agent capable of altering the genetic material in a living organism.
- **Carcinogen:** A substance or agent capable of causing or producing cancer in humans or animals. Authorities/organizations that have evaluated whether or not a substance is a carcinogen are the International Agency for Research on Cancer (IARC), the U.S. National Toxicology Program (NTP) and OSHA.
- **Target Organ Effects:** Chemically-caused effects upon organs and systems such as the liver, kidneys, nervous system, lungs, skin, and eyes from exposure to a material.

To evaluate the potential human effects from exposure to hazardous chemicals, studies on laboratory animals are performed. The terms most commonly used to define the results of the studies are as follows:

- **LD50 (Lethal Dose Fifty)** - The dose of a substance expected to cause the death of 50% of an experimental animal population. This dose may be from oral, dermal or other routes of exposure. The units given for the LD50 are usually milligrams per kilogram body weight of the tested animal (mg/kg).
- **LC50 (Lethal Concentration Fifty)** - A calculated concentration of a substance in air, exposure to which for specified length of time is expected to cause the death of 50% of a laboratory animal population. This concentration is usually in units of milligrams per cubic meter of air (mg/m^3) or milligrams per liter of air (mg/1) and is given for some time period (usually one or four hours).

Other terms occasionally used are:

- **LDLO (Lethal Dose Low)** - The lowest dose of substance introduced by any route other than inhalation reported to have caused death in humans or animals.

- **LCLO (Lethal Concentration Low)** - The lowest concentration of a substance in air that has been reported to have caused death in humans or animals.
- **TDLO (Toxic Dose Low)** - The lowest dose of a substance to which humans or animals have been exposed and reported to produce a toxic effect other than cancer.

3.1.2.1.10 Transportation Information

In the United States, chemical materials are regulated as hazardous by the U.S. Department of Transportation (DOT) and are described in the Hazardous Materials Regulations in the Code of Federal Regulations, 49 Chapter 1 subchapter C are outlined in the LAND portion of Section X.

3.1.2.1.11 Spill and Leak Procedures

Procedural recommendations relative to air, land and water are described. During cleanup of spills or leaks, it may be necessary to use extra personal protective equipment as compared to normal operations.

3.1.2.1.12 Waste Disposal

This section gives guidelines for disposing of a product if it becomes a waste. Recommendations are based upon the physical state and hazardous properties of the material. If the material is designated as hazardous it must be disposed of in a permitted hazardous waste treatment, storage, or disposal facility in accordance with local, state, and federal regulations. If the material is non-hazardous, recommendations for disposal are made depending on the physical state and known characteristics of the material.

3.1.2.1.13 Definition of Terms

This subsection defines common terms found on MSDSs for the reader. The terms are listed alphabetically. If the term you are looking for is not described, other good sources of information are a chemical dictionary or a regular dictionary. Due to the large number of medical terms which might be used on an MSDS, medical terms are not explained here. Advice about the meaning of medical terms can be obtained from someone fully familiar with their use, such as an occupational physician or nurse.

ACGIH - ACGIH stands for American Conference of Governmental Industrial Hygienists. The ACGIH is an association of occupational health professionals employed by government and educational institutions. The Threshold Limit Value (TLV) Committee and Ventilation Committee of the ACGIH publish guidelines which are used worldwide.

Acid, Acidic - See pH.

Active Ingredient - An active ingredient is the part of a product which actually does what the product is designed to do. It is not necessarily the largest or most hazardous part of the product. For example, an insecticidal spray may contain less than 1% pyrethrin, the ingredient which actually kills insects. The remaining ingredients are often called inert ingredients.

Acute - Acute means sudden or brief. Acute can be used to describe either an exposure or a health effect. An acute exposure is a short-term exposure. Short-term means lasting for minutes, hours or days. An acute health effect is an effect that develops either immediately or a short time after an exposure. Acute health effects may appear minutes, hours or even days after an exposure. (See also Chronic).

Aerosol - An aerosol is a collection of very small particles suspended in air. The particles can be liquid (mist) or solid (dust or fume). The term aerosol is also commonly used for a pressurized container (aerosol can) which is designed to release a fine spray of a material such as paint. Inhalation of aerosols is a common route of exposure to many chemicals. Also, aerosols may be fire hazards.

AIHA - AIHA stands for American Industrial Hygiene Association.

Alkali, Alkaline - See pH.

ANSI - ANSI stands for the American National Standards Institute.

Auto-Ignition Temperature - The auto-ignition temperature is the lowest temperature at which a material begins to burn in air in the absence of a spark or flame. Many chemicals will decompose (break down) when heated. The autoignition temperature is the temperature at which the chemicals formed by decomposition begin to burn.

Auto-ignition temperatures for a specific material can vary by one hundred degrees Celsius or more, depending on the test method used. Therefore, values listed on the MSDS may be rough estimates. To avoid the risk of fire or explosion, materials must be stored and handled at temperatures well below the auto-ignition temperature.

Base, Basic - See pH.

Biohazardous Infectious Material - Under the Canadian Controlled Products Regulations, a biohazardous infectious material is a material that contains organisms which can cause disease in humans or animals. For example, a person exposed to a blood sample from someone with hepatitis B may contract the disease. Some jurisdictions require MSDSs for products which contain biohazardous infectious materials.

BOD - BOD stands for biological oxygen demand.

Boiling Point - The boiling point is the temperature at which the material changes from a liquid to a gas. Below the boiling point, the liquid can evaporate to form a vapor. As the material approaches the boiling point, the change from liquid to vapor is rapid and vapour concentrations in the air can be extremely high. Airborne gases and vapors may pose fire, explosion and health hazards. Sometimes, the boiling point of a mixture is given as a range of temperatures. This is because the different ingredients in a mixture can boil at different temperatures. If the material decomposes (breaks down) without boiling, the temperature at which it decomposes may be given with the abbreviation "dec." Some of the decomposition chemicals may be hazardous. (See also Thermal Decomposition Products).

Carcinogen, Carcinogenic, Carcinogenicity - A carcinogen is a substance which can cause cancer. Carcinogenic means able to cause cancer. Carcinogenicity is the ability of a substance to cause cancer. Under the Canadian Controlled Products Regulations, materials are identified as carcinogens if they are recognized as carcinogens by the American Conference of Governmental Industrial Hygienists (ACGIH), or the International Agency for Research on Cancer (IARC). Under the U.S. OSHA Hazard Communication (HAZCOM) Standard, materials are identified as carcinogens on MSDSs if they are listed as either carcinogens or potential carcinogens by IARC or the U.S. National Toxicology Program (NTP), if they are regulated as carcinogens by OSHA, or if there is valid scientific evidence in man or animals demonstrating a cancer causing potential. The lists of carcinogens published by the IARC, ACGIH and NTP include known human carcinogens and some materials which cause cancer in animal experiments. Certain chemicals may be listed as suspect or possible carcinogens if the evidence is limited or so variable that a definite conclusion cannot be made.

Cas Registry Number - The CAS Registry Number is a number assigned to a material by the Chemical Abstracts Service (CAS) of the American Chemical Society (ACS). The CAS number provides a single unique identifier. A unique identifier is necessary because the same material can have many different names. For example, the name given to a specific chemical may vary from one language or country to another. The CAS Registry Number is similar to a telephone number and has no significance in terms of the chemical nature or hazards of the material. The CAS Registry Number can be used to locate additional information on the material, for example, when searching in books or chemical data bases.

CC - Depending on the context, CC can stand for closed cup, cubic centimetres or ceiling concentration.

CCC - CCC stands for Cleveland closed cup, a standard method of determining flash points.

Ceiling (C) - See Exposure Limits for a general explanation.

Chemical Family - The chemical family describes the general nature of the chemical. Chemicals belonging to the same family often share certain physical and chemical properties and toxic effects. However, there may also be important differences. For example, toluene and benzene both belong to the aromatic hydrocarbon family. However, benzene is a carcinogen, but toluene is not.

Chemical Formula - The chemical formula, sometimes called the molecular formula, tells which elements (carbon, hydrogen, oxygen, and so on) make up a chemical. It also gives the number of atoms of each element in one unit or molecule of the chemical. The chemical formula can be used to confirm the identity of ingredients or to indicate the presence of a potentially hazardous element.

For example, zinc yellow has the chemical formula $ZnCrO_4$, which shows that it contains not only zinc (Zn) but also chromium (Cr).

Chemical Name - The chemical name is a proper scientific name for an ingredient of a product. For example, the chemical name of the herbicide 2,4-D is 2,4-dichlorophenoxyacetic acid. The chemical name can be used to obtain additional information.

Chemical Reactivity - Chemical reactivity is the ability of a material to undergo a chemical change. A chemical reaction may occur under conditions such as heating, burning, contact with other chemicals, or exposure to light. Undesirable effects such as pressure buildup, temperature increase or formation of other hazardous chemicals may result. (See also Dangerously Reactive Material and Reactive Flammable Material).

Chronic - Chronic means long-term or prolonged. It can describe either an exposure or a health effect. A chronic exposure is a long-term exposure. Long-term means lasting for months or years. A chronic health effect is an adverse health effect resulting from long-term exposure or a persistent adverse health effect resulting from a short-term exposure. The Canadian Controlled Products Regulations describe technical criteria for identifying materials which cause chronic health effects. (See also Acute).

CNS - CNS stands for central nervous system.

COC - COC stands for Cleveland open cup, a standard method of determining flash points.

COD - COD stands for chemical oxygen demand.

Coefficient of Oil/Water Distribution - The coefficient of oil/water distribution, also called the partition coefficient (abbreviated as P), is the ratio of the solubility of a chemical in an oil to its solubility in water. The P value is typically presented as a logarithm of P (log P). It indicates how easily a chemical can be absorbed into or stored in the body. The P value is also used to help determine the effects of the chemical on the environment.

Combustible - Combustible means able to burn. Broadly speaking, a material is combustible if it can catch fire and burn. However, in many jurisdictions, the term combustible is given a specific regulatory meaning. (See Combustible Liquid). The terms combustible and flammable both describe the ability of a material to burn. Commonly, combustible materials are less easily ignited than flammable materials.

Combustible Liquid - Under the Canadian Controlled Products Regulations, a combustible liquid has a flash point from 37.8 to 93.3 degrees C using a closed cup test. The U.S. OSHA HAZCOM Standard uses a similar definition. This range of flash points is well above normal room temperature. Combustible liquids are, therefore, less of a fire hazard than flammable liquids. If there is a possibility that a combustible liquid will be heated to a temperature near its flash point, appropriate precautions must be taken to prevent a fire or explosion.

Compressed Gas - A compressed gas is a material which is a gas at normal room temperature and pressure but is packaged as a pressurized gas, pressurized liquid or refrigerated liquid. Regardless of whether a compressed gas is packaged in an aerosol can, a pressurized cylinder or a refrigerated container, it must be stored and handled very carefully. Puncturing or damaging the container or allowing the container to become hot may result in an explosion.

Controlled Products - A controlled product is defined as a material, product or substance which is imported or sold and meets the criteria for one or more of the following classes:

- Class A - Compressed Gas
- Class B - Flammable and Combustible Material:
  - Division 1 - Flammable Gas
  - Division 2 - Flammable Liquid
  - Division 3 - Combustible Liquid
  - Division 4 - Flammable Solid
  - Division 5 - Flammable Aerosol
  - Division 6 - Reactive Flammable Material
- Class C - Oxidizing Material
- Class D - Poisonous and Infectious Material:
  - Division 1 - Material Causing Immediate and Serious Toxic
    - Subdivision A - Very Toxic Material
    - Subdivision B - Toxic Material

  - Division 2 - Material Causing Other Toxic Effects:
    Subdivision A - Very Toxic Material
    Subdivision B - Toxic Material
  - Division 3 - Biohazardous Infectious Material
- Class E - Corrosive Material
- Class F - Dangerously Reactive Material

Corrosive Material - A corrosive material can attack (corrode) metals or human tissues such as the skin or eyes. Corrosive materials may cause metal containers or structural materials to become weak and eventually to leak or collapse. Corrosive materials can burn or destroy human tissues on contact and can cause effects such as permanent scarring or blindness.

CU M or CU.M - This stands for cubic meter.

Dangerously Reactive Material - A dangerously reactive material can react vigorously:

- with water to produce a very toxic gas;
- on its own by polymerization or decomposition; or
- under conditions of shock, or an increase in pressure or temperature.

ANSI defines a dangerously reactive material as one that is able to undergo a violent self-accelerating exothermic chemical reaction with common materials, or by itself.

A dangerously reactive material may cause a fire, explosion or other hazardous condition. It is very important to know which conditions (such as shock, heating or contact with water) may set off the dangerous reaction so that appropriate preventive measures can be taken.

Density - The density of a material is its weight for a given volume. Density is usually given in units of grams per millilitre (g/mL) or grams per cubic centimetre (g/cc). Density is closely related to specific gravity (relative density). The volume of a material in a container can be calculated from its density and weight.

Dilution Ventilation - See General Ventilation.

Embryo - An embryo is an organism in the early stages of its development prior to birth. In humans, the embryo is the developing child from conception to the end of the second month of pregnancy. (See also Fetus/Foetus).

Embryotoxic, Embryotoxicity - Embryotoxic means harmful to the embryo. Embryotoxicity is the ability of a substance to cause harm to the embryo. Embryotoxic effects are included as Target Organ Effects.

Engineering Controls - Engineering controls help reduce exposure to potential hazards either by isolating the hazard or by removing it from the work environment. Engineering controls include mechanical ventilation and process enclosure. They are important because they are built into the work process. Engineering controls are usually preferred to other control measures such as the use of personal protective equipment. Substitution of a less hazardous material or industrial process is the best way to reduce a hazard and is often considered to be a type of engineering control.

Evaporation Rate - The evaporation rate is a measure of how quickly the material becomes a vapour at normal room temperature. Usually, the evaporation rate is given in comparison to certain chemicals, such as butyl acetate, which evaporate fairly quickly. For example, the rate might be given as "0.5 (butyl acetate=1)." This means that, under specific conditions, 0.5 grams of the material evaporates during the same time that 1 gram of butyl acetate evaporates. Often, the evaporation rate is given only as greater or less than 1, which means the material evaporates faster or slower than the comparison chemical. In general, a hazardous material with a higher evaporation rate presents a greater hazard than a similar compound with a lower evaporation rate.

Explosion Data - Explosion data is information on the explosive properties of a material. Quantitative explosion data is seldom available and is usually given in descriptive terms such as low, moderate or high. The following types of information can be used to describe the explosive hazard of a material:

- Sensitivity to mechanical impact. This information indicates whether or not the material will burn or explode on shock (for example, dropping a package) or friction (for example, scooping up spilled material).

- Sensitivity to static discharge. This information indicates how readily the material can be ignited by an electric spark.

Detailed information is available on the properties of commercial explosives. A chemical is identified as explosive if it causes a sudden, almost instantaneous release of pressure, gas and heat when subjected to sudden shock, pressure or high temperature.

Explosive Limits - Explosive limits specify the concentration range of a material in air which will burn or explode in the presence of an ignition source (spark or flame). Explosive limits may also be called flammable limits or explosion limits. The lower explosive limit (LEL), or lower flammable limit (LFL), is the lowest concentration of gas or vapour which will burn or explode if ignited. The upper explosive limit (UEL), or upper flammable limit (UFL), is the highest concentration of gas or vapour which will burn or explode if ignited. From the LEL to the UEL, the mixture is explosive. Below the LEL, the mixture is too lean to burn. Above the UEL, the mixture is too rich to burn. However, concentrations above the UEL are still very dangerous because, if the concentration is lowered (for example, by introducing fresh air), it will enter the explosive range. In reality, explosive limits for a material vary since they depend on many factors such as air temperature. Therefore, the values given on an MSDS are approximate. The explosive limits are usually given as the percent by volume of the material in the air. One percent by volume is 10,000 ppm. For example, gasoline has a LEL of 1.4% and a UEL of 7.6%. This means that gasoline vapours at concentrations of 1.4% to 7.6% (14,000 to 76,000 ppm) are flammable or explosive.

Exposure Limits (or Occupational Exposure Limits (OELs)) - An exposure limit is the concentration of a chemical in the workplace air to which most people can be exposed without experiencing harmful effects. Exposure limits should not be taken as sharp dividing lines between safe and unsafe exposures. It is possible for a chemical to cause health effects, in some people, at concentrations lower than the exposure limit. Exposure limits have different names and different meanings depending on who developed them and whether or not they are legal limits. For example, Threshold Limit Values (TLVs) are exposure guidelines developed by the American Conference of Governmental Industrial Hygienists (ACGIH). They have been adopted by many Canadian governments as their legal limits. Permissible Exposure Limits (PELs) are legal exposure limits in the United States. Sometimes, a manufacturer will recommend an exposure limit for a material. Exposure limits have not been set for many chemicals, for many different reasons. For example, there may not be enough information available to set an exposure limit. Therefore, the absence of an exposure limit does not necessarily mean the material is not harmful. There are three different types of exposure limits in common use:

1. Time-weighted average (TWA) exposure limit is the time-weighted average concentration of a chemical in air for a normal 8-hour work day and 40-hour work week to which nearly all workers may be exposed day after day without harmful effects. Time-weighted average means that the average concentration has been calculated using the duration of exposure to different concentrations of the chemical during a specific time period. In this way, higher and lower exposures are averaged over the day or week.
2. Short-term exposure limit (STEL) is the average concentration to which workers can be exposed for a short period (usually 15 minutes) without experiencing irritation, long-term or irreversible tissue damage, or reduced alertness. The number of times the concentration reaches the STEL and the amount of time between these occurrences can also be restricted.
3. Ceiling (C) exposure limit is the concentration which should not be exceeded at any time. SKIN notation (SKIN) means that contact with the skin, eyes and moist tissues (for example, the mouth) can contribute to the overall exposure. The purpose of this notation is to suggest that measures be used to prevent absorption by these routes; for example, the use of protective gloves. If absorption occurs through the skin, then the airborne exposure limits are not relevant.

Extinguishing Media - Extinguishing media are agents which can put out fires involving the material. Common extinguishing agents are water, carbon dioxide, dry chemical, "alcohol" foam, and halogenated gases (Halons). It is important to know which extinguishers can be used so they can be made available at the worksite. It is also important to know which agents cannot be used since an incorrect extinguisher may not work or may create a more hazardous situation. If several materials are involved in a fire, an extinguisher effective for all of the materials should be used.

Fetotoxic, Fetotoxicity - Fetotoxic means the substance is harmful to the fetus/foetus. Fetotoxicity describes the ability of a substance to harm the fetus. (See also Embryotoxicity, Teratogenicity and Reproductive Effects).

Fetus/Foetus - A fetus is an organism in the later stages of development prior to birth. In humans, it is the unborn child from the end of the second month of pregnancy to birth. (See also Embryo).

First Aid - First aid is emergency care given immediately to an injured person. The purpose of first aid is to minimize injury and future disability. In serious cases, first aid may be necessary to keep the victim alive.

Flammable, Flammability - Flammable means able to ignite and burn readily. Flammability is the ability of a material to ignite and burn readily. (See also Combustible). Under the Canadian Controlled Products Regulations and the U.S. HAZCOM Standard, there are specific technical criteria for identifying flammable materials. (See Flammable Aerosol, Flammable Gas, Flammable Liquid, Flammable Solid and Reactive Flammable Material).

Flammable Aerosol - Under the Canadian Controlled Products Regulations, a material is identified as a flammable aerosol if it is packaged in an aerosol container which can release a flammable material. A flammable aerosol is hazardous because it may form a torch (explosive ignition of the spray) or because a fire fueled by the flammable aerosol may flash back.

Flammable and Combustible Material - A material may be classified as a flammable and combustible material if it meets specific criteria for a flammable gas, flammable liquid, combustible liquid, flammable solid, flammable aerosol or reactive flammable material.

Flammable Gas - A flammable gas is a gas which can ignite readily and burn rapidly or explosively. Under the Canadian Controlled Products Regulations and under the U.S. Hazard Communication Standard, there are certain technical criteria for the identification of materials as flammable gases for the purposes of each regulation. Flammable gases can be extremely hazardous in the workplace; for example:

- If the gas accumulates so that its lower explosive limit (LEL) is reached and if there is a source of ignition, an explosion may occur.
- If there is inadequate ventilation, flammable gases can travel a considerable distance to a source of ignition and flash back to the source of the gas.

Flammable Limits - See Explosive Limits.

Flammable Liquid - A flammable liquid gives off a vapour which can be readily ignited at normal working temperatures. Under the Canadian Controlled Products Regulations, a flammable liquid is a liquid with a flash point (using a closed cup test) below 37.8 degrees C. The U.S. Hazard Communication Standard uses a similar, but not identical, definition. Flammable liquids can be extremely hazardous in the workplace; for example:

- If there is inadequate ventilation, vapours can travel considerable distances to a source of ignition and flash back to the flammable liquid.
- It may be difficult to extinguish a burning flammable liquid with water because water may not be able to cool the liquid below its flash point.

Flammable Solid - A flammable solid is a material which can ignite readily and burn vigorously and persistently. There are certain technical criteria in the Canadian Controlled Products Regulations and in the U.S. OSHA Hazard Communication Standard for the identification of flammable solids for the purposes of each regulation. These criteria are based on ease of ignition and rate of burning.

Flammable solids may be hazardous because heat from friction (for example, surfaces rubbing together) or heat from processing may cause a fire. Flammable solids in the form of a dust or powder may be particularly hazardous because they may explode if ignited.

Flash Back - Flash back occurs when a trail of flammable gas, vapor or aerosol is ignited by a distant spark, flame or other source of ignition. The flame then travels back along the trail of gas, vapour or aerosol to its source. A serious fire or explosion could result.

Flash Point - The flash point is the lowest temperature at which a liquid or solid gives off enough vapour to form a flammable air-vapour mixture near its surface. The lower the flash point, the greater the fire hazard. The flash point is an approximate value and should not be taken as a sharp dividing line between safe and hazardous conditions. The flash point is determined by a variety of test methods which give different results. Two types of methods are abbreviated as OC (open cup) and CC (closed cup).

Freezing Point - See Melting Point.

Fumes - Fumes are very small, airborne, solid particles formed by the cooling of a hot vapour. For example, a hot zinc vapour may form when zinc-coated steel is welded. The vapour then condenses to form fine zinc fume as soon as it contacts the cool surrounding air. Fumes are smaller than dusts and are more easily breathed into the lungs.

Gas - Gas is a material without a specific shape or volume. Gases tend to occupy an entire space uniformly at normal room pressure and temperature. The terms vapour and fume are sometimes confused with gas.

General Ventilation - As used in an MSDS, general ventilation, also known as dilution ventilation, is the removal of contaminated air from the general area and the bringing in of clean air. This dilutes the amount of contaminant in the work environment. General ventilation is usually suggested for non-hazardous materials. (See also Mechanical Ventilation, Local Exhaust Ventilation and Ventilation).

GI - GI stands for gastrointestinal (relating to the stomach and intestines).

Hazard, Hazardous - Hazard is the potential for harmful effects. Hazardous means potentially harmful. The hazards of a material are evaluated by examining the properties of the material, such as toxicity, flammability and chemical reactivity, as well as how the material is used. How a material is used can vary greatly from workplace to workplace and, therefore, so can the hazard. In Canada and the U.S., the term hazardous is used by many different regulatory agencies. Definitions may vary. For example, OSHA defines a hazardous chemical as any chemical which is a physical hazard or a health hazard according to the OSHA Hazard Communication (HAZCOM) criteria.

Hazardous Combustion Products - Hazardous combustion products are chemicals which may be formed when a material burns. These chemicals may be toxic, flammable or have other hazards. The chemicals released and their amounts vary, depending upon conditions such as the temperature and the amount of air (or more specifically, oxygen) available. The combustion chemicals may be quite different from those formed by heating the same material during processing (thermal decomposition products). It is important to know which chemicals are formed by hazardous combustion in order to plan the response to a fire involving the material.

Hazardous Decomposition Products - Hazardous decomposition products are formed when a material decomposes (breaks down) because it is unstable, or reacts with common materials such as water or oxygen (in air). This information should be considered when planning storage and handling procedures.

Hazardous Ingredient - Under the Canadian Hazardous Products Act, a chemical must be listed in the Hazardous Ingredients Section of an MSDS if:

- it meets the criteria for a controlled product;
- it is on the Ingredient Disclosure List;
- there is no toxicological information available; or
- the supplier has reason to believe it might be hazardous.

Certain chemicals may be exempt from disclosure on an MSDS if they meet specific criteria set out in the Hazardous Materials Information Review Act.

Hazardous Polymerization - See Polymerize, Polymerization.

Hepatotoxin - Hepatotoxins are agents that can cause toxic effects on the liver.

Highly Toxic - Under the U.S. OSHA HAZCOM Standard, there are specific criteria for materials which must be identified as toxic. The corresponding term under Canadian WHMIS is "Very Toxic" (criteria are not the same).

HR - HR stands for hour.

IARC - IARC stands for the International Agency for Research on Cancer. IARC evaluates information on the carcinogenicity of chemicals, groups of chemicals and chemicals associated with certain industrial processes. IARC has published lists of chemicals which are generally recognized as human carcinogens, probable human carcinogens or carcinogens in animal tests.

IATA - IATA stands for International Air Transport Association.

IDLH - IDLH stands for Immediately Dangerous to Life or Health. For the purposes of respirator selection, NIOSH defines the IDLH concentration as the airborne concentration that poses a threat of exposure to airborne contaminants when that exposure is likely to cause death or immediate or delayed permanent adverse health effects or prevent escape from such an environment. The purpose of establishing an IDLH exposure concentration is to ensure that the worker can escape from a given contaminated environment in the event of failure of the respiratory protection equipment. In the event of failure of respiratory protective equipment, every effort should be made to exit immediately.

Impervious - On an MSDS, impervious is a term used to describe protective gloves and other protective clothing. If a material is impervious to a chemical, then that chemical cannot readily penetrate through the material or damage the material. Different materials are impervious (resistant) to different chemicals. No single material is impervious to all chemicals. If an MSDS recommends wearing impervious gloves, you need to know the type of material from which the gloves should be made. For example, neoprene gloves are impervious to butyl alcohol but not to ethyl alcohol.

Incompatible Materials - Incompatible materials can react with the product or with components of the product and may:

- destroy the structure or function of a product;
- cause a fire, explosion or violent reaction; or
- cause the release of hazardous chemicals.

Inert Ingredient - An inert ingredient is anything other than the active ingredient of a product. It may be a solvent, colorant, filler or dispersing agent. In some cases, inert ingredients may be hazardous.

Ingestion - Ingestion means taking a material into the body by mouth (swallowing).

Inhalation - Inhalation means taking a material into the body by breathing it in.

Irritancy, Irritation - Irritancy is the ability of a material to irritate the skin, eyes, nose, throat or any other part of the body that it contacts. Signs and symptoms of irritation include tearing in the eyes and reddening, swelling, itching and pain of the affected part of the body. Irritancy is often described as mild, moderate or severe, depending on the degree of irritation caused by a specific amount of the material. Irritancy may also be described by a number on a scale of 0 to 4, where 0 indicates no irritation and 4 means severe irritation. Irritancy is usually determined in animal experiments.

ISO - ISO stands for the International Standards Organization.

KG - KG stands for kilogram

Kow - Kow stands for octanol/water partition coefficient.

LC50 - LC stands for lethal concentration. LC50 is the concentration of a material in air which causes the death of 50% (one half) of a group of test animals. The material is inhaled over a set period of time, usually 1 or 4 hours. The LC50 helps determine the short-term poisoning potential of a material.

LD50 - LD stands for lethal dose. LD50 is the amount of a material, given all at once, which causes the death of 50% (one half) of a group of test animals. The LD50 can be determined for any route of entry, but dermal (applied to skin) and oral (given by mouth) LD50's are most common. The LD50 is one measure of the short-term poisoning potential of a material. (See also LC50).

LCLO - LCLO stands for lowest lethal airborne concentration tested. (See also LC50 and LD50).

LDLO - LDLO stands for lowest lethal dose tested. (See also LC50 and LD50).

LEL - See Explosive Limits.

LFL - See Explosive Limits.

Local Exhaust Ventilation - Local exhaust ventilation is the removal of contaminated air directly at its source. This type of ventilation can help reduce worker exposure to airborne materials more effectively than general ventilation. This is because it does not allow the material to enter the work environment. It is usually recommended for hazardous airborne materials. (See also Mechanical Ventilation and Ventilation).

Lower Explosion Limit - See Explosive Limits.

Lower Explosive Limit - See Explosive Limits.

Lower Flammable Limit - See Explosive Limits.

Material Causing Immediate and Serious Toxic Effects - The Canadian Controlled Products Regulations describe technical criteria for identifying materials which cause immediate and serious toxic effects. These criteria use information such as the LD50 or LC50 for a material. Based on the specific information, a material may be identified as toxic or very toxic in the class Poisonous and Infectious Material.

Material Causing Other Toxic Effects - The Canadian Controlled Products Regulations describe technical criteria for identifying materials which cause toxic effects such as skin or respiratory sensitization, mutagenicity and carcinogenicity. Based on the specific information, a material may be identified as toxic or very toxic in the class Poisonous and Infectious Material.

Means of Extinction - See Extinguishing Media.

Mechanical Ventilation - Mechanical ventilation is the movement of air by mechanical means (for example, a wall fan). There are two kinds of mechanical ventilation: general ventilation and local exhaust ventilation. (See also Ventilation).

Melting Point - The melting point is the temperature at which a solid material becomes a liquid. The freezing point is the temperature at which a liquid material becomes a solid. Usually one value or the other is given on the MSDS. It is important to know the freezing or melting point for storage and handling purposes. For example, a frozen or melted material may burst a container. As well, a change of physical state could alter the hazards of the material.

mg/m^3 - The abbreviation mg/m^3 stands for milligrams (mg) of a material per cubic metre (m^3) of air. It is a unit of metric measurement for concentration (weight/volume). The concentrations of any airborne chemical can be measured in mg/m^3, whether it is a solid, liquid, gas or vapor.

MIN - MIN can stand for minute or minimum.

Miscible - Miscible means able to be mixed. Two liquids are said to be miscible if they are partially or completely soluble in each other. Commonly, the term miscible is understood to mean that the two liquids are completely soluble in each other. (See also Solubility).

Mist - A mist is a collection of liquid droplets suspended in air. A mist can be formed when spraying or splashing a liquid. It can also be formed when a vapour condenses into liquid droplets in the air. (See also Aerosol).

ML - ML stands for millilitres (mL).

mm Hg - The abbreviation mm Hg stands for millimeters (mm) of mercury (Hg). It is a common unit of measurement for the pressure exerted by gases such as air. Normal atmospheric pressure is 760 mm Hg.

Molecular Formula - See Chemical Formula.

Molecular Weight - The molecular weight of a chemical is a number showing how heavy one molecule (or unit) of the chemical is compared to the lightest element, hydrogen, which has a weight of 1. The molecular weight has various technical uses, such as calculating conversions from parts per million (ppm) to milligrams per cubic metre (mg/m3) in air.

Mutagen, Mutagenic, Mutagenicity - A mutagen is a substance which can cause changes in the DNA of cells (mutations). Mutagenic means able to cause mutations. Mutagenicity is the ability of a substance to cause mutations. DNA determines the characteristics that children inherit from their parents. DNA also determines how cells in the body divide or reproduce. A number of mutagenicity tests are used to screen chemicals for possible carcinogenicity or reproductive effects. This is because there is some evidence that mutations may increase the risk of cancer and reproductive problems such as infertility or birth defects. However, mutagenicity test results are not very reliable predictors of these effects. One reason for this is that the human body can repair mutations while most mutagenicity tests cannot. Mutagenicity is included on MSDSs because it is an early indicator of potential hazard, and often there is very little other evidence available on possible carcinogenic or reproductive effects. The Canadian Controlled Products Regulations describes technical criteria for identifying materials which are mutagenic. The U.S. OSHA HAZCOM Standard includes mutagenic effects as reproductive target organ effects.

Natural Ventilation - Natural ventilation is a type of general ventilation which depends on natural instead of mechanical means for air movement. Natural ventilation can depend on the wind or the difference in temperature from one area to another to move air through a building. Therefore, it is unpredictable and unreliable. (See also Local Exhaust Ventilation, Mechanical Ventilation and Ventilation).

Nephrotoxins - Nephrotoxins are agents that can cause toxic effects on the kidney.

Neurotoxins - Neurotoxins are agents that can cause toxic effects on the nervous system.

NFPA - NFPA stands for National Fire Protection Association (U.S.).

NIOSH - NIOSH stands for National Institute for Occupational Safety and Health. NIOSH is a branch of the United States government which undertakes research and develops occupational health and safety standards.

NOEL - NOEL stands for No Observable Effect Level.

NOS - NOS stands for not otherwise specified.

NTP - NTP stands for National Toxicology Program. This program is part of the United States Department of Health and Human Services. The NTP has a large program for testing the potential carcinogenicity of chemicals. It also does many other types of studies on short-term and long-term health effects.

OC - OC stands for open cup.

Odor Threshold - The odor threshold is the lowest concentration of a chemical in air that is detectable by smell. The odor threshold should only be regarded as an estimate. This is because odor thresholds are commonly determined under controlled laboratory conditions using people trained in odor recognition. As well, in the workplace, the ability to detect the odor of a chemical varies from person to person and depends on conditions such as the presence of other odorous materials. Odor cannot be used as a warning of unsafe conditions since workers may become used to the smell (adaptation), or the chemical may numb the sense of smell, a process called olfactory fatigue. However, if the odor threshold for a chemical is well below its exposure limit, odor can be used to warn of a problem with your respirator.

OECD - OECD stands for Organization for Economic Cooperation and Development. The OECD is an international agency which supports programs designed to facilitate trade and development. The OECD has published "Guidelines for Testing of Chemicals." These guidelines contain recommended procedures for testing chemicals for toxic and environmental effects and for determining physical and chemical properties.

OEL - OEL stands for Occupational Exposure Limit. (See Exposure Limits for a general explanation).

OSHA - OSHA stands for Occupational Safety and Health Administration. It is the branch of the United States government which sets and enforces occupational health and safety regulations. For example, OSHA sets the legal exposure limits in the United States, which are called Permissible Exposure Limits (PELs). OSHA also specifies what information must be given on labels and Material Safety Data Sheets for materials which have been classified as hazardous using their criteria.

Oxidizing Agent, Oxidizing Material - An oxidizing agent or material gives up oxygen easily or can readily oxidize other materials. Examples of oxidizing agents are oxygen, chlorine and peroxide compounds. These chemicals will support a fire and are highly reactive. Under the Canadian Controlled Products Regulations and under the U.S. OSHA HAZCOM Standard, there are specific criteria for the classification of materials as oxidizing materials.

Particulates Not Otherwise Classified (PNOC) - Particulates not otherwise classified is a term defined by the ACGIH. It is used to describe particulates for which there is no evidence of specific toxic effects such as fibrosis or systemic effects. These materials are not to be considered inert, however, and can produce general toxic effects depending on the airborne concentration.

Partition Coefficient - See Coefficient of Oil/Water Distribution.

PEL - PEL stands for Permissible Exposure Limit. PELs are legal limits in the United States set by the Occupational Safety and Health Administration (OSHA). (See Exposure Limits for a general explanation).

Pensky-Martens Closed Cup - Pensky-Martens Closed Cup (PMCC) is a specific method for determining flash points.

Personal Protective Equipment - Personal protective equipment is clothing or devices worn to help isolate a person from direct exposure to a hazardous material or situation. Recommended personal protective equipment is often listed on an MSDS. This can include protective clothing, respiratory protection and eye protection. The use of personal protective equipment is the least preferred method of protection from hazardous exposures. It can be unreliable and, if it fails, the person can be left completely unprotected. This is why engineering controls are preferred. Sometimes, personal protective equipment may be needed along with engineering controls. For example, a ventilation system (an engineering control) reduces the inhalation hazard of a chemical, while gloves and an apron (personal protective equipment) reduce skin contact. In addition, personal protective equipment can be an important means of protection when engineering controls are not practical: for example, during an emergency or other temporary conditions such as maintenance operations.

pH - The pH is a measure of the acidity or basicity (alkalinity) of a material when dissolved in water. It is expressed on a scale from 0 to 14. Roughly, pH can be divided into the following ranges:

- pH 0 - 2 Strongly acidic
- pH 3 - 5 Weakly acidic
- pH 6 - 8 Neutral
- pH 9 - 11 Weakly basic
- pH 12 - 14 Strongly basic.

PMCC - See Pensky-Martens Closed Cup.

PNS - PNS stands for peripheral nervous system.

Poisonous And Infectious Material - Under the Canadian Controlled Products Regulations, a poisonous and infectious material is any material which meets the criteria for a material causing immediate and serious toxic effects, a material causing other toxic effects, or a biohazardous infectious material.

Polymer - A polymer is a natural or man-made material formed by combining units, called monomers, into long chains. The word polymer means many parts. Examples of polymers are starch (which has many sugar units), polyethylene (which has many ethylene units) and polystyrene (which has many styrene units). Most man-made polymers have low toxicity, low flammability and low chemical reactivity. In these ways, polymers tend to be less hazardous than the chemicals (monomers) from which they are made.

Polymerize, Polymerization - Polymerization is the process of forming a polymer by combining large numbers of chemical units or monomers into long chains. Polymerization can be used to make some useful materials. However, uncontrolled polymerization can be extremely hazardous. Some polymerization processes can release considerable heat, can generate enough pressure to burst a container or can be explosive. Some chemicals can polymerize on their own without warning. Others can polymerize upon contact with water, air or other common chemicals. Inhibitors are normally added to products to reduce or eliminate the possibility of uncontrolled polymerization. Most MSDSs have a section called "Hazardous Polymerization" which indicates whether hazardous polymerization reactions can occur.

Ppb - ppb stands for parts per billion.

Ppm - The abbreviation ppm stands for parts per million. It is a common unit of concentration of gases or vapour in air. For example, 1 ppm of a gas means that 1 unit of the gas is present for every 1 million units of air. One ppm is the same as 1 minute in 2 years or 1 cent in $10,000.

Process Enclosure - As used on an MSDS, process enclosure means that the operation in which the material is used is completely enclosed. A physical barrier separates the worker from the potential health or fire hazard. Process enclosure is usually recommended if the material is very toxic or flammable.

Pyrophoric - Pyrophoric chemicals are defined in the U.S. OSHA HAZCOM standard as chemicals which will ignite spontaneously in air at a temperature of 54.4 degrees C or below. Regulatory definitions in other jurisdictions may differ.

Reactive Flammable Material - Under the Canadian Controlled Products Regulations, a reactive flammable material is a material which is a dangerous fire risk because it can react readily with air or water. This category includes any material which:

- is spontaneously combustible, that is, a material which can react with air until enough heat builds up that it begins to burn;
- can react vigorously with air under normal conditions without actually catching fire;
- gives off dangerous quantities of flammable gas on reaction with water; or
- becomes spontaneously combustible when it contacts water or water vapour.

Reactive flammable materials must be kept dry and isolated from oxygen (in air) or other oxidizing agents. Therefore, they are often stored and handled in an atmosphere of unreactive gas, such as nitrogen or argon.

Relative Density - See Specific Gravity.

Reproductive Effects - Reproductive effects are problems in the reproductive process which may be caused by a substance. Possible reproductive effects include reduced fertility in the male or female, menstrual changes, miscarriage, embryotoxicity, fetotoxicity, teratogenicity, or harmful effects to the nursing infant from chemicals in breast milk. Most chemicals can cause reproductive effects if there is an extremely high exposure. In these cases, the exposed person would experience other noticeable signs and symptoms caused by the exposure. These signs and symptoms act as a warning of toxicity. Chemicals which cause reproductive effects in the absence of other significant harmful effects are regarded as true reproductive hazards. Very few workplace chemicals are known to be true reproductive hazards.

Reproductive Toxicity - The Canadian Controlled Products Regulations describe technical criteria for identifying materials which have reproductive toxicity. These criteria refer to adverse effects on fertility. (See also Reproductive Effects). Other jurisdictions likely have corresponding criteria, which may differ. Under the U.S. OSHA HAZCOM standard, reproductive toxicity is a target organ effect, and includes mutagens, embryotoxins, teratogens and reproductive toxins.

Respiratory Sensitization - See Sensitization.

RTECS - RTECS stands for Registry of Toxic Effects of Chemical Substances.

SARA - SARA stands for Superfund Amendments and Reauthorization Act of 1986 (U.S.).

SEC - SEC stands for second or section.

Sensitization - Sensitization is the development, over time, of an allergic reaction to a chemical. The chemical may cause a mild response on the first few exposures but, as the allergy develops, the response becomes worse with subsequent exposures. Eventually, even short exposures to low concentrations can cause a very severe reaction. There are two different types of occupational sensitization: skin and respiratory. Typical symptoms of skin sensitivity are swelling, redness, itching, pain, and blistering. Sensitization of the respiratory system may result in symptoms similar to a severe asthmatic attack. These symptoms include wheezing, difficulty in breathing, chest tightness, coughing and shortness of breath. The Canadian Controlled Products Regulations and the U.S. OSHA HAZCOM standard describe technical criteria for identifying materials which are respiratory tract sensitizers or skin sensitizers.

Skin Notation - See Exposure Limits for a general explanation.

Skin Sensitization - See Sensitization.

Solubility - Solubility is the ability of a material to dissolve in water or another liquid. Solubility may be expressed as a ratio or may be described using words such as insoluble, very soluble or miscible. Often, on an MSDS, the "Solubility" section describes solubility in water since water is the single most important industrial solvent. Solubility information is useful for planning spill clean-up and fire fighting procedures.

Solvent - A solvent is a material, usually a liquid, which is capable of dissolving another chemical. Chemicals commonly called solvents can dissolve many different chemicals. Examples of common solvents are water, ethanol, acetone, hexane and toluene.

Specific Gravity - Specific gravity is the ratio of the density of a material to the density of water. The density of water is about 1 gram per cubic centimetre (g/cc). Materials which are lighter than water (specific gravity less than 1.0) will float. Most materials have specific gravities exceeding 1.0, which means they are heavier than water and so will sink. Knowing the specific gravity is important for planning spill clean-up and fire fighting procedures. For example, a light flammable liquid such as gasoline may spread and, if ignited, burn on top of a water surface.

Stability - Stability is the ability of a material to remain unchanged in the presence of heat, moisture or air. An unstable material may decompose, polymerize, burn or explode under normal environmental conditions. Any indication that the material is unstable gives warning that special handling and storage precautions may be necessary.

STEL - STEL stands for Short-Term Exposure Limit. (See Exposure Limits for a general explanation).

STP - STP stands for Standard Temperature and Pressure (0 degrees Celsius and one atmosphere pressure).

Synergistic, Synergism - As used on an MSDS, synergism means that exposure to more than one chemical can result in health effects greater than expected when the effects of exposure to each chemical are added together. Very simply, it is like saying 1 + 1 = 3. When chemicals are synergistic, the potential hazards of the chemicals should be re-evaluated, taking their synergistic properties into consideration.

Synonyms - Synonyms are alternative names for the same chemical. For example, methanol and methyl hydrate are synonyms for methyl alcohol. Synonyms may help in locating additional information on a chemical.

Target Organ Effects - Under the U.S. OSHA HAZCOM Standard, chemicals are identified as having target organ effects if there is statistically significant evidence of an acute or chronic health effect determined in a scientifically valid study. The following agents would be included (note, the list is not all- inclusive): hepatotoxins, agents which damage the lungs (including irritants), agents which act on the hematopoietic system, neurotoxins, nephrotoxins, reproductive toxins (mutagens, embryotoxins, teratogens and reproductive toxins), cutaneous hazards (chemicals which affect the dermal layer of the skin) and eye hazards (chemicals which affect the eye or visual capacity). There are no maximum dose criteria for chronic toxicity studies, as specified in the Canadian Controlled Products Regulations.

TCC - TCC stands for Tagliabue closed cup; a standard method of determining flash points. Generally, this appears in abbreviated form as Tag closed cup.

TCLO - TCLO stands for lowest toxic airborne concentration tested (see also LCLO and LC50).

Teratogen, Teratogenic, Teratogenicity - A teratogen is a substance which can cause birth defects. Teratogenic means able to cause birth defects. Teratogenicity is the ability of a chemical to cause birth defects. Teratogenicity results from a harmful effect to the embryo or the fetus/foetus.

The Canadian Controlled Products Regulations describe technical criteria for identifying materials which have teratogenicity and embryotoxicity. (See also Reproductive Effects). Other jurisdictions may also have defined specific criteria. Under the U.S. OSHA HAZCOM Standard, materials which have teratogenic effects are included under reproductive Target Organ Effects.

Thermal Decomposition Products - Thermal decomposition products are chemicals which may be formed when the material is heated but does not burn. These chemicals may be toxic, flammable or have other hazards. The chemicals released and their amounts vary depending upon conditions such as the temperature. The thermal decomposition products may be quite different from the chemicals formed by burning the same material (hazardous combustion products). It is important to know which chemicals are formed by thermal decomposition because this information is used to plan ventilation requirements for processes where a material may be heated.

TLM - TLM stands for Threshold Limit, median (aquatic toxicity rating).

TLV - TLV stands for Threshold Limit Value. It is the occupational exposure limit established by the American Conference of Governmental Industrial Hygienists (ACGIH). TLV is a registered trademark of ACGIH. TLVs are adopted by some governments as their legal limits. (See Exposure Limits for a general explanation).

TLV-C - TLV-C stands for the ACGIH Threshold Limit Value-Ceiling. See also TLV.

TOC - TOC stands for Tagliabue open cup; a standard method of determining flash points. Generally, this appears in abbreviated form as Tag open cup.

Toxic, Toxicity - Toxic means able to cause harmful health effects. Toxicity is the ability of a substance to cause harmful health effects. Descriptions of toxicity (e.g. low, moderate, severe, etc). depend on the amount needed to cause an effect or the severity of the effect. Under the Canadian Controlled Products Regulations and the U.S. OSHA HAZCOM Standard, there are specific technical criteria for identifying a material as toxic for the purpose of each regulation. (See also Very Toxic and Highly Toxic).

Trade Name - A trade name is the name under which a product is commercially known. Some materials are sold under common names, such as Stoddard solvent or degreaser, or internationally recognized trade names, like Varsol. Trade names are sometimes identified by symbols such as (R) or (TM).

TWA - TWA stands for Time-Weighted Average. (See Exposure Limits for a general explanation).

UEL - See Explosive Limits.

UFL - See Explosive Limits.

uG - uG stands for microgram, a unit of mass.

UN - UN stands for United Nations. See also UN Number.

UN Number - UN number stands for United Nations number. The UN number is a four-digit number assigned to a potentially hazardous material (such as gasoline, UN 1203) or class of materials (such as corrosive liquids, UN 1760). These numbers are used by firefighters and other emergency response personnel for identification of materials during transportation emergencies.

UN (United Nations) numbers are internationally recognized. NA (North American) numbers are used only for shipments within Canada and the United States. PINs (Product Identification Numbers) are used in Canada. UN, NA and PIN numbers have the same uses.

Unstable (reactive) - Under the U.S. OSHA HAZCOM standard, a chemical is identified as unstable (reactive) if in the pure state, or as produced or transported, it will vigorously polymerize, decompose, condense, or will become self-reactive under conditions of shock, pressure or temperature.

Upper Explosion Limit - See Explosive Limits.

Upper Explosive Limit - See Explosive Limits.

Upper Flammable Limit - See Explosive Limits.

Vapor - A vapour is the gaseous form of a material which is normally solid or liquid at room temperature and pressure. Evaporation is the process by which a liquid is changed into a vapor. Sublimation is the process by which a solid is changed directly into the vapor state.

Vapor Density - Vapor density is the weight per unit volume of a pure gas or vapor. On an MSDS, the vapor density is commonly given as the ratio of the density of the gas or vapor to the density of air. The density of air is given a value of 1. Light gases (density less than 1) such as helium rise in air. If there is inadequate ventilation, heavy gases and vapors (density greater than 1) can accumulate in low-lying areas such as pits and along floors.

Vapor Pressure - Vapor pressure is a measure of the tendency of a material to form a vapor. The higher the vapor pressure, the higher the potential vapour concentration. In general, a material with a high vapor pressure is more likely to be an inhalation or fire hazard than a similar material with a lower vapor pressure.

Ventilation - Ventilation is the movement of air. One of the main purposes of ventilation is to remove contaminated air from the workplace. There are several different kinds of ventilation. (See General Ventilation, Local Exhaust Ventilation, Mechanical Ventilation and Natural Ventilation).

Very Toxic - Under the Canadian Controlled Products Regulations, there are specific technical criteria for identifying a very toxic material. There are specific criteria for short-term lethality, long-term toxicity, teratogenicity and embryotoxicity, reproductive toxicity, carcinogenicity, respiratory sensitization and mutagenicity. (See also Toxic). Under the U.S. OSHA HAZCOM Standard, the corresponding term is "highly toxic", which has a specific definition.

VOC - VOC stands for volatile organic compound.

Volatile, Volatility - Volatile means a material can evaporate. Volatility is the ability of a material to evaporate. The term volatile is commonly understood to mean that a material evaporates easily. On an MSDS, volatility is commonly expressed as the "% volatile." The percent volatile can vary from 0% (none of the material will evaporate) to 100% (all of the material will evaporate if given enough time). If a product contains volatile ingredients, there may be a need for ventilation and other precautions to control vapor concentrations.

Water Reactive - Under the U.S. OSHA HAZCOM standard, a chemical is identified as water reactive if it reacts with water to release a gas that is either flammable or presents a health hazard.

3.1.2.2 *Occupational Safety and Health Administration*

The United States Occupational Safety and Health Administration is an agency of the United States Department Labor signed into existence by President Richard Nixon in 1970 under the Occupational Safety and Health Act. The intent of enforcing standards for the workplace is to prevent work-related injuries, illnesses, and deaths.

Standards for approximately 500 chemicals have been designated by OSHA and can be found in Appendix A.[1],[2] Exposure to any of the substances listed in Appendix A is to be limited in accordance with the requirements listed in the tables. OSHA sets enforceable permissible exposure limits (PELs) for airborne chemicals to protect workers against the health effects of exposure to hazardous substances. PELs are regulatory limits on the amount or concentration of a substance in the air. They may also contain a skin designation. OSHA PELs are based on an 8-hour time weighted average (TWA) exposure. Existing PEL values can be found in the Code of Federal Regulations 29, section 1910.1000.

Ceiling values, the exposure limit is preceded by a (c), indicates that at no time can the exposure exceed the exposure limit for that substance. Acceptable ceiling concentrations indicate that exposure to any substance on Table 2 in Appendix A is not to exceed the exposure limit at anytime within 8 hours, except for a time period not exceeding the maximum duration and concentration allowed indicated in the "acceptable maximum peak above the acceptable ceiling concentration for an eight-hour shift. All other chemical exposures are 8-hour time weighted averages. Exposure to these substances can not exceed the exposure limit in any eight hour period.[3]

For example, suppose substance X has a 15ppm TWA, 30ppm ceiling, and 60ppm peak. An employee is exposed to a concentration of this substance above 40ppm but never above 60ppm for only a maximum period of 10 minutes. For the remainder of the eight hour shift, exposure can not exceed 15ppm so that the cumulative exposure does not exceed the weighted average of 15ppm.

The cumulative exposure for an 8-hour work shift is computed as follows:

$$E = (C_a T_a + C_b T_b + \ldots C_n T_n)/8$$

Where, E is the equivalent exposure for the working shift. C is the concentration during any period of time T where the concentration remains constant. T is the duration in hours of the exposure at the concentration C. The value of E shall not exceed the 8-hour time

1. Appendix A Table 1 reference.
2. Appendix A Table 2 reference.
3. http://www.osha.gov/pls/oshaweb/owadisp.show_document?p_table=STANDARDS&p_id=9991

weighted average specified in Subpart Z or 29 CFR Part 1910 for the substance involved.

To illustrate the use of the formula above, assume that Substance Z has an 8-hour time weighted average limit of 100ppm. Suppose an employee is subject to the following exposure:

Two hours exposure at 150ppm
Two hours exposure at 75ppm
Four hours exposure at 50ppm

Substitution into the formula gives:

$$(2 \times 150 + 2 \times 75 + 4 \times 50) \div 8 = 81.25\text{ppm}$$

Since 81.25ppm is less than 100ppm, the 8-hour time weighted average limit, the exposure is acceptable.

OSHA has also established Immediately Dangerous to Life and Health (IDLH) standards. IDLH is the concentration of an airborne chemical which may cause irreversible health effects or death. OSHA standards are broad enough to include oxygen deficient circumstances that are free from air contaminants (Table 3 Appendix A). Usually IDLH values are used to determine selection of proper breathing apparatus that must be available to workers in a given situation.[4]

Exposure limits have been used by industrial hygienists for over 50 years as a means of protecting worker health. By implementing these standards as a part of an occupational health and safety program, they are a primary tool in disease prevention.

3.1.2.3 National Institute for Occupational Safety and Health

The National Institute for Occupational Safety and Health (NIOSH) is the United States federal agency responsible for conducting research and making recommendations for the prevention of work-related injury and illness. NIOSH is part of the Centers for Disease Control and Prevention (CDC) within the U.S. Department of Health and Human Services.

The NIOSH Pocket Guide to Chemical Hazards is intended as a source of general industrial hygiene information for workers, employers, and occupational health professionals. The Pocket

---

4. http://en.wikipedia.org/wiki/

Guide presents key information and data in abbreviated tabular form for 677 chemicals or substance groupings (e.g., manganese compounds, tellurium compounds, inorganic tin compounds, etc.) that are found in the work environment. The industrial hygiene information found in the Pocket Guide should help users recognize and control occupational chemical hazards. The chemicals or substances contained in this revision include all substances for which the National Institute for Occupational Safety and Health (NIOSH) has recommended exposure limits (RELs) and recommended short term exposure limits (STEL).[5] These chemical standards have been complied in Table 4 Appendix A.

For NIOSH RELs, "TWA" indicates a time-weighted average concentration for up to a 10-hour workday during a 40-hour workweek. A short-term exposure limit (STEL) is designated by "ST" preceding the value; unless noted otherwise, the STEL is a 15-minute TWA exposure that should not be exceeded at any time during a workday. A ceiling REL is designated by "C" preceding the value; unless noted otherwise, the ceiling value should not be exceeded at any time.

In 1974, NIOSH (which is responsible for recommending health and safety standards) joined OSHA (whose jurisdictions include promulgation and enforcement activities) in developing a series of occupational health standards for substances with existing PELs. This joint effort was labeled the Standards Completion Program and involved the cooperative efforts of several contractors and personnel from various divisions within NIOSH and OSHA. The Standards Completion Program developed 380 substance-specific draft standards with supporting documentation that contained technical information and recommendations needed for the promulgation of new occupational health regulations.

Acting under the authority of the Occupational Safety and Health Act of 1970 (29 USC Chapter 15) and the Federal Mine Safety and Health Act of 1977 (30 USC Chapter 22), NIOSH develops and periodically revises recommended exposure limits (RELs) for hazardous substances or conditions in the workplace. NIOSH also recommends appropriate preventive measures to reduce or eliminate the adverse health and safety effects of these hazards. To formulate these recommendations, NIOSH evaluates all known and available medical, biological, engineering, chemical, trade, and other information relevant

---

5. http://www.cdc.gov/niosh/npg/pgintrod.html#chemicalname

to the hazard. These recommendations are then published and transmitted to OSHA and the Mine Safety and Health Administration (MSHA) for use in promulgating legal standards.

The purpose for establishing an IDLH value in the Standards Completion Program was to determine the airborne concentration from which a worker could escape without injury or irreversible health effects from an IDLH exposure in the event of the failure of respiratory protection equipment. The IDLH was considered a maximum concentration above which only a highly reliable breathing apparatus providing maximum worker protection should be permitted. In determining IDLH values, NIOSH considered the ability of a worker to escape without loss of life or irreversible health effects along with certain transient effects, such as severe eye or respiratory irritation, disorientation, and in coordination, which could prevent escape. As a safety margin, IDLH values are based on effects that might occur as a consequence of a 30-minute exposure. LEL indicates that the IDLH was based on 10% of the lower explosive limit for safety considerations even though the relevant toxicological data indicated that irreversible health effects or impairment of escape existed only at higher concentrations (see Tables 10A through 10C for recommened exposure levels and IDLH limits).[6]

3.1.2.4 *American Conference of Governmental Industrial Hygienists*

ACGIH is an organization comprised of industrial hygienists and environmental health and safety experts. In 1941 the ACGIH the Threshold Limit Values of Chemical Substances Committee. This group was charged with investigating, recommending, and annually reviewing exposure limits for chemical substances. It became a standing committee in 1944. Two years later, the organization adopted its first list of 148 exposure limits, then referred to as Maximum Allowable Concentrations. The term "Threshold Limit Values (TLVs)" was introduced in 1956. The first Documentation of the Threshold Limit Values was published in 1962 and is now in its seventh edition. Today's list of TLVs® includes 642 chemical substances and physical agents (Table 6A Appendix A).[7]

6. http://www.cdc.gov/niosh/npg/pgintrod.html#chemicalname
7. http://www.acgih.org/about/history.htm

3.1.2.5 Agency for Toxic Substances and Disease Registry

The Comprehensive Environmental Response, Compensation, and Liability Act (CERCLA) [42 U.S.C. 9604 et seq.], as amended by the Superfund Amendments and Reauthorization Act (SARA) [Pub. L. 99 499], requires that the Agency for Toxic Substances and Disease Registry (ATSDR) develop jointly with the U.S. Environmental Protection Agency (EPA), in order of priority, a list of hazardous substances most commonly found at facilities on the CERCLA National Priorities List (NPL) (42 U.S.C. 9604(i).1); prepare toxicological profiles for each substance included on the priority list of hazardous substances, and to ascertain significant human exposure levels (SHELs) for hazardous substances in the environment, and the associated acute, subacute, and chronic health effects (42 U.S.C. 9604(i).3); and assure the initiation of a research program to fill identified data needs associated with the substances (42 U.S.C. 9604(i).5).

The ATSDR Minimal Risk Levels (MRLs) were developed as an initial response to the mandate. Following discussions with scientists within the Department of Health and Human Services (HHS) and the EPA, ATSDR chose to adopt a practice similar to that of the EPA's Reference Dose (RfD) and Reference Concentration (RfC) for deriving substance specific health guidance levels for non neoplastic endpoints. An MRL is an estimate of the daily human exposure to a hazardous substance that is likely to be without appreciable risk of adverse noncancer health effects over a specified duration of exposure. These substance specific estimates, which are intended to serve as screening levels, are used by ATSDR health assessors and other responders to identify contaminants and potential health effects that may be of concern at hazardous waste sites. It is important to note that MRLs are not intended to define clean up or action levels for ATSDR or other Agencies (Table 6C Appendix A).

MRLs are intended to serve as a screening tool to help public health professionals decide where to look more closely. They may also be viewed as a mechanism to identify those hazardous waste sites that are not expected to cause adverse health effects. Most MRLs contain some degree of uncertainty because of the lack of precise toxicological information on the people who might be most sensitive (e.g., infants, elderly, and nutritionally or immunologically compromised) to effects of hazardous substances. ATSDR uses a conservative (i.e., protective) approach to address these uncertainties consistent with the public health principle of prevention.

Although human data are preferred, MRLs often must be based on animal studies because relevant human studies are lacking. In the absence of evidence to the contrary, ATSDR assumes that humans are more sensitive than animals to the effects of hazardous substances that certain persons may be particularly sensitive. Thus the resulting MRL may be as much as a hundredfold below levels shown to be nontoxic in laboratory animals.

Proposed MRLs undergo a rigorous review process. They are reviewed by the Health Effects/MRL Workgroup within the Division of Toxicology and Environmental Medicine; an expert panel of external peer reviewers; the agency wide MRL Workgroup, with participation from other federal agencies, including EPA; and are submitted for public comment through the toxicological profile public comment period. Each MRL is subject to change as new information becomes available concomitant with updating the toxicological profile of the substance. MRLs in the most recent toxicological profiles supersede previously published levels. To date, 130 inhalation MRLs, 219 oral MRLs and 8 external radiation MRLs have been derived. A listing of the current published MRLs by route and duration of exposure is provided in Table 6C Appendix A.

3.1.2.6 *World Health Organization*

The World Health Organization was established in 1948 as a specialized agency of the United Nations serving as the directing and coordinating authority for international health matters and public health. One of WHO's constitutional functions is to provide objective and reliable information and advice in the field of human health, a responsibility that it fulfils in part through its publications programs. Through its publications, the Organization seeks to support national health strategies and address the most pressing public health concerns. Recognizing the need of humans for clean air, in 1987 the WHO Regional Office for Europe published Air quality guidelines for Europe, containing health risk assessments of different chemical air contaminants.[8]

3.1.3 Safe Chemical Exposure Tables

See Appendix A.

8. http://www.euro.who.int/document/e71922.pdf

3.2 Job Hazards Analysis Assessment

3.2.1 Introduction

A job hazard analysis must accompany all project workplans. A job hazard analysis identifies hazards associated with work projects and worksites, and identifies protective equipment, modified work procedures, managerial controls and personal protective equipment needed.

Line officers and staff shall use the job hazard analysis to monitor the safety performance of work supervisors, crews or contractors. It serves as the standard against which actual safety performance is measured.

3.2.2 Responsibility and Tailgate Meetings

The person responsible for preparing project work plans, specified HSE Officers, shall complete a Job Hazard Analysis Form. A recommended Job Hazard Analysis Form is provided in Section 3.2.3. A line officer shall approve the job hazard analysis.

Work supervisors shall discuss the job hazards with crew members prior to beginning new projects or changing work sites. Identify any hazards not noted on the job hazard analysis. Discuss ways to reduce these hazards, including the use of protective equipment. Document these meetings. File documentation with other project work documents when the project is completed. Exhibit 1 is a suggested format for a work supervisor's tailgate safety meeting.

Exhibit 1
WORK SUPERVISOR'S TAILGATE MEETINGS

Instructions
To be completed by first line supervisor or work leader at the worksite prior to beginning job and when the hazards change due to a change in worksite location or other condition. Add any hazards that do not appear on the Job Hazard Analysis Form (see Sec 3.2.3).

Study/Project/Job ________ Work Leader/Supervisor ________

Describe Work:

____

____

____

Identify & List Hazards; Hazard Reducing Work Procedures Discussed With Crew:

____

____

____

Protective Equipment Required by:

____

____

Additional Protective Equipment Needed:

____

____

Start of Project (date) _______ Discussed with Crew (date) _______

To be filed at end of project with other project documents.

Signature of Work Supervisor:

____

Date: _______________

3.2.3 Hazards Assessment

3.2.3.1 Purpose of Hazards Assessment

A hazard assessment is an important element of a personal protective equipment (PPE) program because it produces the information needed to select the appropriate PPE for any hazards present or likely to be present at particular workplaces. Employer must have adequate capacity/capability of determining and evaluating the hazards of a particular workplace.

The hazard assessment itself should be viewed as a performance-oriented provision that simply requires management to use their awareness of workplace hazards to enable them to select the appropriate PPE for the work being performed.

Examples of equipment required to be provided by employers include but are not limited to:

- Welding or wire mesh gloves;
- Respirators;
- Hard hats;
- Specialty glasses and goggles such as those used for laser and ultraviolet radiation protection;
- Specialty foot protection such as metatarsal shoes and lineman's shoes with built-in gaffs [such as those used for climbing];
- Face shields;
- Rubber gloves, blankets, cover-ups; and
- Hot sticks and other live-line tools used by power generation workers.

The use of personal protective equipment (PPE) should be required to reduce employees' exposures to hazards when engineering the administrative controls are not feasible or effective in reducing these exposures to acceptable levels. Employers are required to determine all exposures to hazards in their workplace and determine if PPE should be used to protect their workers.

If PPE is to be used to reduce the exposure of employees to hazards, a PPE program should be initialized and maintained. The program should contain identification and evaluation of hazards in the workplace and define whether PPE is an appropriate control measure; if PPE is to be used, how it is selected, maintained and its use evaluated; training of employees using the PPE; and vigilance of the program to determine its effectiveness in preventing employee injury or illness. Other sections of this book provide general provisions and specific guidance on:

- General safety requirements
- Eye and face protection
- Respiratory protection
- Head protection
- Foot protection
- Hand protection
- Personal protective equipment.

The following is a recommended approach to conducting a hazard assessment for PPE. The approach also helps to assign a risk priority code to the hazard to determine the course of actions needed to take to control the hazard.

Follow the instructions as you conduct the hazard assessment and fill in the hazard assessment form. You can make copies of the form or customize it to fit the needs of the work place. This tool can also serve as written certification that a hazard assessment has been performed. Document the hazard assessment for PPE. Make sure that the blank fields at the bottom of the form (indicated by *) are filled out.

- * Name of your work place
- * Address of the work place where you are doing the hazard assessment
- * Name of person certifying that a workplace hazard assessment was done
- * Date the hazard assessment was done

3.2.3.2 *Instructions and Forms for Job Hazard Analysis Assessment for PPE*

The following is a recommended step by step approach to conducting the job hazard analysis assessment for PPE.

Step 1. Conduct a walk through survey of the work area. For each job/task step, note the presence of any of the hazard types listed in Table 3.1, their sources, and the body parts at risk. Fill out the left side of the hazard assessment form. Collect all the information you can. Note that this is a generalized approach. The table may be expanded or modified as appropriate.

In collecting the information, bear the following in mind:

- Look at all steps of a job and ask the employee if there are any variations in the job that are infrequently done and that you might have missed during your observation.
- For purposes of the assessment, assume that no PPE is being worn by the affected employees even though they may actually be wearing what they need to do the job safely.
- Note all observed hazards. This list in Table 3.1 does not cover all possible hazards that employees may face or for which personal protective equipment may be required. Noisy environments or those which may require respirators must be evaluated with appropriate test equipment to quantify the exposure level when overexposure is suspected.

Table 3.1 List of typical workplace hazards.

Type Hazard	General Description of Hazard Type
Impact	Person can strike an object or be struck by a moving or flying or falling object.
Penetration	Person can strike, be struck by, or fall upon an object or tool that would break the skin.
Crush or Pinch	An object(s) or machine may crush or pinch a body or body part.
Harmful Dust	Presence of dust that may cause irritation, or breathing or vision difficulty. May also have ignition potential.
Chemical	Exposure from spills, splashing, or other contact with chemical substances or harmful dusts that could cause illness, irritation, burns, asphyxiation, breathing or vision difficulty, or other toxic health effects. May also have ignition potential.
Heat	Exposure to radiant heat sources, splashes or spills of hot material, or work in hot environments.
Light (optical) Radiation	Exposure to strong light sources, glare, or intense light exposure which is a byproduct of a process.
Electrical Contact	Exposure to contact with or proximity to live or potentially live electrical objects.
Ergonomic Hazards	Repetitive movements, awkward postures, vibration, heavy lifting, etc.
Environmental Hazards	Conditions in the work place that could cause discomfort or negative health effects.

Step 2. Analyze the hazard. For each job task with a hazard source identified, use the job hazard analysis matrix table (Table 3.2) and review the hazard with the affected employee and supervisor. Fill out the right side of the hazard assessment form:

- Rate the **SEVERITY** of injury that would reasonably be expected to result from exposure to the hazard.

Table 3.2 Job hazard analysis matrix and risk priority factors.

Job Hazard Analysis Matrix						
Severity of Injury		**Probability of an Accident Occurring**				
Level	**Description**	**A Frequent**	**B Several Times**	**C Occasional**	**D Possible**	**E Extremely Improbable**
I	Fatal or Permanent Disability	1	1	1	2	3
II	Severe Illness or Injury	1	1	2	2	3
III	Minor Injury or Illness	2	2	2-3	3	3
IV	No Injury or Illness	3	3	3	3	3

Risk Priority		
Code	**Risk Level**	**Action Required**
1	High	Work activities must be suspended immediately until hazard can be eliminated or controlled or reduced to a lower level.
2	Medium	Job hazards are unacceptable and must be controlled by engineering, administrative, or personal protective equipment methods as soon as possible.
3	Low	No real or significant hazard exists. Controls are not required but may increase the comfort level of employees.

- Assign a **PROBABILITY** of an accident actually happening.
- Assign a **RISK CODE** based upon the intersection of the **SEVERITY** and **PROBABILITY** ratings on the matrix.

Step 3. Take action on the assessment. Depending on the assigned risk level/code (or risk priority), take the corresponding action according to Table 3.2:

- If Risk priority is LOW/3 for a task step, it requires no further action.
- If Risk priority is MEDIUM/1 - select and implement appropriate controls.
- If Risk priority is HIGH/1 - immediately stop the task step until appropriate controls can be implemented.

A high risk priority means that there is a reasonable to high probability that an employee will be killed or permanently disabled doing this task step and/or a high probability that the employee will suffer severe illness or injury.

Step 4. Select PPE: Try to reduce employee exposure to the hazard by first implementing engineering, work practice, and/or administrative controls. If PPE is assigned, it must be appropriately matched to the hazard to provide effective protection, durability, and proper fit to the worker. Note the control method to be implemented in the far right column. Guidance on selection of appropriate PPE is provided in Section 3.3.

Step 5. Certify the hazard assessment:

- Certify on the completed hazard assessment form that the hazard assessment has been performed and that the needed controls have been implemented.
- Incorporate any new PPE requirements that have developed into a written accident prevention program.

Exhibit 2 is an example of a form that can be copied and used to certify the assessment.

Exhibit 2.

Job Hazard Analysis for Personal Protective Equipment (PPE) Assessment Certification Form

Job/Task: ______________________ **Location:** ______________________

Job/Task Step	Hazard Type	Hazard Source	Body Parts At Risk	Severity	Probability	Risk Code	Control Method[1]

Note:

1 Engineering, work practice, and/or administrative hazard controls such as guarding must be used, if feasible, before requiring employees to use personal protective equipment.

Certification of Assessment

* Name of work place: ______________________ * Address: ______________________

*Assessment Conducted By: ______________________ Title: ______________________

*Date(s) of Assessment ______________

Implementation of Controls Approved By: ______________ Title: ______________ Date: ______________

3.2.3.3 *General Provisions for Performing Safety Inspections*

Inspections. Inspections identify and eliminate unsafe acts or conditions before accidents or injuries occur. Two types of planned inspections are made: facility safety and health inspections and workplace and project inspections. Facility safety and health inspections are the focus of this subsection.

Responsibility.

- The Health & Safety Officer (HSO) is responsible for facility safety and health inspections.
- Unit manager shall ensure that workplace and project inspections are done by persons knowledgeable of the work activity and who have had training in the electricity health and safety code standards, occupational safety, the inspection process, and use of inspection report forms.

Frequency of Workplace and Project Inspections. All regularly used facilities shall be inspected at least annually. Workplace and project inspections shall be done at least annually. Inspect high-risk workplaces more frequently.

High-Risk Workplaces. Any facility or operation that offers a high potential for injury, illness, or death due to the activity conducted at the location is a high-risk workplace. Examples are:

- Storage areas of:
  - All types of flammables, including paint.
  - Explosives, chemicals, toxicants, or radioactive materials.
  - Materials stacked in such manner that they could overload structural systems or fall upon employees working in the area.
- Vehicle and equipment maintenance, carpentry, and other shops.
- Any location housing an activity involving powered machinery, equipment, electrical devices, and other mechanical apparatus.

- Loading operations or operations involving frequent backing or overhead handling of material; for example, ware-housing and zone fire caches.
- Operations involving chemical storage or usage.
- All facilities used for housing personnel.
- All energized points and high energy transmission systems.

Inspection Content. Compare actual situations and conditions to accepted standards. Variations of actual conditions from these standards are deficiencies. These may be classified as either unsafe conditions or unsafe acts. These may be imminently dangerous, serious, or non-serious. This textbook serves as the basic source of standards for workplace and project inspections.

Critical Inspection Items. Certain parts of equipment and some operations are critical to safe use or safe completion of a job. When a critical part fails or is absent, the machine, equipment, or operation will fail with a high likelihood of losses. Inspect for the presence and safe operation of these critical items. Some examples are:

- Pop-off valve on hot water heater.
- Anti-flashback couplings on oxygen-acetylene torches.
- Proper storage of flammable liquids.
- Eyewash stations in chemical use areas.

Inspection Checklists. Several types of checklists and inspection forms are available from commercial sources. These can be customized to meet local conditions and improve the quality of inspections. Examples are included in this subsection and may be modified to fit the needs of the facility.

Use of Inspection Forms. Safety and Health Inspection Forms, compliance worksheets, and report narratives are suggested for documenting the inspections. These are designed to assist in documenting notices of unsafe or unhealthful working conditions. Computerized or electronically reproduced forms containing similar information may also be used.

3.2.3.4 Inspection Preparation

Inspectors shall prepare for inspections as follows:

1. If the unit is covered by a bargaining agreement, arrange for a union representative to accompany you on the inspection.
2. Review previous inspection reports, reports of unsafe conditions, accident reports, and the unit's program of work to become familiar with the operations of the site to be inspected.
3. Arrange for the unit manager or designee to accompany you.
4. Prepare inspection equipment.
5. Review applicable safety and health standards.
6. Review inspection checklists and other material.
7. Review material to be covered in the opening conference.

3.2.3.5 Hazard Classification

Hazards found during an inspection shall be classified so that managers can allocate time and dollars for their correction in order of priority based on the degree of danger present. Hazards shall be classified as: imminent danger, serious, and non-serious based on the following criteria.

- ***Imminent danger hazards*** would likely cause death, severe injury or high property losses immediately, or before the hazard can be eliminated through normal procedures. Immediate employee protection and abatement is required. An example is a leaking propane gas cylinder in crew quarters.
- ***Serious hazards*** are those in which there is high probability that serious injury, illness, or extensive property damage would result unless corrective action is taken. Abatement shall be accomplished within 14 days. An example is a broken stair tread.
- ***Non-serious hazards*** are those that could cause injury, illness, or property damage. Abatement shall be accomplished in 30 days. An example is a broken window in a workshop.

3.2.3.6 *Inspection Equipment*

Inspectors shall use test or sampling equipment when required to evaluate workplace conditions. Some basic types are listed below. Some situations and conditions require sophisticated equipment that may not be readily available. Contact the Regional Office or Station headquarters for assistance.

Basic Sampling Equipment

Type or Name	Application
Noise Meter	To measure noise levels of equipment or processes.
Detector Tube Pump	To measure levels of airborne contaminants with appropriate tubes.
Smoke Tubes	To identify air flow currents.
Velometer	To measure air flow such as in laboratory hoods.

Basic Inspection Equipment

Receptacle Tester	Indicates if an electrical receptacle box of 110 Volts is correctly wired.
Tape Measure	
Flashlight	
Clipboard	
Camera with Flash	

3.2.3.7 *Inspection Process*

The inspection process shall be a closed loop system to ensure correction of hazards. The following is a graphic display of a closed loop inspection program.

1. Conducting the Inspection
2. Opening Conference. Discuss the following:
    a. Purpose of the inspection.
    b. Items or areas of special interest.
    c. Involvement of employees or employee representatives.
    d. Inspection standards to be applied.
    e. Time and place for a closing conference.

3. Inspection Procedure. Inspection of a workplace or project must be planned to ensure that all situations and conditions are observed.
4. Plan the route of inspection within a workplace inspecting high hazard areas first. Develop an inspection pattern. Complete the inspection of each area before going on. A random pattern is not effective.
5. Wear required protective equipment.
6. Look for imminent danger and serious hazards, then non-serious hazards.
7. Note unsafe acts of employees, or indicators of unsafe acts, and point these out to the manager's representative.
8. Look for good things as well as deficiencies. Note well-organized areas or unique ways of accomplishing work safely.
9. Look for safety and health program promotional items, such as an active poster program, current safety plans, current bulletin boards, and awards.
10. Search for patterns of deficiencies that affect the entire worksite. An example would be water in all the air compressors. This is a strong indicator that no compressor drain program exists.
11. Don't make assumptions. Ask questions, try it, turn it on, shake it, open it up, look at it. Take your time. Be systematic.
12. Remove equipment and tools that are deficient or suspected of being deficient from service. Use appropriate forms for Ground Equipment Warning Tag, to mark them.
13. Photograph conditions that would be difficult to describe or that rapidly change.

3.2.3.8 Closeout Review

At the conclusion of the inspection, meet with the unit manager or representative. Inform the person of unsafe or unhealthful working conditions found during the inspection. Include other topics such as:

- Positive things found.
- The extent of deficiencies found, such as no air compressor drain program or no fire extinguisher inspection program.

- The list of red-tagged and removed-from-service tools and equipment.
- Noise readings on measured equipment.
- The unit manager's responsibilities to correct the hazards and the hazard classification ratings.
- Period of time for the abatement of the unsafe and unhealthful conditions.

3.2.3.9 *Abatement Plan*

If correction of the noted deficiency cannot be accomplished within 30 days of the inspection, the unit manager shall prepare an abatement plan and submit it to the next higher organizational level.

3.2.4 Hazard Assessment and PPE Checklists

This subsection contains checklists that will assist in assigning personal protective equipment, including, but is not limited to, safety glasses, goggles, hard hats, gloves, safety shoes, and heat or electrically resistant clothing. Electrical protective equipment, respiratory protection, hearing protection, PPE for construction site lasers, and equipment like safety belts, lifelines, lanyards, and safety nets, are addressed in separate checklists.

It is important to note that engineering controls should be the primary method of establishing a safe workplace. Personal protective equipment should only be used where engineering controls are not feasible.

Checklist 1. General Safety

General Requirements	Please Circle
Has a hazard assessment been conducted in the workplace to identify possible hazards that would require the use of PPE?	Y N N/A ??
Is there a written certification of hazard assessment which identifies the workplace evaluated, the person certifying that the evaluation has been performed, and the date(s) of the hazard assessment?	Y N N/A ??
Based on the hazards identified, has PPE been selected for all appropriate individuals?	Y N N/A ??

Checklist 1 (cont.) General Safety

General Requirements	Please Circle
Have individuals involved been informed of the PPE selection decisions?	Y N N/A ??
If PPE is necessary to prevent injury or impairment by exposure to chemical hazards, radiological hazards, or mechanical irritants through absorption, inhalation or physical contact, is it provided?	Y N N/A ??
Has the selected PPE been fitted to appropriate individuals?	Y N N/A ??
Is PPE maintained in a sanitary and reliable condition?	Y N N/A ??
Do appropriate individuals use the PPE selected?	Y N N/A ??
Is defective or damaged PPE removed from service immediately? (shall not be used)	Y N N/A ??

Training	
Has each individual who is required to use PPE been provided with training?	Y N N/A ??
Has training on PPE included all of the following elements: when PPE is necessary; what PPE is necessary; how to properly don, doff, adjust, and wear PPE; the limitations of the PPE; and the proper care, maintenance, useful life and disposal of the PPE.	Y N N/A ??
Have the trained individuals demonstrated an understanding of the training and the ability to use PPE properly before being allowed to perform work requiring the use of PPE?	Y N N/A ??
Are individuals retrained when there is reason to believe that they do not have the understanding or skill to use PPE properly?	Y N N/A ??
Is retraining conducted whenever changes in the workplace or changes in types of PPE make previous training obsolete?	Y N N/A ??

Checklist 1 (cont.) General Safety

Training	Please Circle
Is there written certification for each person who has received PPE training that includes the following: a statement indicating the person understood the training; the name of the person trained; the date(s) of the training; and the subject of the certification?	Y N N/A ??

Head, Foot, and Hand Protection	
Are protective helmets used wherever there is the possible danger of head injury from impact, or from falling or flying objects, or from electrical shock and burns?	Y N N/A ??
Do protective helmets that are used in the workplace that were purchased prior to July 5, 1994 meet the Standard Safety Requirements for Industrial Head Protection, ANSI Z89.1-1969?	Y N N/A ??
Do protective helmets that are used in the workplace that were purchased after July 5, 1994 meet the Standard for Personnel Protection – Protective Headwear for Industrial Workers – Requirements, ANSI Z89.1-1986?	Y N N/A ??
Is protective footwear used wherever there is the danger of foot injuries due to falling or rolling objects, or objects piercing the sole, and where feet are exposed to electrical hazards?	Y N N/A ??
Does protective footwear that is used in the workplace that was purchased prior to July 5, 1994 meet the requirements of the Standard for Men's Safety-Toe Footwear, ANSI Z41.1-1967?	Y N N/A ??
Does protective footwear that is used in the workplace that was purchased after July 5, 1994 meet the requirements of the Standard for Personal Protection--Protective Footwear, ANSI Z41-1991?	Y N N/A ??

Checklist 1 (cont.) General Safety

Head, Foot, and Hand Protection	Please Circle
Are appropriate protective gloves used wherever there is the danger to hands of exposure to hazards such as those from skin absorption of harmful substances, severe cuts or lacerations, severe abrasions, punctures, chemical burns, thermal burns, and harmful temperature extremes?	Y N N/A ??

Eye and Face Protection	
Are individuals issued and required to wear appropriate eye protective devices while participating or observing activities which present a potential eye safety hazard? Note: Eye potential hazards include: caustic or explosive chemicals or materials, hot liquids or solids, molten materials, welding operations of any type, repairing or servicing of vehicles, heat treatment or tempering of metals, the shaping of solid materials and laser device operation and experimentation.	Y N N/A ??
Do all protective eye and face devices purchased after July 5, 1994 comply with Z87.1-1989? Note: Regular prescription eye glasses do not meet this requirement. Goggles or other protective glasses meeting the Standard must be worn over-top prescription eye glasses.	Y N N/A ??

Posting Requirements	
Are all lab or shop entrances, areas and equipment requiring the use of PPE devices posted with a sign indicating this requirement?	Y N N/A ??

Checklist 2

Chemical Safety Checklist - The objective of this surveillance is to ensure that practices for handling, storing, and using chemicals provide effective protection for the health and safety of employees.

Surveillance Guideline
CHEMICAL SAFETY

Surveillance No.:______________
Facility: ______________________
Date Completed:______________

Activity 1: Evaluate the general hazard communication program, including documentation and training, MSDS availability, worker exposure control, etc.

Yes No N/A

___ ___ ___ 1. Is there a written hazard communication program which addresses labeling and other forms of warning, material data safety sheets (MSDS), and employee information and training?

___ ___ ___ 2. Is there a list of hazardous chemicals known to be present that also provides a reference to appropriate MSDSs?

___ ___ ___ 3. Is a hazard communication program available to all employees?

___ ___ ___ 4. Are MSDSs readily available during each work shift to employees in their workplace for each hazardous chemical?

___ ___ ___ 5. Do employees receive effective information and training on hazardous chemicals in their work area or with which they work?

- Does the training include the physical and health hazards of the chemicals including exposure limits?
- Does the training include measures employees can take to protect themselves from the chemical hazards (procedures, work practices, and PPE)?
- Does the training include the details of the hazard communication program, including an explanation of

the labeling system, MSDSs, and how employees can obtain and use hazard information?

___ ___ ___ 6. Have the employees been informed of the location and availability of the written hazard communication program, the required list of chemicals, and the location of MSDSs?

___ ___ ___ 7. Do employers ensure that employees do not exceed Permissible Exposure Levels (PEL) or American Conference of Governmental Industrial Hygienist Threshold Limit Values (TLV), whichever is more restrictive?

___ ___ ___ 8. Do employers measure employee exposure to chemicals if it is believed that action levels or PELs may have been exceeded?

___ ___ ___ 9. Do employers notify employees of monitoring results within 15 workdays of receipt of such information?

___ ___ ___ 10. Do employers have a Chemical Hygiene Plan for laboratory use of chemicals (laboratory use - laboratory scale, multiple chemicals/procedures, non-production chemical use, standard laboratory protective practices)?

- Is the Chemical Hygiene Plan capable of protecting employees from health hazards associated with hazardous chemicals?
- Does the Chemical Hygiene Plan provide provision for keeping exposures below limits?
- Is the Chemical Hygiene Plan readily available to employees?
- Does the Chemical Hygiene Plan identify standard operating procedures relevant to safety and health considerations for handling chemicals?
- Does the Chemical Hygiene Plan provide criteria used to determine and implement control measures to reduce employee exposure the hazardous chemicals?
- Does the Chemical Hygiene Plan provide requirements that fume hoods and other protective equipment function properly?
- Does the Chemical Hygiene Plan contain provisions for employee information and training relative to chemical hazards?

___ ___ ___ 13. Do training records substantiate that workers have received required training on hazard recognition and control related to chemical safety?

Activity 2: Observe chemical storage locations and storage practices.

Yes No N/A

___ ___ ___ 14. Is each container containing a hazardous chemical labeled, tagged or marked with the identity of the chemical and appropriate hazard warnings? (Labels are not required on portable containers into which chemicals are transferred from labeled containers and which are intended only for the immediate use of the employee who made the transfer.)

___ ___ ___15. Are the labels on containers legible and prominently displayed?

___ ___ ___ 16. Do containers and portable tanks containing flammable and combustible liquids meet UN hazard warning requirements?

___ ___ ___ 17. Is each portable tank containing flammable/combustible liquid provided with a means of venting to limit internal pressure to less than 68.94 kPa?

___ ___ ___ 18. Do flammable storage cabinets contain less than 227.1 liters of Class I or Class II liquids, or less than 454.1 liters of Class III liquids?

___ ___ ___ 19. Are flammable storage cabinets conspicuously labeled "Flammable Keep Fire Away"?

___ ___ ___20. Does the flammable storage cabinet meet the design requirements?

___ ___ ___ 21. Do flammable storage rooms meet the requirements of safe room design, quantity of liquid stored, ventilation, storage spacing requirements, electrical wiring, etc.?

___ ___ ___ 22. Are suitable fire control devices, such as fire extinguishers, available at locations where flammable or combustible liquids are stored?

___ ___ ___ 23. Are incompatible chemicals/materials segregated to prevent accidental contact with one another?

___ ___ ___ 24. Is the minimum amount of hazardous chemical necessary for work stored in the open in the laboratory work area?

___ ___ ___ 25. Are safety showers and/or an eyewash stations provided within the work area for immediate emergency use?

Activity 3: Observe work activities involving handling or use of chemicals.

Yes No N/A

___ ___ ___ 26. Are users of flammable/combustible liquids familiar with the hazard classification of the liquid?

___ ___ ___ 27. Is bonding used when transferring Class I flammable liquids between two conductive containers of greater volume than 4.2 liters?

___ ___ ___ 28. Has the maintenance work package or work activity been reviewed by representatives of the safety and/or environmental organizations for appropriate safety and environmental controls?

___ ___ ___ 29. Was an appropriate job safety analysis performed for the work activity, identifying the hazards and implementing controls?

___ ___ ___ 30. Do the work instructions address use of specific personnel protective equipment such as safety glasses with side shields, rubber gloves, aprons, etc.?

___ ___ ___ 31. Do the work instructions include procedures in sufficient detail to ensure the safety and health of the workers?

___ ___ ___ 32. Were potential chemical hazards discussed with workers during the pre-job brief?

___ ___ ___ 33. Are workers using the personal protective equipment specified in the work instructions or procedures while performing the activity?

___ ___ ___ 34. Do workers follow instructions/procedures prescribed in the work package?

___ ___ ___ 35. Can workers describe the chemical hazards associated with the work they are performing?

___ ___ ___ 36. Were workers provided the opportunity to review and understand the MSDSs for chemicals associated with the activity?

___ ___ ___ 37. Can workers describe the correct response to an emergency involving chemical hazards?

NOTE

The Facility Representative should avoid interrupting maintenance personnel performing work on systems containing hazardous chemicals. The Facility Representative should wait for opportune times to conduct business with Facility Operators.

OTHER:

NOTES/COMMENTS:

PERSONNEL CONTACTED:

PROCEDURES REVIEWED:

FINDINGS:
Finding No.:
Description:

OBSERVATIONS:
Observation No.:
Description:

FOLLOW-UP ITEMS:

CONTRACTOR MANAGEMENT DEBRIEFED AND RESULTS:

Signature: ______________________________ Date: _____/_____/_____
Facility Representative

3.2.5 Bibliography

Cheremisinoff, N. P., 1996Safety Management Practices for Hazardous Materials, Marcel Dekker Publishers, N.Y.

DOE 5480.4, *Environmental Protection, Safety and Health Protection Standards*

DOE O 440.1A, *Worker Protection Management for DOE Federal and Contractor Employees*

DOE G 440.1-1, *Worker Protection Management for DOE Federal and Contractor Employees Guide*

DOE G 440.1-3, *Occupational Exposure Assessment Implementing Guide*

DOE-HDBK-1100-96, *Chemical Hazard Analysis*

DOE-HDBK-1101-96, *Process Safety Management for Highly Hazardous Chemicals*

29 CFR 1910.106, *Flammable and Combustible Liquids*

29 CFR 1910.119, *Process Safety Management of Highly Hazardous Chemicals*

29 CFR 1910.1200, *Hazard Communication*

29 CFR 1910.1450, *Occupational Exposures to Hazardous Chemicals in Laboratories*

29 CFR 1910.132, Personal Protective Equipment, *General Requirements*

48 CFR 970.5204-2, *Integration of Environment, Safety, and Health Into Work Planning and Execution*

10 CFR 830.120, *Quality Assurance*

National Fire Protection Association, Standard 45, *Fire Protection for Laboratories Using Chemicals*

American Conference of Governmental Industrial Hygienists, *Threshold Limit Values for Chemical Substances and Physical Agents and Biological Exposure Indices*

3.3 Personal Protective Equipment

3.3.1 Introduction

The use of personal protective equipment (PPE) should be required to reduce employee exposure to hazards when engineering and administrative controls are not feasible or effective in reducing these exposures to acceptable levels. Employers are required to determine if PPE should be used to protect their workers.

If PPE is to be used, a PPE program should be implemented. This program should address the hazards present; the selection, maintenance, and use of PPE; the training of employees; and monitoring of the program to ensure its ongoing effectiveness.

3.3.2 General Provisions

Hazards exist in every workplace in many different forms: sharp edges, falling objects, flying sparks, chemicals, noise and a myriad of other potentially dangerous situations. Employers should protect their employees from workplace hazards that can cause injury. Controlling a hazard at its source is the best way to protect employees.

Depending on the hazard or workplace conditions, the use of engineering or work practice controls to manage or eliminate hazards to the greatest extent possible is recommended. For example, building a barrier between the hazard and the employees is an engineering control; changing the way in which employees perform their work is a work practice control.

When engineering, work practice and administrative controls are not feasible or do not provide sufficient protection, employers are required to provide personal protective equipment (PPE) to their employees and ensure its use. PPE is equipment worn to minimize exposure to a variety of hazards. Examples of PPE include such items as gloves, foot and eye protection, protective hearing devices (earplugs, muffs) hard hats, respirators and full body suits.

This subsection will help both employers and employees do the following:

- Understand the types of PPE,
- Know the basics of conducting a "hazard assessment" of the workplace,
- Select appropriate PPE for a variety of circumstances,
- Understand what kind of training is needed in the proper use and care of PPE.

The information in this guide is general in nature and does not address all workplace hazards or PPE requirements. The information, methods and procedures in this subsection are based on a number of references noted in the bibliography and may be considered general guidance. The employer should perform a job hazards assessment before deciding on the levels of protection required and in selecting specific PPE to match work environment conditions.

3.3.3 Requirements of a PPE Program

The objective of a PPE program is to ensure the greatest possible protection for employees in the workplace. Cooperative efforts of both employers and employees will help in establishing and maintaining a safe and healthful work environment.

Employers are responsible for:

- Performing a "job hazard assessment" of the workplace to identify and control physical and health hazards.
- Identifying and providing appropriate PPE for employees.
- Training employees in the use and care of the PPE.
- Maintaining PPE, including replacing worn or damaged PPE.
- Periodically reviewing, updating and evaluating the effectiveness of the PPE program.

Employees should:

- Properly wear PPE,
- Attend training sessions on PPE,
- Care for, clean and maintain PPE, and
- Inform a supervisor of the need to repair or replace PPE.

A PPE Program should be designed based on comprehensive safety and health program which should be unto itself based on identification of physical and health hazards in the workplace. This process is known as a "hazard assessment" or a "job hazards assessment."

Potential hazards may be physical or health-related and a comprehensive hazard assessment should identify hazards in both categories. Examples of physical hazards include moving objects, fluctuating temperatures, high intensity lighting, rolling or pinching objects, electrical connections and sharp edges. Examples of health hazards include overexposure to harmful dusts, chemicals or radiation.

The hazard assessment should begin with a walk-through survey of the facility to develop a list of potential hazards in the following basic hazard categories:

- Impact,
- Penetration,
- Compression (roll-over),
- Chemical,
- Heat/cold,
- Harmful dust,

- Light (optical) radiation, and
- Biologic.

In addition to noting the basic layout of the facility and reviewing any history of occupational illnesses or injuries, things to look for during the walk-through survey include:

- Sources of electricity.
- Sources of motion such as machines or processes where movement may exist that could result in an impact between workers and equipment.
- Sources of high temperatures that could result in burns, eye injuries or fire.
- Types of chemicals used in the workplace.
- Sources of harmful dusts.
- Sources of light radiation, such as welding, brazing, cutting, furnaces, heat treating, high intensity lights, etc.
- The potential for falling or dropping objects.
- Sharp objects that could poke, cut, stab or puncture.
- Biologic hazards such as blood or other potentially infected material.

When the walk-through is complete, the employer should organize and analyze the data so that it may be efficiently used in determining the proper types of PPE required at the worksite. The employer should become aware of the different types of PPE available and the levels of protection offered. It is a good idea to select PPE that will provide a level of protection greater than the minimum required to protect employees from hazards.

The workplace should be periodically reassessed for any changes in conditions, equipment or operating procedures that could affect occupational hazards. This periodic reassessment should also include a review of injury and illness records to spot any trends or areas of concern and taking appropriate corrective action. The suitability of existing PPE, including an evaluation of its condition and age, should be included in the reassessment.

Documentation of the hazard assessment is required through a written certification that includes the following information:

- Identification of the workplace evaluated;
- Name of the person conducting the assessment;
- Date of the assessment; and

- Identification of the document certifying completion of the hazard assessment.

3.3.4 Guidelines for PPE Selection

All PPE clothing and equipment should be of safe design and construction, and should be maintained in a clean and reliable fashion. Employers should take the fit and comfort of PPE into consideration when selecting appropriate items for their workplace.

3.3.5 Worker Training

Employers are required to train each employee who must use PPE. Employees must be trained to know at least the following:

- When PPE is necessary.
- What PPE is necessary.
- How to properly put on, take off, adjust and wear the PPE.
- The limitations of the PPE.
- Proper care, maintenance, useful life and disposal of PPE.

Employers should make sure that each employee demonstrates an understanding of the PPE training as well as the ability to properly wear and use PPE before they are allowed to perform work requiring the use of the PPE. If an employer believes that a previously trained employee is not demonstrating the proper understanding and skill level in the use of PPE, that employee should receive retraining. Other situations that require additional or retraining of employees include the following circumstances:

- changes in the workplace or in the type of required PPE that make prior training obsolete.

The employer must document the training of each employee required to wear or use PPE by preparing a certification containing the name of each employee trained, the date of training and a clear identification of the subject of the certification.

3.3.6 Eye and Face Protection

Employees can be exposed to a large number of hazards that pose danger to their eyes and face. Many occupational eye injuries occur because workers are not wearing any eye protection while others result from wearing improper or poorly fitting eye protection. Employers must be sure that their employees wear appropriate eye and face protection and that the selected form of protection is appropriate to the work being performed and properly fits each worker exposed to the hazard.

3.3.6.1 Prescription Lenses

Everyday use of prescription corrective lenses will not provide adequate protection against most occupational eye and face hazards, so employers must make sure that employees with corrective lenses either wear eye protection that incorporates the prescription into the design or wear additional eye protection over their prescription lenses. It is important to ensure that the protective eyewear does not disturb the proper positioning of the prescription lenses so that the employee's vision will not be inhibited or limited.

Employees who wear contact lenses must wear eye or face PPE when working in hazardous conditions.

3.3.6.2 Eye Protection for Exposed Workers

Eye protection should be routinely considered for use by carpenters, electricians, machinists, mechanics, millwrights, plumbers and pipefitters, sheetmetal workers and tinsmiths, assemblers, sanders, grinding machine operators, sawyers, welders, laborers, chemical process operators and handlers, and timber cutting and logging workers. Workers in other job categories should decide whether there is a need for eye and face PPE through a hazard assessment.

Examples of potential eye or face injuries include:

- Dust, dirt, metal or wood chips entering the eye from activities such as chipping, grinding, sawing, hammering, the use of power tools or even strong wind forces.
- Chemical splashes from corrosive substances, hot liquids, solvents or other hazardous solutions.

- Objects swinging into the eye or face, such as tree limbs, chains, tools or ropes.
- Radiant energy from welding, harmful rays from the use of lasers or other radiant light (as well as heat, glare, sparks, splash and flying particles).

3.3.6.3 *Types of Protection*

Selecting the most suitable eye and face protection for employees should take into consideration the following elements:

- Ability to protect against specific workplace hazards.
- Should fit properly and be reasonably comfortable to wear.
- Should provide unrestricted vision and movement.
- Should be durable and cleanable.
- Should allow unrestricted functioning of any other required PPE.

The eye and face protection selected for employee use must identify the manufacturer. Any new eye and face protective devices must comply with product safety standards of performance such as ANSI Z87.1-1989 or be at least as effective s these standard requires. The reader should examine product literature closely to ensure that equipment states performance being equivalent or exceeding an product performance standard.

3.3.6.4 *Welding Operations*

Table 3.3 shows the minimum protective shades for a variety of welding, cutting and brazing operations in general industry practice.

Table 3.3 Filter lenses for protection against radiant energy.

Welding Operation	Shade Number
Shielded metal-arc welding 1.6-, 2.4-, 3.2-, 4.0-mm diameter electrodes	10
Gas-shielded arc welding (nonferrous) 1.6-, 2.4-, 3.2-, 4.0-mm diameter electrodes	11

Table 3.3 (cont.) Filter lenses for protection against radiant energy.

Welding Operation	Shade Number
Gas-shielded arc welding (ferrous) 1.6-, 2.4-, 3.2-, 4.0-mm diameter electrodes	12
Shielded metal-arc welding 4.8-, 5.6-, 6.4-mm diameter electrodes 7.9-, 9.5-mm diameter electrodes	14
Atomic hydrogen welding	10–14
Carbon-arc welding	14
Soldering	2
Torch brazing	3 or 4
Light cutting, up to 25.4 mm	3 or 4
Medium cutting, 25.4 to 152.4 mm	4 or 5
Heavy cutting, more than 152.4 mm	5 or 6
Gas welding (light), up to 3.2-mm	4 or 5
Gas welding (medium), 3.2- to 12.7-mm	5 or 6
Gas welding (heavy), more than 12.7-mm	6 or 8

Source: 29 CFR 1926.102(b).1.

3.3.6.5 *Laser Operations*

Laser retinal burns can be painless, so it is essential that all personnel in or around laser operations wear appropriate eye protection. Table 3.4 below lists maximum power or energy densities and appropriate protection levels for optical densities 5 through 8.

Table 3.4 Recommended laser safety glass.

Intensity, CW Maximum Attenuation Power Density (watts/cm^2)	Optical Density Attenuation	
	(O.D.)	Factor
10^{-2}	5	10^5
10^{-1}	6	10^6
1.0	7	10^7
10.0	8	10^8

Source: 29 CFR 1926.102(b).1.

3.3.7 Head Protection

Protecting employees from potential head injuries is a key element of any safety program. Employers must ensure that their employees wear head protection if any of the following apply:

- Objects might fall from above and strike them on the head;
- They might bump their heads against fixed objects, such as exposed pipes or beams; or
- There is a possibility of accidental head contact with electrical hazards.

Examples of occupations in which employees should be required to wear head protection include construction workers, carpenters, electricians, linemen, plumbers and pipefitters, timber and log cutters, welders, among many others. Whenever there is a danger of objects falling from above, such as working below others who are using tools or working under a conveyor belt, head protection must be worn. Hard hats must be worn with the bill forward to protect employees properly.

Protective helmets or hard hats should do the following:

- Resist penetration by objects.
- Absorb the shock of a blow.
- Be water-resistant and slow burning.
- Have clear instructions explaining proper adjustment and replacement of the suspension and headband.

Hard hats must have a hard outer shell and a shock-absorbing lining that incorporates a headband and straps that suspend the shell from 2.54 cm to 3.18 cm away from the head. This type of design provides shock absorption during an impact and ventilation during normal wear. In the U.S, OSHA standards recommend that protective headgear must meet ANSI standard Z89.1-1986 (Protective Headgear for Industrial Workers) or provide an equivalent level of protection. The reader should review product literature carefully to ensure that protective headgear meets ANSI or similar product performance standards.

3.3.7.1 *Types of Hard Hats*

Hard hats are divided into three industrial classes:

- Class A hard hats provide impact and penetration resistance along with limited voltage protection (up to 2,200 volts).
- Class B hard hats provide the highest level of protection against electrical hazards, with high-voltage shock and burn protection (up to 20,000 volts). They also provide protection from impact and penetration hazards by flying/falling objects.
- Class C hard hats provide lightweight comfort and impact protection but offer no protection from electrical hazards.

3.3.7.2 *Size and Care*

Never drill holes, paint or apply labels to protective headgear as this may reduce the integrity of the protection. Do not store headgear in direct sunlight, such as on the rear window shelf of a car, since sunlight and extreme heat can damage them.

3.3.8 Foot and Leg Protection

If an employee's feet may be exposed to electrical hazards, nonconductive footwear should be worn. On the other hand, workplace exposure to static electricity may necessitate the use of conductive footwear.

Examples of situations in which an employee should wear foot and/or leg protection include:

- When heavy objects such as barrels or tools might roll onto or fall on the employee's feet;
- Working with sharp objects such as nails or spikes that could pierce the soles or uppers of ordinary shoes;
- Exposure to molten metal that might splash on feet or legs;
- Working on or around hot, wet or slippery surfaces; and
- Working when electrical hazards are present.

3.3.8.1 Special Purpose Shoes

Workers exposed to electrical hazards must never wear conductive shoes.

Special Note: Nonconductive footwear must not be used in explosive or hazardous locations.

3.3.8.2 Protective Footwear Care

Workers should follow the manufacturers' recommendations for cleaning and maintenance of protective footwear.

3.3.9 Hand and Arm Protection

If a workplace hazard assessment reveals that employees face potential injury to hands and arms that cannot be eliminated through engineering and work practice controls, employers must ensure that employees wear appropriate protection.

3.3.9.1 Protective Gloves Types

The variety of potential occupational hand injuries makes selecting the right pair of gloves challenging. It is mandatory that employees use gloves provided by the employer.

3.3.9.2 Fabric and Coated Fabric Gloves

Fabric and coated fabric gloves are made of cotton or other fabric to provide varying degrees of protection.

Fabric gloves protect against dirt, slivers, chafing and abrasions. They do not provide sufficient protection for use with rough, sharp or heavy materials. Adding a plastic coating will strengthen some fabric gloves.

Coated fabric gloves are normally made from cotton flannel with napping on one side. By coating the unnapped side with plastic, fabric gloves are transformed into general-purpose hand protection offering slip-resistant qualities. These gloves are used for tasks ranging from handling bricks and wire to chemical laboratory containers. When selecting gloves to protect against chemical exposure hazards, always check with the manufacturer or review the manufacturer's product literature to determine the gloves' effectiveness against specific workplace chemicals and conditions.

3.3.9.3 *Chemical- and Liquid-Resistant Gloves*

Chemical-resistant gloves are made with different kinds of rubber: natural, butyl, neoprene, nitrile and fluorocarbon (viton); or various kinds of plastic: polyvinyl chloride (PVC), polyvinyl alcohol and polyethylene. These materials can be blended or laminated for better performance. As a general rule, the thicker the glove material, the greater the chemical resistance but thick gloves may impair grip and dexterity, having a negative impact on safety. Examples of chemical-resistant gloves include:

- **Butyl gloves** are made of a synthetic rubber and protect against a wide variety of chemicals, such as peroxide, rocket fuels, highly corrosive acids (nitric acid, sulfuric acid, hydrofluoric acid and red-fuming nitric acid), strong bases, alcohols, aldehydes, ketones, esters and nitrocompounds. Butyl gloves also resist oxidation, ozone corrosion and abrasion, and remain flexible at low temperatures. Butyl rubber does not perform well with aliphatic and aromatic hydrocarbons and halogenated solvents.
- **Natural (latex) rubber gloves** are comfortable to wear, which makes them a popular general-purpose glove. They feature outstanding tensile strength, elasticity and temperature resistance. In addition to resisting abrasions caused by grinding and polishing, these gloves protect workers' hands from most water solutions of acids, alkalis, salts and ketones. Latex gloves have caused allergic reactions in some individuals and may not be appropriate for all employees. Hypoallergenic gloves, glove liners and powderless gloves are possible alternatives for workers who are allergic to latex gloves.
- **Neoprene gloves** are made of synthetic rubber and offer good pliability, finger dexterity, high density and tear resistance. They protect against hydraulic fluids, gasoline, alcohols, organic acids and alkalis. They generally have chemical and wear resistance properties superior to those made of natural rubber.
- **Nitrile gloves** are made of a copolymer and provide protection from chlorinated solvents such as

trichloroethylene and perchloroethylene. Although intended for jobs requiring dexterity and sensitivity, nitrile gloves stand up to heavy use even after prolonged exposure to substances that cause other gloves to deteriorate. They offer protection when working with oils, greases, acids, caustics and alcohols but are generally not recommended for use with strong oxidizing agents, aromatic solvents, ketones and acetates.

The following (Table 3.5) is a materials selection chart which will guide the reader in the selection of gloves based on the chemicals required to be handled on the job.

Table 3.5 Material chemical resistance guide.

Chemical	Neoprene	Latex/ Rubber	Butyl	Nitrile
Acetaldehyde*	VG	G	VG	G
Acetic acid	VG	VG	VG	VG
Acetone*	G	VG	VG	P
Ammonium hydroxide	VG	VG	VG	VG
Amy acetate*	F	P	F	P
Aniline	G	F	F	P
Benzaldehyde*	F	F	G	G
Benzene*	P	P	P	F
Butyl acetate	G	F	F	P
Butyl alcohol	VG	VG	VG	VG
Carbon disulfide	F	F	F	F
Carbon tetrachloride*	F	P	P	G
Castor oil	F	P	F	VG
Chlorobenzene*	F	P	F	P
Chloroform*	G	P	P	F

Table 3.5 (cont.) Material chemical resistance guide.

Chemical	Neoprene	Latex/ Rubber	Butyl	Nitrile
Chloronaphthalene	F	P	F	F
Chromic acid (50%)	F	P	F	F
Citric acid (10%)	VG	VG	VG	VG
Cyclohexanol	G	F	G	VG
Dibutyl phthalate*	G	P	G	G
Diesel fuel	G	P	P	VG
Diisobutyl ketone	P	F	G	P
Dimethylformamide	F	F	G	G
Dioctyl phthalate	G	P	F	VG
Dioxane	VG	G	G	G
Epoxy resins, dry	VG	VG	VG	VG
Ethyl acetate*	G	F	G	F
Ethyl alcohol	VG	VG	VG	VG
Ethyl ether*	VG	G	VG	G
Ethylene dichloride*	F	P	F	P
Ethylene glycol	VG	VG	VG	VG
Formaldehyde	VG	VG	VG	VG
Formic acid	VG	VG	VG	VG
Freon 11	G	P	F	G
Freon 12	G	P	F	G
Freon 21	G	P	F	G
Freon 22	G	P	F	G
Furfural*	G	G	G	G
Gasoline, leaded	G	P	F	VG

Table 3.5 (cont.) Material chemical resistance guide.

Chemical	Neoprene	Latex/ Rubber	Butyl	Nitrile
Glycerin	VG	VG	VG	VG
Hexane	F	P	P	G
Hydrazine (65%)	F	G	G	G
Hydrochloric acid	VG	G	G	G
Hydrofluoric acid (48%)	VG	G	G	G
Hydrogen peroxide (30%) G	G	G	G	G
Hydroquinone	G	G	G	F
Isooctane	F	P	P	VG
Kerosene	VG	F	F	VG
Ketones	G	VG	VG	P
Lacquer thinners	G	F	F	P
Lactic acid (85%)	VG	VG	VG	VG
Lauric acid (36%)	VG	F	VG	VG
Lineolic acid	VG	P	F	G
Linseed oil	VG	P	F	VG
Maleic acid	VG	VG	VG	VG
Methyl alcohol	VG	VG	VG	VG
Methylamine	F	F	G	G
Methyl bromide	G	F	G	F
Methyl chloride*	P	P	P	P
Methyl ethyl ketone*	G	G	VG	P
Methyl isobutyl ketone*	F	F	VG	P
Methyl metharcrylate	G	G	VG	F

Table 3.5 (cont.) Material chemical resistance guide.

Chemical	Neoprene	Latex/ Rubber	Butyl	Nitrile
Monoethanolamine	VG	G	VG	VG
Morpholine	VG	VG	VG	G
Naphthalene	G	F	F	G
Napthas, aliphatic	VG	F	F	VG
Napthas, aromatic	G	P	P	G
Nitric acid*	G	F	F	F
Nitric acid, red and white fuming	P	P	P	P
Nitromethane (95.5%)*	F	P	F	F
Nitropropane (95.5%)	F	P	F	F
Octyl alcohol	VG	VG	VG	VG
Oleic acid	VG	F	G	VG
Oxalic acid	VG	VG	VG	VG
Palmitic acid	VG	VG	VG	VG
Perchloric acid (60%)	VG	F	G	G
Perchloroethylene	F	P	P	G
Petroleum distillates (naphtha)	G	P	P	VG
Phenol	VG	F	G	F
Phosphoric acid	VG	G	VG	VG
Potassium hydroxide	VG	VG	VG	VG
Propyl acetate	G	F	G	F
Propyl alcohol	VG	VG	VG	VG
Propyl alcohol (iso)	VG	VG	VG	VG

Table 3.5 (cont.) Material chemical resistance guide.

Chemical	Neoprene	Latex/ Rubber	Butyl	Nitrile
Sodium hydroxide	VG	VG	VG	VG
Styrene	P	P	P	F
Styrene (100%)	P	P	P	F
Sulfuric acid	G	G	G	G
Tannic acid (65)	VG	VG	VG	VG
Tetrahydrofuran	P	F	F	F
Toluene*	F	P	P	F
Toluene diisocyanate (TDI)	F	G	G	F
Trichloroethylene*	F	F	P	G
Triethanolamine (85%)	VG	G	G	VG
Tung oil	VG	P	F	VG
Turpentine	G	F	F	VG
Xylene*	P	P	P	F

(Key: VG: Very Good; G: Good; F: Fair; P: Poor (not recommended). Chemicals marked with an asterisk (*) are for limited service only).

U.S. Government publication titled Personal Protective Equipment, U.S. Department of Labor, Occupational Safety and Health Administration, OSHA 3151-12R, 2003

3.3.9.4 Protective Gloves Care

Protective gloves should be inspected before each use to ensure that they are not torn, punctured or made ineffective in any way.

3.3.10 Protective Equipment for the Body

Workers who face possible bodily injury of any kind that cannot be eliminated through engineering, work practice or administrative controls, must wear appropriate body protection while performing their jobs.

3.3.11 Hearing Protection

Employee exposure to excessive noise depends upon a number of factors, including:

- The loudness of the noise as measured in decibels (dB).
- The duration of each employee's exposure to the noise.
- Whether employees move between work areas with different noise levels.
- Whether noise is generated from one or multiple sources.

Generally, the louder the noise, the shorter the exposure time before hearing protection is required. For instance, employees may be exposed to a noise level of 85 dB for 8 hours per day (unless they experience a Standard Threshold Shift) before hearing protection is required. On the other hand, if the noise level reaches 115 dB hearing protection is required if the anticipated exposure exceeds 15 minutes.

If engineering and work practice controls do not lower employee exposure to workplace noise to acceptable levels, employees must wear appropriate hearing protection.

Manufacturers of hearing protection devices must display the device's NRR on the product packaging.

Some types of hearing protection include:

- **Single-use earplugs** are made of waxed cotton, foam, silicone rubber or fiberglass wool. They are self-forming and, when properly inserted, they work as well as most molded earplugs.
- **Pre-formed or molded earplugs** must be individually fitted by a professional and can be disposable or reusable. Reusable plugs should be cleaned after each use.
- **Earmuffs** require a perfect seal around the ear. Glasses, facial hair, long hair or facial movements such as chewing may reduce the protective value of earmuffs.

3.3.12 Respiratory Protection

In the control of those occupational diseases caused by breathing air contaminated with harmful dusts, fogs, fumes, mists, gases, smokes, sprays, or vapors, the primary objective of a respiratory protection program is to prevent atmospheric contamination. This should be accomplished as far as feasible by accepted engineering control measures (for example, enclosure or confinement of the operation, general and local ventilation, and substitution of less toxic materials). When effective engineering controls are not feasible, or while they are being instituted, appropriate respirators should be used.

3.3.12.1 Definitions

The following U.S. OSHA definitions are important terms used in the respiratory protection standard in this subsection.

Air-purifying respirator means a respirator with an air-purifying filter, cartridge, or canister that removes specific air contaminants by passing ambient air through the air-purifying element.

Assigned protection factor (APF) means the workplace level of respiratory protection that a respirator or class of respirators is expected to provide to employees when the employer implements a continuing, effective respiratory protection program as specified by this section.

Atmosphere-supplying respirator means a respirator that supplies the respirator user with breathing air from a source independent of the ambient atmosphere, and includes supplied-air respirators (SARs) and self-contained breathing apparatus (SCBA) units.

Canister or cartridge means a container with a filter, sorbent, or catalyst, or combination of these items, which removes specific contaminants from the air passed through the container.

Demand respirator means an atmosphere-supplying respirator that admits breathing air to the facepiece only when a negative pressure is created inside the facepiece by inhalation.

Emergency situation means any occurrence such as, but not limited to, equipment failure, rupture of containers, or failure of control equipment that may or does result in an uncontrolled significant release of an airborne contaminant.

Employee exposure means exposure to a concentration of an airborne contaminant that would occur if the employee were not using respiratory protection.

End-of-service-life indicator (ESLI) means a system that warns the respirator user of the approach of the end of adequate respiratory protection, for example, that the sorbent is approaching saturation or is no longer effective.

Escape-only respirator means a respirator intended to be used only for emergency exit.

Filter or air purifying element means a component used in respirators to remove solid or liquid aerosols from the inspired air.

Filtering facepiece (dust mask) means a negative pressure particulate respirator with a filter as an integral part of the facepiece or with the entire facepiece composed of the filtering medium.

Fit factor means a quantitative estimate of the fit of a particular respirator to a specific individual, and typically estimates the ratio of the concentration of a substance in ambient air to its concentration inside the respirator when worn.

Fit test means the use of a protocol to qualitatively or quantitatively evaluate the fit of a respirator on an individual. (See also Qualitative fit test QLFT and Quantitative fit test QNFT.)

Helmet means a rigid respiratory inlet covering that also provides head protection against impact and penetration.

High efficiency particulate air (HEPA) filter means a filter that is at least 99.97% efficient in removing monodisperse particles of 0.3 micrometers in diameter.

Hood means a respiratory inlet covering that completely covers the head and neck and may also cover portions of the shoulders and torso.

Immediately dangerous to life or health (IDLH) means an atmosphere that poses an immediate threat to life, would cause irreversible adverse health effects, or would impair an individual's ability to escape from a dangerous atmosphere.

Interior structural firefighting means the physical activity of fire suppression, rescue or both, inside of buildings or enclosed structures which are involved in a fire situation beyond the incipient stage. (See 29 CFR 1910.155)

Loose-fitting face piece means a respiratory inlet covering that is designed to form a partial seal with the face.

Maximum use concentration (MUC) means the maximum atmospheric concentration of a hazardous substance from which an employee can be expected to be protected when wearing a respirator, and is determined by the assigned protection factor of the respirator or class of respirators and the exposure limit of the hazardous substance. The MUC can be determined mathematically by multiplying the assigned protection factor specified for a respirator by the required permissible exposure limit, short-term exposure limit, or ceiling limit. When no exposure limit is available for a hazardous substance, an employer must determine an MUC on the basis of relevant available information and informed professional judgment.

Negative pressure respirator (tight fitting) means a respirator in which the air pressure inside the facepiece is negative during inhalation with respect to the ambient air pressure outside the respirator.

Oxygen deficient atmosphere means an atmosphere with oxygen content below 19.5% by volume.

Positive pressure respirator means a respirator in which the pressure inside the respiratory inlet covering exceeds the ambient air pressure outside the respirator.

Powered air-purifying respirator (PAPR) means an air-purifying respirator that uses a blower to force the ambient air through air-purifying elements to the inlet covering.

Pressure demand respirator means a positive pressure atmosphere-supplying respirator that admits breathing air to the facepiece when the positive pressure is reduced inside the facepiece by inhalation.

Qualitative fit test (QLFT) means a pass/fail fit test to assess the adequacy of respirator fit that relies on the individual's response to the test agent.

Quantitative fit test (QNFT) means an assessment of the adequacy of respirator fit by numerically measuring the amount of leakage into the respirator.

Respiratory inlet covering means that portion of a respirator that forms the protective barrier between the user's respiratory tract and an air-purifying device or breathing air source, or both. It may be a facepiece, helmet, hood, suit, or a mouthpiece respirator with nose clamp.

Self-contained breathing apparatus (SCBA) means an atmosphere-supplying respirator for which the breathing air source is designed to be carried by the user.

Service life means the period of time that a respirator, filter or sorbent, or other respiratory equipment provides adequate protection to the wearer.

Supplied-air respirator (SAR) or airline respirator means an atmosphere-supplying respirator for which the source of breathing air is not designed to be carried by the user.

Tight-fitting face piece means a respiratory inlet covering that forms a complete seal with the face.

User seal check means an action conducted by the respirator user to determine if the respirator is properly seated to the face.

3.3.12.2 *Establishing a Respiratory Protection Program*

Employers should develop and implement a written respiratory protection program with required worksite-specific procedures and elements for required respirator use. The program must be administered by a suitably trained program administrator.

In any workplace where respirators are necessary to protect the health of the employee or whenever respirators are required by the employer, the employer shall establish and implement a written respiratory protection program with worksite-specific procedures. The program must be updated as necessary to reflect those changes in workplace conditions that affect respirator use. The employer shall include in the program the following provisions of this section, as applicable:

- Procedures for selecting respirators for use in the workplace;
- Medical evaluations of employees required to use respirators;
- Fit testing procedures for tight-fitting respirators;
- Procedures for proper use of respirators in routine and reasonably foreseeable emergency situations;
- Procedures and schedules for cleaning, disinfecting, storing, inspecting, repairing, discarding, and otherwise maintaining respirators;
- Procedures to ensure adequate air quality, quantity, and flow of breathing air for atmosphere-supplying respirators;
- Training of employees in the respiratory hazards to which they are potentially exposed during routine and emergency situations;
- Training of employees in the proper use of respirators, including putting on and removing them, any limitations on their use, and their maintenance; and
- Procedures for regularly evaluating the effectiveness of the program.

Where respirator use is not required an employer may provide respirators at the request of employees or permit employees to use their own respirators, if the employer determines that such respirator use will not in itself create a hazard.

In addition, the employer must establish and implement those elements of a written respiratory protection program necessary to ensure that any employee using a respirator voluntarily is medically able to use that respirator, and that the respirator is cleaned, stored, and maintained so that its use does not present a health hazard to the user. Exception: Employers are not required to include in a written respiratory protection program those employees whose only use of respirators involves the voluntary use of filtering face pieces (dust masks).

The employer shall designate a program administrator who is qualified by appropriate training or experience that is commensurate with the complexity of the program to administer or oversee the respiratory protection program and conduct the required evaluations of program effectiveness.

The employer shall provide respirators, training, and medical evaluations at no cost to the employee.

The employer must evaluate respiratory hazard(s) in the workplace, identify relevant workplace and user factors, and base respirator selection on these factors. This includes specifying appropriately protective respirators for use in IDLH atmospheres, and limits the selection and use of air-purifying respirators.

The employer shall identify and evaluate the respiratory hazard(s) in the workplace; this evaluation shall include a reasonable estimate of employee exposures to respiratory hazard(s) and an identification of the contaminant's chemical state and physical form. Where the employer cannot identify or reasonably estimate the employee exposure, the employer shall consider the atmosphere to be IDLH.

The employer shall select respirators from a sufficient number of respirator models and sizes so that the respirator is acceptable to, and correctly fits, the user.

The employer must select a respirator for employee use that maintains the employee's exposure to the hazardous substance, when measured outside the respirator, at or below the MUC.

Employers must not apply MUCs to conditions that are immediately dangerous to life or health (IDLH); instead, they must use respirators listed for IDLH conditions in paragraph (d).1 of this standard.

When the calculated MUC exceeds the IDLH level for a hazardous substance, or the performance limits of the cartridge or canister, then employers must set the maximum MUC at that lower limit.

The respirator selected shall be appropriate for the chemical state and physical form of the contaminant.

Medical evaluation. Using a respirator may place a physiological burden on employees that varies with the type of respirator worn, the job and workplace conditions in which the respirator is used, and the medical status of the employee. Accordingly, this paragraph specifies the minimum requirements for medical evaluation that employers must implement to determine the employee's ability to use a respirator.

The employer shall provide a medical evaluation to determine the employee's ability to use a respirator, before the employee is fit tested or required to use the respirator in the workplace. The employer may discontinue an employee's medical evaluations when the employee is no longer required to use a respirator.

The medical evaluation shall obtain the information requested by the questionnaire.

The employer shall ensure that a follow-up medical examination is provided for an employee who gives a positive response to any question. The follow-up medical examination shall include any medical tests, consultations, or diagnostic procedures that the examiner deems necessary to make a final determination.

The employer shall provide the employee with an opportunity to discuss the questionnaire and examination results with the medical examiner.

Fit testing. This is an important part of a program. It requires that, before an employee may be required to use any respirator with a negative or positive pressure tight-fitting facepiece, the employee must be fit tested with the same make, model, style, and size of respirator that will be used. The employer shall ensure that employees using a tight-fitting facepiece respirator pass an appropriate qualitative fit test (QLFT) or quantitative fit test (QNFT).

Continuing respirator effectiveness. Appropriate surveillance shall be maintained of work area conditions and degree of employee exposure or stress. When there is a change in work area conditions or degree of employee exposure or stress that may affect respirator effectiveness, the employer shall reevaluate the continued effectiveness of the respirator.

The employer shall ensure that employees leave the respirator use area:

- To wash their faces and respirator facepieces as necessary to prevent eye or skin irritation associated with respirator use; or
- If they detect vapor or gas breakthrough, changes in breathing resistance, or leakage of the facepiece; or

- To replace the respirator or the filter, cartridge, or canister elements.

If the employee detects vapor or gas breakthrough, changes in breathing resistance, or leakage of the facepiece, the employer must replace or repair the respirator before allowing the employee to return to the work area.

Maintenance and care of respirators. The recommended respiratory protection program requires the employer to provide for the cleaning and disinfecting, storage, inspection, and repair of respirators used by employees.

- **Cleaning and disinfecting.** The employer shall provide each respirator user with a respirator that is clean, sanitary, and in good working order. The respirators shall be cleaned and disinfected at the following intervals:
  a. Respirators issued for the exclusive use of an employee shall be cleaned and disinfected as often as necessary to be maintained in a sanitary condition;
  b. Respirators issued to more than one employee shall be cleaned and disinfected before being worn by different individuals;
  c. Respirators maintained for emergency use shall be cleaned and disinfected after each use; and
  d. Respirators used in fit testing and training shall be cleaned and disinfected after each use.
- **Storage.** The employer shall ensure that respirators are stored as follows:
  a. All respirators shall be stored to protect them from damage, contamination, dust, sunlight, extreme temperatures, excessive moisture, and damaging chemicals, and they shall be packed or stored to prevent deformation of the facepiece and exhalation valve.
  b. In addition emergency respirators shall be:
     i. Kept accessible to the work area;
     ii. Stored in compartments or in covers that are clearly marked as containing emergency respirators; and

iii. Stored in accordance with any applicable manufacturer instructions.

- **Inspection.** The employer shall ensure that respirators are inspected as follows:
  a. All respirators used in routine situations shall be inspected before each use and during cleaning;
  b. All respirators maintained for use in emergency situations shall be inspected at least monthly and in accordance with the manufacturer's recommendations, and shall be checked for proper function before and after each use; and
  c. Emergency escape-only respirators shall be inspected before being carried into the workplace for use.
  d. The employer shall ensure that respirator inspections include the following:
     i. A check of respirator function, tightness of connections, and the condition of the various parts including, but not limited to, the facepiece, head straps, valves, connecting tube, and cartridges, canisters or filters; and
     ii. A check of elastomeric parts for pliability and signs of deterioration.
     iii. Section, self-contained breathing apparatus shall be inspected monthly. Air and oxygen cylinders shall be maintained in a fully charged state and shall be recharged when the pressure falls to 90% of the manufacturer's recommended pressure level. The employer shall determine that the regulator and warning devices function properly.
     iv. For respirators maintained for emergency use, the employer shall:
        (1) Certify the respirator by documenting the date the inspection was performed, the name (or signature) of the person who made the inspection, the findings, required remedial action, and a serial number or other means of identifying the inspected respirator; and
        (2) Provide this information on a tag or label that is attached to the storage compartment for the respirator, is kept with the respirator,

or is included in inspection reports stored as paper or electronic files. This information shall be maintained until replaced following a subsequent certification.

- **Repairs.** The employer shall ensure that respirators that fail an inspection or are otherwise found to be defective are removed from service, and are discarded or repaired or adjusted in accordance with the following procedures:
  a. Repairs or adjustments to respirators are to be made only by persons appropriately trained to perform such operations and shall use only the respirator manufacturer's approved parts designed for the respirator;
  b. Repairs shall be made according to the manufacturer's recommendations and specifications for the type and extent of repairs to be performed; and
  c. Reducing and admission valves, regulators, and alarms shall be adjusted or repaired only by the manufacturer or a technician trained by the manufacturer.

Identification of filters, cartridges, and canisters. The employer shall ensure that all filters, cartridges and canisters used in the workplace are labeled and color coded with the NIOSH approval label and that the label is not removed and remains legible.

Training and Recordkeeping. The employer shall provide effective training to employees who are required to use respirators. The training must be comprehensive, understandable, and recur annually, and more often if necessary.

The employer shall ensure that each employee can demonstrate knowledge of at least the following:

- Why the respirator is necessary and how improper fit, usage, or maintenance can compromise the protective effect of the respirator;
- What the limitations and capabilities of the respirator are;
- How to use the respirator effectively in emergency situations, including situations in which the respirator malfunctions;

- How to inspect, put on and remove, use, and check the seals of the respirator;
- What the procedures are for maintenance and storage of the respirator;
- How to recognize medical signs and symptoms that may limit or prevent the effective use of respirators; and
- The general requirements of this section.

Training shall be conducted in a manner that is understandable to the employee. The employer shall provide the training prior to requiring the employee to use a respirator in the workplace.

Retraining shall be administered annually, and when the following situations occur:

- Changes in the workplace or the type of respirator render previous training obsolete;
- Inadequacies in the employee's knowledge or use of the respirator indicate that the employee has not retained the requisite understanding or skill; or
- Any other situation arises in which retraining appears necessary to ensure safe respirator use.

Program evaluation and documentation. The employer is required to conduct evaluations of the workplace to ensure that the written respiratory protection program is being properly implemented, and to consult employees to ensure that they are using the respirators properly. The employer shall conduct evaluations of the workplace as necessary to ensure that the provisions of the current written program are being effectively implemented and that it continues to be effective.

The employer shall regularly consult employees required to use respirators to assess the employees' views on program effectiveness and to identify any problems. Any problems that are identified during this assessment shall be corrected. Factors to be assessed include, but are not limited to:

- Respirator fit (including the ability to use the respirator without interfering with effective workplace performance);
- Appropriate respirator selection for the hazards to which the employee is exposed;

- Proper respirator use under the workplace conditions the employee encounters; and
- Proper respirator maintenance.

Recordkeeping. This section requires the employer to establish and retain written information regarding medical evaluations, fit testing, and the respirator program. This information will facilitate employee involvement in the respirator program, and assist the employer in auditing the adequacy of the program.

3.3.12.3 *Levels of Protection*

Respirators for IDLH atmospheres - The employer shall provide the following respirators for employee use in IDLH atmospheres:

- A full face piece pressure demand SCBA certified by NIOSH for a minimum service life of thirty minutes, or
- A combination full face piece pressure demand supplied-air respirator (SAR) with auxiliary self-contained air supply.

For all IDLH atmospheres, the employer shall ensure that:

- One employee or, when needed, more than one employee is located outside the IDLH atmosphere;
- Visual, voice, or signal line communication is maintained between the employee(s) in the IDLH atmosphere and the employee(s) located outside the IDLH atmosphere;
- The employee(s) located outside the IDLH atmosphere are trained and equipped to provide effective emergency rescue;
- The employer or designee is notified before the employee(s) located outside the IDLH atmosphere enter the IDLH atmosphere to provide emergency rescue;
- The employer or designee authorized to do so by the employer, once notified, provides necessary assistance appropriate to the situation;
- Employee(s) located outside the IDLH atmospheres are equipped with: Pressure demand or other positive

pressure SCBAs, or a pressure demand or other positive pressure supplied-air respirator with auxiliary SCBA; and either appropriate retrieval equipment for removing the employee(s) who enter(s) these hazardous atmospheres where retrieval equipment would contribute to the rescue of the employee(s) and would not increase the overall risk resulting from entry; or equivalent means for rescue where retrieval equipment is not required.

Procedures for interior structural firefighting. This is a special IDLH situation. The U.S. OSHA protocol requires:

- At least two employees enter the IDLH atmosphere and remain in visual or voice contact with one another at all times;
- At least two employees are located outside the IDLH atmosphere; and
- All employees engaged in interior structural firefighting use SCBAs.

One of the two individuals located outside the IDLH atmosphere may be assigned to an additional role, such as incident commander in charge of the emergency or safety officer, so long as this individual is able to perform assistance or rescue activities without jeopardizing the safety or health of any firefighter working at the incident. Nothing in this section is meant to preclude firefighters from performing emergency rescue activities before an entire team has assembled.

All oxygen-deficient atmospheres shall be considered IDLH.

Respirators for atmospheres that are not IDLH - The employer shall provide a respirator that is adequate to protect the health of the employee and ensure compliance with all other OSHA statutory and regulatory requirements, under routine and reasonably foreseeable emergency situations.

For protection against gases and vapors, the employer shall provide:

- An atmosphere-supplying respirator, or
- An air-purifying respirator, provided that the respirator is equipped with an end-of-service-life indicator (ESLI) certified by NIOSH for the contaminant.

For protection against particulates, the employer shall provide:

- An atmosphere-supplying respirator; or
- An air-purifying respirator equipped with a filter such as a high efficiency particulate air (HEPA) filter, or an air-purifying respirator equipped with a filter; or
- For contaminants consisting primarily of particles with mass median aerodynamic diameters (MMAD) of at least 2 micrometers, an air-purifying respirator equipped with any filter certified for particulates.

Assigned Protection Factors (APFs) - Employers must use the assigned protection factors listed in Table 3.6 below to select a respirator that meets or exceeds the required level of employee

Table 3.6 Assigned protection factors.

Type of Respirator[1],[2]	Quarter Mask	Half Mask	Full Facepiece	Helmet/ Hood	Loose-Fitting Facepiece
1. Air-Purifying Respirator	5	10[3]	50	—	—
2. Powered Air-Purifying Respirator (PAPR)	—	50	1,000	25/1,000[4]	25
3. Supplied-Air Respirator (SAR) or Airline Respirator					
• Demand mode	—	10	50	—	—
• Continuous flow mode	—	50	1,000	25/1,000[4]	25
• Pressure-demand or other positive-pressure mode	—	50	1,000	—	—

Table 3.6 (cont.) Assigned protection factors.

Type of Respirator[1],[2]	Quarter Mask	Half Mask	Full Facepiece	Helmet/ Hood	Loose-Fitting Facepiece
4. Self-Contained Breathing Apparatus (SCBA)					
• Demand mode	—	10	50	50	—
• Pressure-demand or other positive-pressure mode (e.g., open/ closed circuit)	—	—	10,000	10,000	—

Notes:

1. Employers may select respirators assigned for use in higher workplace concentrations of a hazardous substance for use at lower concentrations of that substance, or when required respirator use is independent of concentration.

2. The assigned protection factors in Table are only effective when the employer implements a continuing, effective respirator program including training, fit testing, maintenance, and use requirements.

3. This APF category includes filtering facepieces and half masks with elastomeric facepieces.

4. The employer must have evidence provided by the respirator manufacturer that testing of these respirators demonstrates performance at a level of protection of 1,000 or greater to receive an APF of 1,000. This level of performance can best be demonstrated by performing a WPF or SWPF study or equivalent testing. Absent such testing, all other PAPRs and SARs with helmets/hoods are to be treated as loose-fitting facepiece respirators, and receive an APF of 25.

5. These APFs do not apply to respirators used solely for escape. For escape respirators used in association with specific substances employers must refer to the appropriate substance-specific standards in that subpart

protection. When using a combination respirator (e.g., airline respirators with an air-purifying filter), employers must ensure that the assigned protection factor is appropriate to the mode of operation in which the respirator is being used.

3.3.12.4 *Rules for Breathing Air Quality and Use*

This subsection requires the employer to provide employees using atmosphere-supplying respirators (supplied-air and SCBA) with breathing gases of high purity. The employer shall ensure that compressed air, compressed oxygen, liquid air, and liquid oxygen used for respiration accords with the following specifications:

- Compressed and liquid oxygen shall meet the requirements for medical or breathing oxygen; and
- Compressed breathing air shall meet at least the requirements for Grade D breathing air described in ANSI/Compressed Gas Association Commodity Specification for Air, G-7.1-1989, to include:
  a. Oxygen content (v/v) of 19.5-23.5%;
  b. Hydrocarbon (condensed) content of 5 milligrams per cubic meter of air or less;
  c. Carbon monoxide (CO) content of 10 ppm or less;
  d. Carbon dioxide content of 1,000 ppm or less; and
  e. Lack of noticeable odor.

The employer shall ensure that compressed oxygen is not used in atmosphere-supplying respirators that have previously used compressed air.

The employer shall ensure that oxygen concentrations greater than 23.5% are used only in equipment designed for oxygen service or distribution.

The employer shall ensure that cylinders used to supply breathing air to respirators meet the following requirements:

- Cylinders are tested and maintained as prescribed in the Shipping Container Specification Regulations of the Department of Transportation (49 CFR part 173 and part 178);
- Cylinders of purchased breathing air have a certificate of analysis from the supplier that the breathing air meets the requirements for Grade D breathing air; and
- The moisture content in the cylinder does not exceed a dew point of -45.6 deg.C at 1 atmosphere pressure.

The employer shall ensure that compressors used to supply breathing air to respirators are constructed and situated so as to:

- Prevent entry of contaminated air into the air-supply system;
- Minimize moisture content so that the dew point at 1 atmosphere pressure is 5.56 deg.C below the ambient temperature;
- Have suitable in-line air-purifying sorbent beds and filters to further ensure breathing air quality. Sorbent beds and filters shall be maintained and replaced or refurbished periodically following the manufacturer's instructions.
- Have a tag containing the most recent change date and the signature of the person authorized by the employer to perform the change. The tag shall be maintained at the compressor.

For compressors that are not oil-lubricated, the employer shall ensure that carbon monoxide levels in the breathing air do not exceed 10 ppm.

For oil-lubricated compressors, the employer shall use a high-temperature or carbon monoxide alarm, or both, to monitor carbon monoxide levels. If only high-temperature alarms are used, the air supply shall be monitored at intervals sufficient to prevent carbon monoxide in the breathing air from exceeding 10 ppm.

The employer shall ensure that breathing air couplings are incompatible with outlets for nonrespirable worksite air or other gas systems. No asphyxiating substance shall be introduced into breathing air lines.

3.3.13 Bibliography

Eye and Face Protection: ANSI Z87.1-1989 (USA Standard for Occupational and Educational Eye and Face Protection).

ANSI Standard Z89.1-1986 (Protective Headgear for Industrial Workers)

Foot Protection: ANSI Z41.1-1991.

A.M. Best Co. (AMB). 1990. Bests Safety Directory. AMB: Odwick, NJ.

Hering, S.V., Ed. 1989. Air Sampling Instruments for Evaluation of Atmospheric Contaminants, American Conference of Governmental Industrial Hygienists: Cincinnati, Ohio.

Handbook of Industrial Toxicology and Hazardous Materials, Cheremisinoff, N. P. , Marcel Dekker Publishers, NY, NY. , 1999
Chemical Protective Clothing Performance Index, 2nd Edition, Krister Forsberg, Lawrence H. Keith, ISBN: 978-0-471-32844-5, September 1999
Industrial Solvents Handbook, 2nd Edition, N. P. Cheremisinoff, Marcel Dekker Publishers, New York and Basel, 2003.
Handbook of Hazardous Chemical Properties, N. P. Cheremisinoff, Butterworth-Heinemann, Oxford, UK 2000.
Handbook of Pollution and Hazardous Materials Compliance, N. P. Cheremisinoff, Noyes Publications, Park Ridge, N.J., 1996.
Noise Control in Industry, N. P. Cheremisinoff, Noyes Publications, Park Ridge, N.J., 1996.
Materials Selection Deskbook, N. P. Cheremisinoff, Noyes Publications, Park Ridge, N.J., 1996.
Safety Management Practices for Hazardous Materials, N. P. Cheremisinoff, Marcel Dekker Publishers, N.Y., 1996.
Handbook of Emergency Response to Toxic Chemical Releases: A Guide to Compliance, N. P. Cheremisinoff, Noyes Publications, Park Ridge, N.J., 1995.
Environmental and Health & Safety Management: A Guide to Compliance, N. P. Cheremisinoff, Noyes Publications, Park Ridge, N.J., 1995.
Dangerous and Toxic Properties of Industrial and Consumer Products, N. P. Cheremisinoff, et. al., Marcel Dekker Publishers, Inc., N.Y., 1994.
The following are links to WWW sites on safety equipment:
http://www.occupationalhazards.com/zone/342/OHRespiratory_Protection.aspx
http://www.osha.gov/SLTC/etools/respiratory/
http://www.cdc.gov/niosh/npptl/topics/protclothing/
http://www.osha.gov/dts/osta/otm/otm_viii/otm_viii_1.html

3.4 First Aid and Resuscitation

3.4.1 Introduction[12]

First aid is emergency care provided for injury or sudden illness before professional emergency medical treatment is available. The first aid provider in the workplace is someone who is trained and

12. Occupational Safety and Health Administration Best Practices Guide: Fundamentals of a Workplace First aid Program.

experienced in the delivery of initial medical emergency procedures, using a limited amount of equipment to perform a primary assessment and intervention while awaiting arrival of emergency medical service (EMS) personnel.

A workplace first aid program is part of a comprehensive safety and health management system consisting of the following four essential elements:

- Management leadership and employee involvement;
- Worksite evaluation and analysis;
- Hazard and risk prevention and control; and
- Safety and health training.

Additionally, the fundamental elements for a first aid program at the workplace should include:

- Identifying and assessing the workplace risks displaying the potential to cause worker injury or illness.
- Designing and implementing a workplace first aid program that:
  - aims to limit the outcome of accidents or exposures; and
  - includes sufficient quantities of appropriate and readily accessible first aid supplies and first aid equipment.
- Assigning and training first aid providers who:
  - receive first aid training suitable to the specific workplace; and
  - receive periodic refresher courses on first aid skills and knowledge.
- Familiarizing all workers with the first aid program, including what workers should do if a coworker is injured or ill. Establishing the policies and program in writing will assist with implementing this and other program elements.
- Providing for scheduled evaluation and changing of the first aid program to keep the program current and applicable to emerging risks in the workplace, including regular assessment of the adequacy of the first aid training course.

3.4.2 First Aid Assessments[13]

First aid assessments must be conducted in each area and department to determine the appropriate first aid facilities and suitably trained people. The assessment should be documented in consultation with the first aider and a health and safety representative or expert with consideration to:

- Size, design, and layout of the facility or workplace;
- The number and distribution of employees including arrangements such as shift work, after hours and field trips;
- The nature of hazards and the severity of the risk;
- The location of the workplace; and
- Known occurrences of accidents or illness.

3.4.3 Elements of a First Aid Training Program[14]

There are a number of elements to include when planning a first aid training program for a particular workplace including:

1. Teaching Methods
Training programs should consider and include the following principles:

- Basing the curriculum on a consensus of scientific evidence where available;
- Having trainees develop "hands-on" skills through the use of mannequins and partner practice;
- Having proper first aid supplies and equipment available;
- Exposing trainees to acute injury and illness settings as well as to the appropriate response through the use of visual aids;
- Including a course information resource for reference both during and after training;

13. La Trobe University Occupational Health and Safety Manual, December 1999.
14. Occupational Safety and Health Administration Best Practices Guide: Fundamentals of a Workplace First Aid Program.

- Allowing adequate time for emphasis on frequently occurring situations;
- Emphasizing skills training and confidence-building over classroom lectures; and
- Emphasizing quick response to first aid situations.

2. Preparing to Respond to a Health Emergency

The training program should include instruction or discussion in the following:

- Prevention as a strategy in reducing and limiting fatalities, illnesses and injuries;
- Interacting with the existing local Emergency Management System (if available);
- Maintaining a current list of emergency telephone numbers (police, fire, ambulance, poison control) easily accessible by all employees;
- Understanding the legal aspects of providing first aid care,
- Understanding the effects of stress, fear of infection, panic; how they interfere with performance; and what to do to overcome these barriers to action;
- Learning the importance of universal precautions and body substance isolation to provide protection from blood-borne pathogens and other potentially infectious materials. Learning about personal protective equipment – gloves, eye protection, masks, and respiratory barrier devices. Appropriate management and disposal of blood-contaminated sharps and surfaces; and awareness of blood-borne Pathogens standard.

3. Evaluating the Scene and the Victim(s)

An effective training program should include instruction in the following areas:

- Assessing the scene for safety, number of injured, and nature of the event;
- Assessing the toxic potential of the environment and the need for respiratory protection;

- Establishing the presence of a confined space and the need for respiratory protection and specialized training to perform a rescue;
- Prioritizing care when there are several injured;
- Assessing each victim for responsiveness, airway patency (blockage), breathing, circulation, and medical alert tags;
- Taking a victim's history at the scene, including determining the mechanism of injury;
- Performing a logical head-to-toe check for injuries;
- Stressing the need to continuously monitor the victim;
- Emphasizing early activation of the Emergency Management System and emergency medical personnel;
- Indications for and methods of safely moving and rescuing victims; and
- Repositioning ill/injured victims to prevent further injury.

4. Responding to Life-Threatening Emergencies

The training program should be designed or adapted for the specific worksite and may include first aid instruction in the following:

- Establishing responsiveness;
- Establishing and maintaining an open and clear airway;
- Performing rescue breathing;
- Treating airway obstruction in a conscious victim;
- Performing cardiopulmonary resuscitation (CPR);
- Using an automated external defibrillator (AED);
- Recognizing the signs and symptoms of shock and providing first aid for shock due to illness or injury;
- Assessing and treating a victim who has an unexplained change in level of consciousness or sudden illness;
- Controlling bleeding with direct pressure;
- Dealing with a variety of poisonings including:
  - Ingested poisons: alkali, acid, and systemic poisons. Role of the Poison Control Center;
  - Inhaled poisons: carbon monoxide; hydrogen sulfide; smoke; and other chemical fumes, vapors, and gases. Assessing the toxic potential of the environment and the need for respirators;

  - Knowledge of the chemicals at the worksite and of first aid and treatment for inhalation or ingestion; and
  - Effects of alcohol and illicit drugs so that the first aid provider can recognize the physiologic and behavioral effects of these substances.
- Recognizing asphyxiation and the danger of entering a confined space without proper respiratory protection. Additional training is required if first aid personnel will assist in the rescue from the confined space.
- Responding to wide-range of medical emergencies including:
  - Chest pain;
  - Stroke;
  - Breathing problems;
  - Anaphylactic reaction;
  - Hypoglycemia in diabetics taking insulin;
  - Seizures;
  - Pregnancy complications;
  - Abdominal injury;
  - Reduced level of consciousness; and
  - Impaled object.

5. Responding to Non-Life-Threatening Emergencies

The training program should be designed for the specific worksite and to include first aid instruction for the management of the following:

- Wounds
  - Assessment and first aid for wounds including abrasions, cuts, lacerations, punctures, avulsions, amputations and crush injuries;
  - Principles of wound care, including infection precautions; and
  - Principles of body substance isolation, universal precautions and use of personal protective equipment.
- Burns
  - Assessing the severity of a burn;
  - Recognizing whether a burn is thermal, electrical, or chemical and how to administer the appropriate first aid; and

  - Reviewing corrosive chemicals at a specific worksite, along with administering the appropriate first aid.
- Temperature Extremes
  - Exposure to cold, including frostbite and hypothermia; and
  - Exposure to heat, including heat cramps, heat exhaustion and heat stroke.
- Musculoskeletal Injuries
  - Fractures;
  - Sprains, strains, contusions and cramps;
  - Head, neck, back and spinal injuries; and
  - Appropriate handling of amputated body parts.
- Eye injuries
  - First aid for eye injuries; and
  - First aid for chemical burns.
- Oral Injuries
  - Lip and tongue injuries; broken and missing teeth; and
  - The importance of preventing aspiration of blood and/or teeth.
- Bites and Stings
  - Human and animal bites; and
  - Bites and stings from insects; instruction in first aid treatment of anaphylactic shock.

3.4.4 Periodic Program Updates[15]

The first aid program should be reviewed periodically to determine if it continues to address the current needs of the specific workplace. Training, supplies, equipment and first aid policies should be added or modified to account for changes in workplace safety and health hazards, worksite locations and worker schedules since the last program review. The first aid training program should be kept up-to-date with the latest first aid techniques and knowledge. Outdated training and reference materials should be replaced or removed from the workplace.

15. Occupational Safety and Health Administration Best Practices Guide: Fundamentals of a Workplace First Aid Program.

3.4.5 First Aiders[16]

First aiders should be selected from those staff members whose duties normally do not remove them from the workplace. First aiders should be on-site at the workplace as often as possible.

Additionally, prospective first aiders should be:

- Staff who show evidence of enthusiasm and a ability to deal with injury and illness;
- Able to relate well to and communicate clearly with other staff;
- In reasonable health;
- Able to exercise sound judgment especially regarding the need to involve other support services;
- Able to be called away from their ordinary work and respond in short notice; and
- Appointed to the first aider role via their own free will.

3.4.6 First Aid Training and Certification[17]

First aiders assume the initial treatment of people suffering injury and illness. The treatment provided by first aiders should be consistent with their training and competency. When in doubt, a first aider should recommend that an injured employee seek medical advice. A first aider should not be responsible for ongoing care.

In order to be effective, a workplace first aid in the training plan must cover the following issues and principles:

- Principals of First aid, Aims of first aid and priorities of care;
- Expired Air Resuscitation (EAR);
- Cardio Pulmonary resuscitation (CPR);
- Hygiene, infection control and medical waste removal;
- Reports and legalities;
- First aid kits;
- Practical decision making;

16. La Trobe University Occupational Health and Safety Manual, December 1999.
17. Ibid.

- Function of respiratory and circulatory systems;
- Cardiac emergencies;
- Secondary assessment & altered conscious state;
- Burns Fractures; theory and practical;
- Soft Tissue Injury; theory and practical;
- Head Injuries;
- Spinal Injuries;
- Eye Injuries;
- Respiratory emergencies; Asthma; hyperventilation;
- Diabetes, Stroke, Epilepsy,
- Allergic reactions;
- Poisoning;
- Moving a patient; and
- Material Safety Data Sheets and how to read them.

3.4.7 First Aid for Electric Shock[18]

Shock is a common occupational hazard associated with working with electricity. A person who has stopped breathing is not necessarily dead but is in immediate danger. Life is dependent on oxygen, which is breathed into the lungs and then carried by the blood to every body cell. Since body cells cannot store oxygen and since the blood can hold only a limited amount (and only for a short time), death will surely result from continued lack of breathing.

However, the heart may continue to beat for some time after breathing has stopped, and the blood may still be circulated to the body cells. Since the blood will, for a short time, contain a small supply of oxygen, the body cells will not die immediately. For a very few minutes, there is some chance that the person's life may be saved.

The process by which a person who has stopped breathing can be saved is called artificial ventilation (respiration). The purpose of artificial respiration is to force air out of the lungs and into the lungs, in rhythmic alternation, until natural breathing is reestablished. Records show that seven out of ten victims of electric shock were revived when artificial respiration was started in less than three minutes. After three minutes, the chances of revival decrease rapidly.

Artificial ventilation should be given only when the breathing has stopped. Do not give artificial ventilation to any person who

18. Electricians Toolbox Etc., http://www.elec-toolbox.com/Safety/safety.htm

is breathing naturally. You should not assume that an individual who is unconscious due to electrical shock has stopped breathing. To tell if someone suffering from an electrical shock is breathing, place your hands on the person's sides at the level of the lowest ribs. If the victim is breathing, you will usually be able to feel movement.

Once it has been determined that breathing has stopped, the person nearest the victim should start the artificial ventilation without delay and send others for assistance and medical aid. The only logical, permissible delay is that required to free the victim from contact with the electricity in the quickest, safest way. This step must be done with enormous care; otherwise, the first aid responder may be injured.

In the case of portable electric tools, lights, appliances, equipment, or portable outlet extensions, the victim should be freed from contact with the electricity by turning off the supply switch or by removing the plug from its receptacle. If the switch or receptacle cannot be quickly located, the suspected electrical device may be pulled free of the victim. Other persons arriving on the scene must be clearly warned not to touch the suspected equipment until it is de-energized.

The injured person should be pulled free of contact with stationary equipment (such as a bus bar) if the equipment cannot be quickly de-energized or if the survival of others relies on the electricity and prevents immediate shutdown of the circuits. This can be done quickly and easily by carefully applying the following procedures:

1. Protect yourself with dry insulating material.
2. Use a dry board, belt, clothing, or other available non-conductive material to free the victim from electrical contact. Do NOT touch the victim until the source of electricity has been removed.

Once the victim has been removed from the electrical source, it should be determined whether the person is breathing. If the person is not breathing, a method of artificial respiration is used.

Cardiopulmonary Resuscitation (CPR)

CPR has been developed to provide aid to a person who has stopped breathing and suffered a cardiac arrest. The techniques are relatively easy to learn.

Note: A heart that is in fibrillation cannot be restricted by closed chest cardiac massage. A special device called a defibrillator is available in some medical facilities and ambulance services.

Muscular contractions are so severe with 200 milliamperes and over that the heart is forcibly clamped during the shock. This clamping prevents the heart from going into ventricular fibrillation, making the victim's chances for survival better.

3.4.8 Skills Update

First aid responders tend to experience long intervals between learning certain skills such as CPR and AED and actually using them. Numerous studies have shown a retention rate of only 6-12 months of these crucial, life-saving skills. Skill reviews and practice sessions are encouraged at least every 6 months for CPR and AED skills. Instructor-led retraining for life-threatening emergencies should occur at least annually. Retraining for non-life-threatening response should also occur periodically.

3.4.9 First Aid Supplies[19]

It is important for the employer to give a specific person at the workplace the responsibility for choosing the types and amounts of first aid supplies and for maintaining those supplies. The supplies must be adequate, should address the kinds of injuries that commonly occur, and must be stored in an area where they are readily available for emergency access. An automated external defibrillator (AED) should be considered when selecting first aid supplies and equipment.

Employers who have unique or revolving first aid needs should consider upgrading their first aid kits. Consultation with the local fire and rescue service or emergency medical professionals and experts may be beneficial. By assessing the specific needs of their workplaces, employers can ensure the availability of adequate first aid supplies. Employers should periodically reassess the demand for these supplies and adjust their inventories accordingly.

19. Occupational Safety and Health Administration Best Practices Guide: Fundamentals of a Workplace First Aid Program.

3.4.10 Bibliography

Accident Compensation Act 1985, Victorian Consolidated Legislation, http://www.austlii.edu.au/au/legis/vic/consol_act/aca1985204/.

American Red Cross. First Aid: Responding to Emergencies. Third Edition. 2001.

ANSI Z308.1-2003. Minimum Requirements for Workplace First Aid Kits. Arlington VA:

American Heart Association, Heartsaver First Aid with CPR and AED, Publication 70-2562, Dallas: American Heart Association, 2002.

Electricians Toolbox Etc., http://www.elec-toolbox.com/Safety/safety.htm.

La Trobe University, Occupational Health and Safety Manual, December 1999.

Occupational Health and Safety Act of 2004, prepared by the Legislative Counsel Office, http://www.gov.pe.ca/law/statutes/pdf/o-01_01.pdf.

Occupational Safety and Health Administration, Best Practices Guide: Fundamentals of a Workplace First aid Program, OSHA 3317-06N, 2006.

Workplace Health and Safety Act 1995, Code of Practice, First Aid in the Workplace, http://www.legislation.qld.gov.au/LEGISLTN/CURRENT/W/WorkplHSaA95.pdf.

3.5 Fire Protection, Evacuation, First Responder and Emergency Planning

3.5.1 Flammability Properties

3.5.1.1 General Information

The majority of information organized in this subsection has been taken from the Handbook of Industrial Toxicology and Hazardous Materials for which permission has been granted.

Two main categories of liquids are flammable and combustible, and are determined mainly by the liquid's flash point. Both categories of liquids will burn, but it is into which of these two categories the liquid belongs that determines its relative fire hazard. Flammable liquids are generally considered the more hazardous of the two categories mainly because they release ignitable vapors.[20]

---

20. Cheremisinoff, N. P., Handbook of Industrial Toxicology and Hazardous Materials, Marcel Dekker Publishers, New York, New York, 1999. Both the Publisher and Author hereby grant the Government of Nigeria the use materials reprinted in this section of the Nigerian Electricity Safety Standards.

A flammable material is any liquid having a flash point below 37.8°C. The U.S. National Fire Protection Agency (NFPA) expands this definition by including the stipulation that the vapor cannot exceed 275.79 kPa at a liquid temperature of 37.8°C, with the theory being that such liquids are capable of releasing vapor at a rate sufficient to be ignitable. Since this aspect of the definition relating to vapor pressure has little fire-ground application it is often ignored. However, it is important to note that if the heat from a fire raises the liquid temperature to a temperature above the liquid's flash point, it will automatically increase the vapor pressure inside a closed container. Any other source of sufficient heat will produce the same result.

Within the combustible liquid category are those materials with a flash point above 37.8°C, Combustible liquids are considered less hazardous than flammable liquids because of their higher flash points. However, this statement can be misleading since there are circumstances when it is not a valid assumption. It is possible for certain combustible liquids to be at their flash point when a hot summer sun has been striking their metal container for some time. Additionally, during the transportation of some combustible products, the product is either preheated or a heat source is maintained to make the product more fluid than it would be at atmospheric temperatures. One reason this is done is to facilitate transportation or pumping; i.e. to aid with the movement of a material that is very viscous, such as asphalt or tar. Also, some materials classed as combustible solids will be heated to their melting point. Naphthalene is one example of this treatment. Naphthalene might be heated to a temperature above its melting point, which is about 80°C. Despite its fairly high ignition temperature (almost 526.7°C), it would not be unreasonable to surmise that a spill of liquid naphthalene could present a serious fire hazard. Fortunately, with naphthalene, quick action with adequate amounts of water applied as spray streams should cool and solidify it, thus greatly minimizing the fire risk.

It is important to note that a combustible liquid at or above its flash point will behave in the same manner that a flammable liquid would in a similar emergency. As an example, No.2 fuel oil when heated to a temperature of 65.6°C can be expected to act or react in the same way gasoline would at 10°C. In most instances, however, to reach this elevated temperature will require the introduction of an external heat source. Some common examples of combustible petroleum liquids are given in Table 3.7.

Table 3.7 Examples of petroleum liquids that are combustible.

Product	Flash Point (°C)
Kerosene	37.8+
Fuel Oils	37.8–60
Diesel Oil	54.4
Lubricating Oil	148.9
Asphalt	204.4

It is important to note that the extinguishing techniques, controlling actions, or fire-prevention activities implemented can differ greatly depending upon which of the two categories the liquid falls in. To have the ability to categorize a liquid correctly when it is not so identified, it is only necessary to know its flash point. By definition, the flash point of a liquid determines whether a liquid is flammable or combustible.

The categories of liquids are further subdivided into classes according to the flash point plus the boiling point of certain liquids. These divisions are summarized in Table 3.8, which shows that flammable liquids fall into Class 1, and combustible liquids into Classes 2 and 3. The products that are at the low end (37.8°C) of the Class 2 combustible-liquid group might be thought of as borderline cases. These could act very much like flammable liquids if atmospheric temperatures were in the same range. It is not a common industry practice to identify either stationary or portable (mobile) liquid containers by the class of liquid it contains. The usual practice is to label either "Flammable" or "Combustible" and include the required United Nations hazard placard.

Flash point is the temperature a liquid must be at before it will provide the fuel vapor required for a fire to ignite. A more technical definition for flash point is: The lowest temperature a liquid may be at and still have the capability of liberating flammable vapors at a sufficient rate that, when united with the proper amounts of air, the air-fuel mixture will flash if a source of ignition is presented. The amounts of vapor being released at the exact flash-point temperature will not sustain the fire and, after flashing across the liquid surface, the flame will go out. It must be remembered that at the

Table 3.8 Classes of flammable and combustible liquids.

Class	Flash Point (°C)	Boiling Point (°C)
1	Below 37.8	—
1A	Below 22.8	Below 37.8
1B	Below 22.8	At or above 37.8
1C	22.8-37.2	—
2	37.8-59.4	—
3	60 or above	Below 37.8
3A	60-92.8	At or above 137.8
3B	93.3 or above	—

flash-point temperature, the liquid is releasing vapors and, as with other ordinary burnable materials, it is the vapors that burn. The burning process for both ordinary combustible solids and liquids requires the material to be vaporized. It may also be in the form of a very fine mist, which will be instantly vaporized if a source of heat is introduced. It is not the actual solid or the liquid that is burning, but the vapors being emitted from it. For this reason, when we speak of a fuel we are referring to the liberated vapor. It is an accepted phenomenon, assuming sufficient amounts of air to be present, that the greater the volume of released vapor, the larger the fire will be.

The technical literature sometimes refers to the "fire point", which in most instances is just a few degrees above the flash point temperature, and is the temperature the liquid must be before the released vapor is in sufficient quantity to continue to burn, once ignited. However, because a flash fire will normally ignite any Class "A" combustible present in the path of the flash, it is reasonable to accept the flash point as being the critical liquid temperature in assessing a fire hazard. Any of the other combustibles ignited by the flash fire that is, wood, paper, cloth, etc., once burning, could then provide the additional heat necessary to bring the liquid to its fire point.

A crucial objective upon arrival of the first responding fire forces is to determine if the liquid present is a product that is vaporizing at the time or, if it is not, and what condition may be present that is capable of providing the required heat to cause the liquid to reach its flash point. This information would have a direct influence on the selection of control and/or extinguishing activities. An emergency involving a petroleum liquid, which is equal to or above flash point, means that a fuel source consisting of flammable vapors will be present. This, in turn, means the responding fire-fighting forces will be faced with either a highly hazardous vapor cloud condition or with a fire if ignition has occurred before arriving at the scene. Conversely, if it is a liquid at a temperature below its flash point, then fuel would not be immediately available to burn.

A source of air or more specifically, oxygen must be present. A reduction in the amount of available air to below ideal quantities causes the fire to diminish. Moreover, reduce the fuel quantity available and the fire will also diminish in size. Almost all extinguishing techniques developed are methods of denying the fire one or both of these requirements. By cooling a material below its flash point, vapor production is halted, thus removing the fuel from the fire. When utilizing a smothering-type extinguishing agent, the principle involved consists of altering the air-fuel mixture. When the vapor is no longer in its explosive range, the fire dies, either due to insufficient fuel or a lack of oxygen.

The flash point tells us the conditions under which we can expect the fuel vapor to be created, but it is the explosive range which tells us that a certain mixture of fuel vapor and air is required for the vapor to become ignitable. The terms flammable limit and combustible limit are also used to describe the explosive range. These three terms have identical meanings and can be used interchangeably. This information is reported as the lower explosive limits (LEL) and the upper explosive limits (UEL). The values that are reported for the LEL and UEL are given as a percentage of the total volume of the air-fuel mixture. The area between the LEL and the UEL is what is known as the explosive range. The figures given for the amount of fuel vapor required to place a substance within its explosive range are shown as a percentage of the total air-fuel mixture. To compute how much air is required to achieve this mixture, subtract the listed percentage from 100 percent: the remainder will be the amount of air needed. Even though it is only the oxygen contained in the air that the fire consumes, flammable ranges are shown as air-fuel ratios because it is the air that is so readily available. Any air-fuel mixture in which the vapor is above the UEL, or any air-fuel

Table 3.9 Explosive range of gasoline.

	LEL(%)	UEL (%)
Gasoline vapor	1.5	7.6
Air	98.5	92.4
Total volume	100	100

mixture in which the vapor is below the LEL, will not burn. Using gasoline as an example, the explosive range can be computed as follows, in Table 3.9.

This example helps to illustrate that large volumes of air are required to burn gasoline vapors. The explosive ranges for the different grades of gasoline, or even those of most other petroleum liquids, are such that average explosive-range figures that are suitable for use by the fire fighter would be the LEL at 1 percent and the UEL at about 7 percent. The vapor content of a contaminated atmosphere may be determined through the use of a combustible gas-detecting instrument, referred to as an explosimeter.

If a fire involving a petroleum liquid does occur, an extinguishing technique that may be appropriate is the altering of the air-fuel mixture. One technique utilized will necessitate the use of an extinguishing agent such as a foam with the capability of restricting the air from uniting with the vapor. Another technique is to prevent the liquid from having the ability to generate vapor. Usually this is a cooling action and is accomplished with water spray streams. In both cases, extinguishment is accomplished as a result of altering the air-fuel mixture to a point below the LEL for the specific liquid.

Flammability, the tendency of a material to burn, can only be subjectively defined. Many materials that we normally do not consider flammable will burn, given high enough temperatures. Neither can flammability be gauged by the heat content of materials. Fuel oil has a higher heat content than many materials considered more flammable because of their lower flashpoint. In fact, flashpoint has become the standard for gauging flammability.

3.5.1.2 *Flammability Designation*

The most common systems for designating flammability are the Department of Transportation (DOT) definitions, the National Fire

Protection Association's (NFPA) system, and the Environmental Protection Agency's (EPA) Resource Conservation and Recovery Act's (RCRA) definition of ignitable wastes, all of which use flashpoint in their schemes. The NFPA diamond, which comprises the backbone of the NFPA Hazard Signal System, uses a four-quadrant diamond to display the hazards of a material. The top quadrant (red quadrant) contains flammability information in the form of numbers ranging from zero to four. Materials designated as zero will not burn. Materials designated as four rapidly or completely vaporize at atmospheric pressure and ambient temperature, and will burn readily (flashpoint <22.8°C and boiling point < 37.8°C). The NFPA defines a flammable liquid as one having a flashpoint of 93.3°C or lower, and divides these liquids into five categories:

1. Class IA: liquids with flashpoints below 22.8°C and boiling points below 37.8°C. An example of a Class 1A flammable liquid is n-pentane (NFPA Diamond: 4).
2. Class IB: liquids with flashpoints below 22.8°C and boiling points at or above 37.8°C. Examples of Class IB flammable liquids are benzene, gasoline, and acetone (NFPA Diamond: 3).
3. Class IC: liquids with flashpoints at or above 22.8°C and below 37.8°C. Examples of Class IC flammable liquids are turpentine and n-butyl acetate (NFPA Diamond: 3).
4. Class II: liquids with flashpoints at or above 37.8°C but below 60°C, Examples of Class II flammable liquids are kerosene and camphor oil (NFPA Diamond: 2).
5. Class III: liquids with flashpoints at or above 60°C but below 93.3°C. Examples of Class III liquids are creosote oils, phenol, and naphthalene. Liquids in this category are generally termed combustible rather than flammable (NFPA Diamond: 2). The DOT system designates those materials with a flashpoint of 37.8°C or less as flammable, those between 37.8°C and 93.3°C as combustible, and those with a flashpoint of greater than 93.3°C as nonflammable. EPA designates those wastes with a flashpoint of less than 60°C as ignitable hazardous wastes.

The elements required for combustion are few – a substrate, oxygen, and a source of ignition. The substrate, or flammable material, occurs in many classes of compounds but most often is organic. Generally, compounds within a given class exhibit increasing heat contents with increasing molecular weights (MW).

These designations serve as useful guides in storage, transport, and spill response. However, they do have limitations. Since these designations are somewhat arbitrary, it is useful to understand the basic concepts of flammability.

Other properties specific to the substrate that are important in determining flammable hazards are the auto-ignition temperature, boiling point, vapor pressure, and vapor density. Auto-ignition temperature (the temperature at which a material will spontaneously ignite) is more important in preventing fire from spreading (e.g., knowing what fire protection is needed to keep temperatures below the ignition point) but can also be important in spill or material handling situations. For example, gasoline has been known to spontaneously ignite when spilled onto an overheated engine or manifold. The boiling point and vapor pressure of a material are important not only because vapors are more easily ignited than liquids, but also because vapors are more readily transportable than liquids (they may disperse, or when heavier than air, flow to a source of ignition and flash back). Vapors with densities greater than one do not tend to disperse but rather to settle into sumps, basements, depressions in the ground, or other low areas, thus representing active explosion hazards.

Oxygen, the second requirement for combustion, is generally not limiting. Oxygen in the air is sufficient to support combustion of most materials within certain limits. These limitations are compound specific and are called the explosive limits in air. The upper and lower explosive limits (UEL and LEL) of several common materials are given in Table 3.10.

The source of ignition may be physical (such as a spark, electrical arc, small flame, cigarette, welding operation, or a hot piece of equipment), or it may be chemical in nature, such an exothermic reaction. In any case, when working with or storing flammables, controlling the source of ignition is often the easiest and safest way to avoid fires or explosions.

Once a fire has started, control of the fire can be accomplished in several ways: through water systems (by reducing the temperature), carbon dioxide or foam systems (by limiting oxygen), or through removal of the substrate (by shutting off valves or other controls).

Table 3.10 Explosive limits of hazardous materials.

Compound	LEL %	UEL %	Flashpoint°C	Vapor Density
Acetone	2.15	13	-20	2.0
Acetylene	2.50	100	Gas	0.9
Ammonia,	16	25	Gas	0.6
Benzene	1.30	7.1	–11.1	7.8
Carbon monoxide	12.4	74	Gas	1.0
Gasoline	1.4	7.6	–42.8	3-4
Hexane	1.1	7.5	–21.7	3.0
Toluene	1.2	7.1	4.4	3.1
Vinyl chloride	3.6	33	Gas	2.2
p-xylene	1.0	6.0	32.2	3.7

Chapter 4 provides detailed discussion on the theories of fire and specific information on hydrocarbons, as well as chemical specific fire characteristics.

3.5.2 Ignition Temperature

Consider the emergency situation where there is a spill of gasoline. We may immediately conclude that two of the requirements for a fire exist. First, the gasoline, which would be at a temperature above its flash point, will be releasing flammable vapors; thus a source of fuel will be present. Moreover, there is ample air available to unite with the fuel thus there is the potential for the mixture to be in its explosive range. The only remaining requirement needed to have a fire is a source of heat at or above the ignition temperature of gasoline. Technically speaking, all flammable vapors have an exact minimum temperature that has the capability of igniting the specific air-vapor mixture in question. This characteristic is referred to as the ignition temperature and could range from as low as 148.9°C for the vapor from certain naphthas to over 482.2°C for asphaltic material vapor. Gasoline vapor is about halfway between, at 315.6°C. A rule of thumb for the ignition temperature of petroleum-liquid vapors is 260°C. This figure may appear low for several of the hydrocarbon vapors, but it is higher than that of most ordinary combustibles, and is close enough to the actual ignition temperatures of the products most frequently present at emergency scenes to give a suitable margin of safety.

In emergency situations, it is best to take conservative approaches by assuming that all heat sources are of a temperature above the ignition temperature of whatever liquid may be present. This approach is not an overreaction when it is realized that almost all the normally encountered spark or heat sources are well above the ignition temperature of whatever petroleum liquid might be present. Among the more common sources of ignition would be smoking materials of any kind (cigarettes, cigars, etc.), motor vehicles, and equipment powered by internal combustion engines; also electrically operated tools or equipment, as well as open-flame devices such as torches and flares. The removal of any and all potential ignition sources from the area must be instituted immediately and methodically. The operation of any motor vehicle, including diesel-powered vehicles, must not be permitted within the immediate vicinity of either a leak or spill of a flammable liquid. The probability of a spark from one of the many possible sources on a motor vehicle is always present. Also, under no circumstances should motor vehicles be allowed to drive through a spill of a petroleum product.

Ignition sources are not necessarily an external source of heat; it could be the temperature of the liquid itself. Refineries and chemical plants frequently operate processing equipment that contains a liquid above its respective ignition temperature. Under normal operating conditions, when the involved liquid is totally contained within the equipment, no problems are presented because the container or piping is completely filled with either liquid or vapor. If full and totally enclosed, it means there can be no air present; thus an explosive or ignitable mixture cannot be formed. If the enclosed liquid, which in certain stages of its processing, may be above the required ignition temperature should be released to the atmosphere, there is a possibility that a vapor-air mixture could be formed and hence, ignition could occur. This type of ignition is referred to as auto-ignition. Auto-ignition is defined as the self-ignition of the vapors emitted by a liquid heated above its ignition temperature and that, when escaping into the atmosphere, enter into their explosive range. Some typical ignition temperatures for various petroleum liquids are 315.6°C for gasoline, 287.8°C for naphtha and petroleum ethers, 210°C for kerosene, and 385°C for methanol.

From the above discussions, the important elements that are responsible for a fire are:

- Fuel in the form of a vapor that is emitted when a liquid is at or above its flash point temperature.
- Air that must combine with the vapor in the correct amounts to place the mixture in the explosive range.
- Heat, which must be at least as hot as the ignition temperature, must then be introduced.

3.5.3 Flammability Limits

Petroleum liquids have certain characteristics that can exert an influence on the behavior of the liquid and or vapor that is causing the problem. For this reason, these features may have a bearing on the choice of control practices or extinguishing agents under consideration. These characteristics include the weight of the vapor, the weight of the liquid, and whether the liquid will mix readily with water. The specific properties of importance are vapor density, specific gravity, and water solubility. Before discussing these important physical properties, let's first examine the data in Table 3.11 which

Table 3.11 Limits of flammability of gases and vapors, % in air.

Gas or Vapor	LEL	UEL
Hydrogen	4.00	75.0
Carbon monoxide	12.5	74.0
Ammonia	15.5	26.60
Hydrogen sulfide	4.30	45.50
Carbon disulfide	1.25	44.0
Methane	5.30	14.0
Ethane	3.00	12.5

Table 3.11 (cont.) Limits of flammability of gases and vapors, % in air.

Gas or Vapor	LEL	UEL
Propane	2.20	9.5
Butane	1.90	8.5
Iso-butane	1.80	8.4
Pentane	1.50	7.80
Iso-pentane	1.40	7.6
Hexane	1.20	7.5
Heptane	1.20	6.7
Octane	1.00	3.20
Nonane	0.83	2.90
Decane	0.67	2.60
Dodecane	0.60	
Tetradecane	0.50	
Ethylene	3.1	32.0
Propylene	2.4	10.3
Butadiene	2.00	11.50
Butylene	1.98	9.65
Amylene	1.65	7.70
Acetylene	2.50	81.00
Allylene	1.74	
Benzene	1.4	7.1
Toluene	1.27	6.75
Styrene	1.10	6.10
o-Xylene	1.00	6.00

Table 3.11 (cont.) Limits of flammability of gases and vapors, % in air.

Gas or Vapor	LEL	UEL
Naphthalene	0.90	
Antluacene	0.63	
Cyclo-propane	2.40	10.4
Cyclo-hexene	1.22	4.81
Cyclo-hexane	1.30	8.0
Methyl cyclo-hexane	1.20	
Gasoline-regular	1.40	7.50
Gasoline-73 octane	1.50	7.40
Gasoline-92 octane	1.50	7.60
Gasoline-100 octane	1.45	7.50
Naphtha	1.10	6.00

lists the flammability limits of some common gases and liquids. Two general conclusions can be drawn. First, the lower the material's LEL, obviously the more hazardous. Also note that there are some materials that have wide explosive ranges. This aspect is also significant from a fire standpoint. As an example, comparing hydrogen sulfide to benzene, although the LEL for H2S is more than 3 times greater, its explosive range is 7 times wider. This would suggest that H2S is an extremely hazardous material even though its LEL is relatively high. In fact, H2S fires are generally so dangerous that the usual practice is to contain and allow burning to go to completion rather than to fight the fire.

3.5.4 Vapor Density

Vapor density is a measure of the relative weight of vapor compared to the weight of air. Published data on the characteristics of petroleum products usually include the vapor density. The value of unity has been arbitrarily assigned as the weight of air. Hence any vapor that is reported to have a density of greater than 1 is heavier than air, and any vapor with a density of less than 1 is lighter than air. Vapors weighing more than 1 will usually flow like water, and those weighing less will drift readily off into the surrounding atmosphere. Even heavier-than-air flammable petroleum-liquid vapor can be carried along with very slight air currents. It may spread long distances before becoming so diluted with enough air as to place it below the lower explosive limit (LEL), at which time it would become incapable of being ignited. There are catastrophic incidences that have occurred whereby ignitable air-vapor mixtures have been detected as far as one-half mile from the vapor source. For this reason, while responding to a spill or leak, we must consider environmental and topographical features of the surroundings, such as wind direction, the slope of the ground, any natural or artificial barriers that may channel the liquid or vapors. It is critical in a non-fire incident such as a spill or leak to determine the type of petroleum liquid present and its source. Information about the material's vapor density enables us to make reasonable predictions as to the possible behavior of the emitting vapor. These factors may influence the route of approach, the positioning of firefighting apparatus and personnel, the need for and the route of evacuation, and the boundaries of the potential problem area. It is essential that no apparatus or other motor vehicles or personnel be located in the path that a vapor cloud will most likely follow.

As a rule of thumb one should approach a hydrocarbon spill (non-fire situation) under the assumption that the liquid is vaporizing (the vapors will be invisible) and that the liberated vapors are heavier than air unless proven otherwise. The expected conduct of a heavier-than-air vapor is for it to drop and spread at or below ground level much as a liquid would. The big difference is that a liquid will be visible and its boundaries well defined. One can expect that the invisible heavier-than-air vapor will settle and collect in low spots such as ditches, basements, sewers, etc. As the vapor travels, it will be mixing with the air, thus some portions of the cloud may be too rich to burn, other sections too lean, and still others well within the explosive range. Some typical vapor densities for petroleum products are 3 to 4 for gasoline, 2.5 for naphtha, and 1.1 for methanol. For comparison, the vapor density for hydrogen gas is 0.1.

3.5.5 Specific Gravity

The property of specific gravity indicates a petroleum liquid's weight relative to the weight of an equal volume of water. The specific gravity of water is assigned the value of unity as a reference point. Hence, other liquids are evaluated relative to water; those lighter than water have a value less than unity, whereas those that are heavier have a value that is greater than unity. In general, petroleum products are lighter than water; as a result, they can be expected to float on and spread over the water's surface. The exceptions to this are thick, viscous materials such as road tars and heavy "bunker" fuel.

With their low specific gravity, most petroleum liquids, if spilled onto a pool of water, have a tendency to spread quickly across the water's surface. Unless the fluid contacts an obstacle, the oil will continue to spread until it is of microscopic thickness. For this reason, a relatively small amount of oil floating on water is capable of covering a large area of the surface. In a spill situation, knowledge of the liquid's specific gravity can help determine which one of several tactics will be implemented to mitigate the spread of contamination or to eliminate a fire hazard. Knowing the specific gravity will help determine the following:

- If it will be possible to use only water, or must we apply a different agent for purposes of smothering the fire.
- How great the probability is that the burning liquid could result in the involvement of exposures because of it floating on any water that is applied.
- Can the displacement of the fuel by water be considered an effective technique to control a leak from a container?

Some specific gravities of common petroleum liquids are 1 to 1.1 for asphalt, 0.8 for gasoline, and 0.6 for naphtha.

3.5.6 Water Solubility

Another important property is water solubility, which may be described as the ability of a liquid to mix with water. Since most petroleum products are lighter than water, and even if they are well mixed with water, they will separate into a layer of water and a layer of the hydrocarbon. Exceptions to this are polar solvents such as methanol and other alcohols. These types of materials will readily mix with water and can even become diluted by it.

Information on solubility is important in an emergency situation because a petroleum fire will require the application of a regular-type foam or an alcohol-resistant-type foam for extinguishment or for vapor-suppression purposes. For use on a water-soluble liquid, a good-quality alcohol-resistant foam is generally applied.

3.5.7 Responding to Fires

In an emergency situation involving a flammable liquid, the product can be expected to behave as follows:

- When accidentally released from its container it almost always results in a fire response.
- If a fire does occur, flammable liquids prove to be virtually impossible to extinguish by cooling with water.
- If the liquid is contained, the confined space will consist of a vapor-rich mixture.
- After extinguishment, there is still the strong possibility for a reflash owing to the continued production of vapor.

It is important to remember that during an emergency the escaping flammable liquids are low-flash products and, as a result, are releasing vapor at the usual atmospheric temperatures. These materials are, therefore, very susceptible to ignition. Because of this, they are generally encountered as an event requiring fire-control procedures. Also, for the same reasons, and the frequent need for large quantities of chemical extinguishing agents, they can present difficult extinguishment problems. The fact that the temperature at which many flammable liquids release vapor is well below the temperature of the water that is being used for fire-control purposes means that extinguishment by the cooling method is not feasible. This does not mean water cannot be used. In fact, the use of water serves an important function in spite of its limitations. However, to obtain extinguishment other tactics utilizing a different agent and techniques are needed. Exactly what agent and what technique will be dictated by the size of the fire, the type of storage container or processing equipment involved, and the fire-fighting resources that are available. When dealing with a low flash point flammable liquid, the probability of a reflash occurring after initial extinguishment is achieved is a high probability.

The probability of an ignitable air-fuel mixture existing inside a closed storage container of a flammable liquid is normally low. Flammable liquids have the capacity of generating vapor below the commonly encountered atmospheric temperatures; thus, the space between the tank top and the liquid surface will most usually contain a vapor-rich atmosphere (i.e., conditions will be above the UEL). As the vapor being liberated drives the air from the container, the vapor-rich mixture being above the UEL cannot be ignited; moreover, if a fire should occur outside the tank or vessel, it will not propagate a flame back into the tank. The major exceptions would be those instances where the product had a flash point temperature about the same as the prevailing atmospheric temperature. Another example would be when a tank containing a low vapor pressure product with a flash point in the same range as the prevailing atmospheric temperature is suddenly cooled. A thunderstorm accompanied by a downpour of rain could cause the tank to breathe in air if the liquid is cooled below the temperature at which it is capable of emitting vapor (the boiling point).

In contrast to flammable liquids, an emergency situation that involves a combustible liquid will have a much different behavior. The expected behavior of a combustible product would be for:

- the liquid to present no significant vapor problem;
- a fire to be readily extinguished by cooling the liquid with water; and
- the atmosphere above the liquid level to be below the LEL of any confined product.

Most combustible liquids do not present a vapor problem if accidentally released into the atmosphere. The probability of a fire, therefore, is considerably less than it would be if the spill was of a flammable material. If, however, the combustible liquid is at a temperature higher than its flash point, then it can be expected to behave in the identical manner of a flammable liquid. One major difference between the two in a fire situation is that the potential exists for cooling the combustible liquid below its flash point by the proper application of water (generally applied in the form of water spray). In the event the liquid is burning, and if the fire forces are successful in achieving the required reduction in liquid temperature, then vapor production will cease and the fire will be extinguished because of a lack of vapor fuel. Unless this reduction in liquid temperature can be brought about, the fire will necessitate the same control considerations a low-flash liquid fire would.

With a fire in a storage tank containing a combustible liquid, normally the application of a foam blanket is the only practical method to achieve extinguishment. This is normally done when the entire exposed surface is burning and is necessitated by the fact that the sheer size of the fire makes it very difficult to apply water spray in the amounts and at the locations needed. Also, the volume of oil that has become heated is such that the large quantities of water needed to cool the liquid would introduce the possibility of overflowing the tank. Another problem is that the water requirements for both protecting exposures and attempting extinguishment might be far greater than what is available. These factors alone would cause the emergency forces to treat all refined product tank fires alike and, regardless of the flash point of the liquid, initiate the required steps to apply a foam blanket to the liquid surface. In addition there is always concern for any reaction the water may cause when contacting an oil or hydrocarbon heated above the boiling point of water (212°F). In general, because a spill is generally shallow, spilled or splashed combustible liquids do not present the type of problems that a large storage tank does. It is reasonable to expect to be able to extinguish shallow pools or surface spills with water spray.

In the case of a storage tank with a combustible product, the atmosphere inside the container will normally consist of air. On occasion, there will be detectable odors such as that associated with fuel oil. These odors, which are a good indicator of the presence of a combustible product, are not considered fuel vapor. During a fire situation, flame impingement or radiant heat on a container could cause the liquid to become heated to the temperature at which it would emit vapors. Should this occur, and the vapors being generated then start to mix with the air already in the tank, at some instant the space in the container would then contain a mixture that was within the explosive range of the product; if ignition occurred, a forceful internal explosion could result.

The case of crude oil is somewhat unique compared to fires with refined petroleum products. Burning crude oil has the capability of developing a "heat wave". Crude oil has a composition of different fractions of petroleum products. In a manner similar to a refinery operation which distills, or heats crude oil to separate it into the various usable products - such as gasoline and asphalt, a fire accomplishes the same effect. As crude oil burns it releases the fractions that have lower flash points first, and these are burned. The heavier fractions sink down into the heated oil. This movement of light fractions up to the fire and heavier, heated fractions down into the crude produces the phenomenon known as a heat wave. Crude oil has the same basic characteristics as other flammable liquids. There are different grades of crude oil produced from various geographic locations throughout the world, with some crudes having more heavy-asphalt type material than others. Some have greater quantities of sulfur (creating the problem of poisonous hydrogen sulfide gas generation during a fire) and still others more light, gaseous fractions; however, a common characteristic among them is that all will have varying amounts of impurities and some entrained water.

When liquids of this type burn, creating a heat wave which is comprised of the higher boiling point components plus whatever impurities may be present in the product, radiant heat from the flame heats the liquid surface, the light products boil-off, thus creating the vapor that is burning. The remaining hot, heavier materials transfer their heat down into the oil. As it is formed, this heat wave, or layer of heated crude oil components, can reach temperatures as high as 315.6°C and spread downward at a rate of from 30.5 to 45.7 centimeters per hour faster than the burn-off rate of the crude oil. This would mean that with a crude oil burn-off of 30.5 centimeters per hour, at the end of two hours the heat wave would be somewhere between 61.0 and 91.4 centimeters thick. Once this heat wave is created, the chances of extinguishing a crude-oil tank fire, unless it is of small size, are poor, and any water or foam applied could result in a "slopover" of burning oil.

Crude oil normally contains some entrained water and/or an emulsion layer of water and oil. In addition, crude-oil storage tanks will have some accumulations of water on the uneven tank bottoms. In a fire, when a heat wave is formed and comes in contact with any water, a steam explosion will occur, thus agitating the hot oil above it with great force. The evolution of the steam explosion can be attributed to the reaction of water to high temperatures. When water is heated to its boiling point of 100°C, water vapor or steam is generated. Steam that is produced expands approximately 1700 times in volume over the volume of the water that boils away. If a heat wave of a temperature well above 100°C contacts any water entrained in the oil, or some of the bottom water, which is usually in larger quantities, the instantaneous generation of steam will act like a piston, causing the oil to be flung upward with considerable violence and force. This reaction is so strong that it causes the oil to overflow the tank shell. This sudden eruption is known as a boilover. When the hot oil and steam reaction takes place, the oil is made frothy, this in turn further increases its volume. The reaction resulting from the heat wave contacting entrained water can be expected to be of lesser activity than from contact with bottom water. The reason for this difference is that the quantities of water converted to steam in a given spot are usually less.

Another phenomenon associated with a crude oil fire is slopover. Basically, the same principles that are responsible for a boilover are the cause of a slopover. The fundamental difference is that in a slopover the reaction is from water that has entered the tank since the start of the fire. Usually this introduction is the result of firefighting activities. A slopover will occur at some point after the heat wave has been formed. Either the water from the hose streams or, after the bubbles collapse, the water in the foam will sink into the oil, contacting the heat wave, where it is converted to steam, and the agitation of the liquid surface spills some amount of oil over the tank rim.

The proper extinguishing agents suitable for petroleum liquid emergencies must be capable of performing the identical functions as those agents used in the combating of structural-type fires. There will be times when circumstances dictate the use of a cooling agent, whereas at other times it will be a smothering agent, and on some occasions, both agents will be necessary. There are a variety of agents capable of accomplishing each of these objectives, as well as being appropriate to combat Class B-type fires. Agents suitable for use on Class B-type fires include halon, carbon dioxide (COD), dry chemicals, foam, and aqueous film forming foam. A description of the major types of fire fighting agents is given below.

3.5.8 Fire Fighting Agents

3.5.8.1 Water

For chemical fires water can be used for the dissipation of vapors, for the cooling of exposed equipment, for the protection of personnel, for control purposes, and, for actual extinguishment. For the cooling down of exposed equipment such as pipelines, pinups, or valves, it is recommended that water be applied at the minimum flow rate of 3.8 liters per minute per 10 square feet of exposed surface area. Some general guidelines to consider when applying water are as follows:

- All areas of any piping, containers, etc., that are exposed to the fire's heat or flame should be kept wetted during the course of the fire.
- The use of water streams to push and move the burning liquid away from exposed equipment is recommended provided that it can be done safely.
- The rate of flow from any hose stream should not be less than 378.4 l/min regardless of its purpose.
- The use of spray streams is recommended whenever possible.
- Back-up lines for lines in active service should be provided and they should be at least of the same capacity as the attack line.
- Any equipment being protected is cool enough if water applied to it no longer turns to steam.

With storage tanks or processing equipment exposed to fire or radiant heat, the cooling of any metal above the liquid level inside the vessel is critical. Metal surfaces that have a constant film of water flowing over them will not reach a surface temperature above the boiling point of water. This temperature is well below that which would subject the metal to loss of integrity because of softening.

Water flows employed to cool exposed vertical storage tanks can be calculated using as a requirement one hose line (flowing 756.9 l/min) per 3.0 meters of tank diameter. Assuming an average tank height of 15.2 feet, this would give a water flow capability in excess of the recommended rate. On the fire ground, wind conditions, personnel deficiencies (fatigue, lack of experience or training, etc.), strewn feathering, and so forth have historically resulted in not all the water that is flowing actually doing its intended function. The rule of thumb of one line flowing a minimum of 756.9 l/min for each 3.0 meters of tank diameter provides a suitable safety margin to overcome the loss of water not reaching its target. This flow is only required on the side or sides of the tank being heated; therefore, if a 30.5-meter-diameter tank is receiving heat on just one-half of its circumference, it would require five hose streams of 756.9 l/min, each applied to the heated area for cooling purposes. It must be anticipated that these minimum flows will need to be maintained for a time period of at least 60 minutes. Tank truck incidents in which the fire burned for several hours are not unusual. It should be appreciated that a relatively minor fire on a tank truck or rail car could require in excess of 75,686 liters of water for control and/or extinguishment. It is imperative that as early into the event as possible, an accurate assessment of water flow requirements should be made and flow rates adjusted accordingly. First responders must be constantly alert for any indication of an increase in the internal pressure of a container. Such an increase or any visible outward distortion of the tank shell would be an indication that additional water flows are required. These warning signs, which would be an indication of increased internal pressures, could justify the immediate use of unmanned monitors or hoseholders, larger size nozzles, and the pulling back of all personnel to a safe location.

In preparing emergency response plans for petroleum liquid spills or fires, it should be taken into consideration that the required water rates could be needed for long periods of time. Built into the plans must be provisions for an uninterrupted supply at a suitable volume. The rates stipulated in the foregoing do not include amounts of water that may be needed for the protection of fire-fighting personnel who are involved in activities such as rescue work or valve closing and block off operations. If these or other water-consuming activities are required, additional water must be provided. Of equal importance to the amount of water being used is that water be used in the right place. In general, to application of water should have as its objective one or more of the following goals:

- The cooling of the shell of any container that is being subjected to high heat levels. This is most effectively accomplished by applying the water to the uppermost portions of the container and allowing it to cascade down the sides.
- The cooling of any piece of closed-in equipment containing a liquid or gas and exposed to high heat levels. This is most effectively accomplished by applying the water spray over the entire area being heated.
- The protection of any part of a container, piping, or item of processing equipment receiving direct time

contact. This is most effectively accomplished with a very narrow spray pattern or even a straight stream directed at the point of flame contact.

- The cooling of steel supports of any container or pipe rack that may be subjected to high heat levels. This is most effectively accomplished by the application of narrow spray streams to the highest part of the support being heated and permitting the water to run down the vertical length of the support.

3.5.8.2 Foam

The application of a foam blanket is the only means available to the fire forces for the extinguishment of large petroleum storage tank fires. The foam blanket extinguishes by preventing vapor, rising from the liquid surface, from uniting with the surrounding air and forming a flammable mixture. Although the water in a foam does provide some incidental cooling action, this is considered of more importance for cooling heated metal parts, thus reducing the possibility of re-ignition, than as an extinguishing factor. A good-quality foam blanket of several inches in thickness has also been proved effective as a vapor suppressant on low flash-point liquids. Foam may also be used to suppress vapor, hence the layer of foam will be instrumental both in preventing ignition and reducing the contamination of the surrounding atmosphere. Since foam is still water, even if in a different form from that usually used, it may conduct electricity; consequently, its use on live electrical equipment is not recommended.

There are basically two methods of foam application to fires. The first involves the application of chemical foam, which is generated from the reaction of a powder with water. This type of foam has been replaced largely by a technique that involves the formation of foam bubbles when a foaming agent and water are expanded by the mechanical introduction of air. This type, which is not a chemical reaction, is referred to as mechanical foam. Another name for the same material is air-foam. There are a variety of foam concentrates designed to fit different hazards. These include regular protein-based foams, fluoroprotein foams, aqueous film forming types, alcohol-resistant types, as well as foams that are compatible with dry chemical powders and those that will not freeze at below-zero temperatures. Of the many types, the most suitable for general all-around petroleum the use would be either a good-quality fluoroprotein or a good-quality aqueous film forming foam (AFFF). Foam liquids are also available in a wide range of concentrates, from 1 percent to 10 percent. The 3 percent and 6 percent protein and fluoroprotein types are usually employed as low-expansion agents with an expansion ratio of about 8 to 1. That is, for each 378.4 liters of foam solution (water/concentrate mix) to which air is properly introduced, it will then develop approximately 3,027.4 liters of finished foam. Foam concentrates of other than 3 percent or 6 percent generally are either high-expansion (as high as 1,000 to 1) or alcohol-resistant types.

In foam applications, the manufacturer will provide a percentage rating of a concentrate, which identifies the quantity of concentrate required to be added to water to achieve a correct solution mixture. For each 378.4 liters of solution flowing, a concentrate rated as 3 percent would mean 3 gallons of concentrate per 367.1 liters of water, whereas a 6 percent concentrate would mix with 355.7 liters of water. This readily explains why only half as much space is required to store or transport the amount of 3 percent concentrate needed to generate a given quantity of foam than would be needed for a 6 percent concentrate to make the same volume of foam.

Once the application of foam is initiated, it must be applied as gently as possible in order to develop a good vapor-tight blanket on the liquid surface. Any agitation of the foam blanket or of the burning liquid surface will serve to prolong the operation and to waste foam supplies. Water streams cannot be directed into the foam blanket or across the foam streams because the water will dilute and break down the foam. To be assured that all metal surfaces are cool enough and a good, thick (10.2 centimeters or more) blanket of foam has been applied, continue application for a minimum of five minutes after all visible fire is extinguished.

One of the agents considered suitable for the extinguishment of petroleum-liquid fires is aqueous film forming foam. This is a liquid concentrate that contains a fluorocarbon surfactant to help float and spread the film across the petroleum-liquid surface and is commonly referred to as "A Triple F". AFFF concentrates of 1, 3, or 6 percent are available, all with about an 8 to 1 expansion ratio. This material is one of the mechanical-type foaming agents. The same kind of air-aspirating nozzles and proportioners that are used for protein-based foams are usable with AFFF concentrates. The primary advantage of AFFF over other foaming agents is its ability to form a thin aqueous film that travels ahead of the usual foam bubbles. This film has the ability to flow rapidly across the burning liquid surface, thus extinguishing the tire by excluding the air as it moves across the surface. The regular foam bubbles formed and flowing behind the film have good securing qualities, which serve to prevent rehashing from occurring. As with all types of foams, care must be exercised that the foam blanket, once formed, is not disturbed. Water streams should not be directed into the foam blanket or onto the same target a foam stream is aimed at. Water will dilute the foam below the needed concentration and, simultaneously, the force of the stream will destroy the foam's blanketing effect. The blanket must be maintained until all flames are extinguished, all heated metal surfaces cooled, and other sources of ignition removed from the vicinity.

3.5.8.3 *Alcohol-Resistant Foams*

Foams that are suitable for water-soluble polar solvents are formulated to produce a bubble that is stable in those fuels and tends to mix and unite with water. Fuels of this type dissolve the water contained in regular foam very rapidly resulting in the collapse of the bubbles. The breakdown is so fast and complete with regular protein or fluoroprotein based foams that unless the rate at which the foam is being applied is well above the recommended rate, the blanket will not form at all. Alcohol-type foam concentrates are most commonly available at strengths of 3 percent, 6 percent, or 10 percent. Because of the possibility of breakdown, regular foams are not considered suitable for polar solvent-type fires. The exception would be a fire in a container of fairly small diameter or a shallow spill, either of which would allow for the possibility of applying foam at sufficient rates to the point of overwhelming the tire.

3.5.8.4 *High Expansion Foams*

High-expansion foam includes foaming agents with the expansion ratio between the solution and the foam bubbles of from 20 to 1 to as high as 1,000 to 1. This agent has been found suitable when combating certain types of Class A and Class B fires. Originally developed to help fight fires inside mines, it is most effective when used in confined areas. Extinguishment is accomplished both by the smothering action of the foam blanket and the cooling action obtained from the water as the bubbles break down. Light, fluffy bubbles break apart and are easily dispersed by even relatively moderate wind currents. Bubbles formed at ratios greater than about 400 to 1 are most likely to be adversely affected by regular air movement as well as the thermal updrafts created by the fire. In an effort to overcome the susceptibility of the bubbles to wind currents, medium expansion foams have been introduced, which have expansion ratios ranging from about 20:1 to 200:1. High-expansion foam concentrates require special foam generators both for proportioning the liquid with water, and aspirating the mixture. Many high expansion foam-dispensing devices have a discharge range of only a few feet; thus, they must be operated fairly close to the area being blanketed.

3.5.8.5 *Other Extinguishing Agents*

Other extinguishing agents that are suitable for use on fires involving petroleum liquids include dry chemical powders, carbon dioxide gas, and halon gases. Each of these agents, while being capable of extinguishing Class B fires, usually is available in either hand-held extinguishers or the larger wheeled or trailer-mounted portable units.

In some petroleum refineries or chemical plants, an on-site fire brigade that is equipped with an apparatus capable of dispensing large volumes of dry chemical, or a vehicle with a large-capacity carbon dioxide (CO_2) cylinder is common practice.

3.5.8.6 *Carbon Dioxide*

Carbon Dioxide (CO_2) is used in fighting electrical fires. It is non-conductive and, therefore, the safest to use in terms of electrical safety. It also offers the least likelihood of damaging equipment. However, if the discharge horn of a CO_2 extinguisher is allowed to accidentally touch an energized circuit, the horn may transmit a shock to the person handling the extinguisher.

The very qualities which cause CO_2 to be a valuable extinguishing agent also make it dangerous to life. When it replaces oxygen in the air to the extent that combustion cannot be sustained, respiration also cannot be sustained. Exposure of a person to an atmosphere of high concentration of CO_2 will cause suffocation.

3.5.9 Electrical Fire Prevention

To prevent electrical fires, pay special attention to the following areas.

- Against short circuit: Electrical short circuit has three situations: grounding short-circuit, short-circuit between the lines, and completely short-circuit.
  - To prevent short circuits: 1) Install electrical equipment according to the circuit voltage, current strength and the use of nature, correct wiring. In acidic, hot or humid places, it is necessary to use acid with anti-corrosion, high temperature and moisture-resistant wires. Traverse firmly installed to prevent shedding and should not be tied into wire or wires tightly knotting hung on the wire or nails. 2) Mobile power tools wire, a good layer of protection to prevent the mechanical damage, loss. 3) No bare wire inserted in the socket on the end. 4) Power master switch, the switch should be installed for the use of the current strength of the insurance unit, and conduct periodic current operation to eliminate hidden perils.
- Prevent overloads: Various conductors have certain load, when the load current intensity over Traverse, Traverse temperature surge, insulating layer can lead to fire. Prevent circuit overload by: 1) Ensure that all electrical equipment are in strict accordance with the electrical safety codes matching the corresponding wires, and the correct installation, are not allowed to randomly indiscriminate access. 2) Check to see that the appropriate use of load is induced on wires. Make sure that electrical equipment do not exceed load limitations, and that production levels and needs have separation, control use. 3) To prevent single-phase three-phase motor running, it is necessary to install three-phase switching single-phase power distribution boards running lights. 4) Circuit master switch, the switch should be installed in traffic safety and wire line Fusing easy for the insurance.
- Anti-contact resistance: When a conductor with another wire or wires and switches, protective devices, meters and electrical connections are exposed, they form a resistance, called contact resistance. If the contact resistance is too large, then when the current is passed it will result in heating to the point where the wire insulation layers will catch on fire, metal wire fuse, and sparks are generated. If there is any source of fuel

in close proximity then the potential for a fire exists. To avoid this: 1) All connections should be secured and layers should be thoroughly clean of grease, dirt and oxide films. 2) Apply 6-102 mm cross-sectional area of wire, welding method of connecting. 102 mm cross section above the wire should be connected by wiring tablets. 3) Regular line connections should be made and any connections found to be tap loose or hot should be immediately corrected.

- Check electrical equipment regularly for sparks or arcing. These conditions can trigger spontaneous combustion of flammable gases or cause dust explosions. To prevent electrical equipment from contributing to fires: 1) Conduct regular inspections of insulation resistance and monitor the quality of insulation layers. 2) Inspect and correct naked wires and metal contacts to prevent short circuit. 3) Install explosion-proof seals or isolated lighting fixtures, switches and protective devices.
- Lighting fires: 1) Lamp fire such as bulbs, including incandescent, iodine-tungsten lamp, and high pressure mercury lamps generate high temperature surfaces. The greater the power along with continuous service, the higher temperatures will be experienced. Thermal radiation from hot surfaces can be a source of spontaneous combustion if fuels and combustible materials are stored nearby. Fire Prevention bulbs should pay attention to make sure: no paper lamp door, or the use of paper, cloth lamp; likely to be hit in place, the lamps should have solid metal enclosures; they should not be closer than 30 cm from cloth, paper or clothing, and should be isolated from flammable materials. 2) Fluorescent fire: Ballasts can be the cause for such fires. Attention should be given to the installation of ventilation in order to provide cooling the fluorescent and to prevent leakage. Ballasts should be installed from the bottom up, not down. During installation, care should be given to prevent Asphalt seals from spilling and resulting in burning on the ballast. Ballasts may be installed with high temperature alarms that automatically cut off the power supply when temperature becomes excessive or smoke is detected.

3.5.10 Firefighting Guidance

3.5.10.1 Types

The types of firefighting are:

- **Structural firefighting** is fighting fires involving buildings and other structures. Most city firefighting is structural firefighting. Structural firefighting is what most people think of when they think of what their local "fire department" does.
- **Wildland firefighting** is fighting forest fires, brush or bush fires, and fires in other undeveloped areas.
- **Proximity firefighting** is fighting fires in situations where the fire produces a very high level of heat, such as aircraft and some chemical fires. Proximity firefighting is not the same thing as structural firefighting and in some (but not all) cases takes place outdoors.
- **Entry firefighting** is a highly specialized form of firefighting involving the actual direct entry of firefighters into a fire with a very high level of heat. It is not the same thing as proximity firefighting, although it is used in many of the same situations (such as flammable liquid fires).

Three elements necessary for a fire to burn: oxygen, fuel, and a source of heat. Remove any one of the three and the fire can no longer burn. All firefighting tactics are based on removing one or more of the three elements in the fire triangle. Fire fighting involves a range of methods and equipment for fighting fires, depending on the circumstances and availability.

3.5.10.2 Firefighting Agents and Extinguishers

Water: The standard method is to pour water onto the fire. This reduces the heat, hopefully to the point that the fire can no longer burn. However, water is not suitable for liquid fuel fires, because it is likely to spread the fuel and make the fire larger, nor for electrical fires, because of the risk of electric shock.

Foam: Foam is put onto fires fuelled by liquid fuel. It works by preventing oxygen getting to the fire, but does not spread the fuel around.

Powder: Powder is used on electrical fires, as it does not conduct electricity. Like foam, it works by cutting off the oxygen supply to the fire.

Gas: Non-combustible gases are sometimes used in fire-suppression systems installed in buildings. They work by being heavier than air, thus displacing the oxygen. However, because the oxygen is displaced from the building, humans are unable to survive for long either, so gas systems are only used in special circumstances.

Fire beaters and rakes: Fire beaters and rakes are low-tech ways to fight small fire in scrub. They work by spreading the fire out, thus reducing the concentration of heat, and by removing the source of fuel from the fire.

Extinguishers: Fire extinguishers are canisters of water, foam, or powder which can be carried by one person to the fire. They are of limited capacity, but can be used to stop a fire before it becomes large.

Vehicles: Various vehicles are used by fire fighters, including trucks which pump water from a nearby supply onto a fire, and tankers which carry water to the site of the fire when no local supply is available.

Fire breaks: For fires in forested areas, bulldozers or other heavy equipment may be used to cut a clear path through the forest in front of an advancing fire, or such fire breaks may be permanently maintained in forested areas. Fire breaks provide an area with little fuel content, for the purpose of the fire stopping when it reaches the break. However, strong winds might carry sparks from the fire across the fire break to further forest on the other side. Even so, it may slow down the fire enough for fire crews to cope with the rest.

Back-burning: Instead of constructing fire breaks, an area ahead of an advancing fire might be set alight and burnt in a controlled manner to remove the fuel from that area, thus impeding the advance of the main fire.

For small fires, a fire extinguisher can be used. Remember P.A.S.S. (Pull - Aim - Squeeze - Sweep). Fire extinguishers are rated for the types of fires they are effective against. Class A: ordinary flammable solids, such as wood and paper; Class B: flammable liquids, such as grease, oil, and gasoline; Class C: electrical fires. Class D: flammable metals.

Fire extinguishers are divided into four categories, based on different types of fires. Each fire extinguisher also has a numerical rating that serves as a guide for the amount of fire the extinguisher can handle. The higher the number, the more fire-fighting power. The following is a quick guide to help choose the right type of extinguisher.

- Class A extinguishers are for ordinary combustible materials such as paper, wood, cardboard, and most plastics. The numerical rating on these types of extinguishers indicates the amount of water it holds and the amount of fire it can extinguish.
- Class B fires involve flammable or combustible liquids such as gasoline, kerosene, grease and oil. The numerical rating for class B extinguishers indicates the approximate number of square feet of fire it can extinguish.
- Class C fires involve electrical equipment, such as appliances, wiring, circuit breakers and outlets. Never use water to extinguish class C fires - the risk of electrical shock is far too great! Class C extinguishers do not have a numerical rating. The C classification means the extinguishing agent is non-conductive.
- Class D fire extinguishers are commonly found in a chemical laboratory. They are for fires that involve combustible metals, such as magnesium, titanium, potassium and sodium. These types of extinguishers also have no numerical rating, nor are they given a multi-purpose rating - they are designed for class D fires only.

Some fires may involve a combination of these classifications.

- **Water extinguishers** or APW extinguishers (air-pressurized water) are suitable for **class A fires only**. Never use a water extinguisher on grease fires, electrical fires or class D fires - the flames will spread and make the fire bigger! Water extinguishers are filled with water and pressurized with oxygen. Again - water extinguishers can be very dangerous in the wrong type of situation. Only fight the fire if you're certain it contains ordinary combustible materials only.
- **Dry chemical extinguishers** come in a variety of types and are suitable for a combination of class A, B and C

Figure 3.1 Fire Extinghishers

fires. These are filled with foam or powder and pressurized with nitrogen.

- **BC -** This is the regular type of dry chemical extinguisher. It is filled with sodium bicarbonate or potassium bicarbonate. The BC variety leaves a mildly corrosive residue which must be cleaned immediately to prevent any damage to materials.
- **ABC -** This is the multipurpose dry chemical extinguisher. The ABC type is filled with mono-ammonium phosphate, a yellow powder that leaves a sticky residue that may be damaging to electrical appliances such as a computer.

Dry chemical extinguishers have an advantage over CO_2 extinguishers since they leave a non-flammable substance on the extinguished material, reducing the likelihood of re-ignition.

- **Carbon Dioxide (CO_2) extinguishers** are used for class B and C fires. CO_2 extinguishers contain carbon dioxide, a non-flammable gas, and are highly pressurized. The pressure is so great that it is not uncommon for bits of dry ice to shoot out the nozzle. They don't work very well on class A fires because they may not be able to displace enough oxygen to put the fire out, causing it to re-ignite.

CO_2 extinguishers have an advantage over dry chemical extinguishers since they don't leave a harmful residue - a good choice for an electrical fire on a computer or other favorite electronic device such as a stereo or TV.

These are only the common types of fire extinguishers. There are many others to choose from. Base your selection on the classification and the extinguisher's compatibility with the items you wish to protect.

3.5.10.3 Vehicles

The "fire truck" and "fire engine" are two different vehicles, and the terms should not be used interchangeably. In general, fire engines are outfitted to pump water at a fire. Trucks do not pump water and are outfitted with ladders and other equipment for use in ventilation of the building, rescue, and related activity. In many jurisdictions it is common to have a ladder (or truck) company and an engine company stationed together in the same firehouse.

3.5.10.4 Firefighting Gear

Basic firefighting gear includes:

- Bunker gear - refers to the set of protective clothing worn by firefighters (Figures 3.2 and 3.3). This is made of a modern fire-resistant material such as Nomex, and has replaced the old rubber coat which was traditionally worn. Nomex is a registered trademark for flame resistant meta-aramid material developed in the early 1960s by DuPont. The original use was for parachutes in the space program. A Nomex hood is a common piece of firefighting equipment. It is placed on the head on top of a firefighter's face mask. The hood protects the portions of the head not covered by the helmet and face mask from the intense heat of the fire.
- SCBA, or self-contained breathing apparatus, is worn in any IDLH (immediately dangerous to life or health)

Figure 3.2 Protective clothing worn by firemen.

Figure 3.3 Protective clothing and equipment used by firemen.

situation, which includes entry into burning buildings (where hot air and smoke inhalation present a danger) or any situation involving hazardous materials exposure. It consists of a breathing mask and a tank of air worn on the back.

- Helmet, boots, gloves.
- Tools: traditional hand tools such as the pike, halligan, pulaski, axe, hydrant wrench; power tools such as saws.
- Hoses and nozzles.
- Other equipment.

"Two in, two out" refers to the rule that firefighters never enter a building or other IDLH situation alone but go in as a pair. Two go in and two come out; also, the two in can refer to the two who enter the building while a backup team of two remains outside the building ready to enter if needed.

To minimize the risk of electrocution, electrical shock, and electricity-related burns while fighting wildland fires, NIOSH recommends that fire departments and fire fighters take the following precautions [IFSTA 1998a,b; NWCG 1998; NFPA 1997; 29 CFR* 1910.332(b); 29 CFR 1910.335(b); Brunacini 1985]:

Fire departments should do the following:

- Keep fire fighters a minimum distance away from downed power lines until the line is de-energized. This minimum distance should equal the span between two poles.
- Ensure that the incident commander conveys strategic decisions related to power line location to all suppression crews on the fireground and continually reevaluates fire conditions.
- Establish, implement, and enforce standard operating procedures (SOPs) that address the safety of fire fighters when they work near downed power lines or energized electrical equipment. For example, assign one of the fireground personnel to serve as a spotter to ensure that the location of the downed line is communicated to all fireground personnel.
- Do not apply solid-stream water applications on or around energized, downed power lines or equipment.
- Ensure that protective shields, barriers, or alerting techniques are used to protect fire fighters from electrical hazards and energized areas. For example, rope off the energized area.
- Train fire fighters in safety-related work practices when working around electrical energy. For example, treat all downed power lines as energized and make fire fighters aware of hazards related to ground gradients.

- Ensure that fire fighters are equipped with the proper personal protective equipment (Nomex® clothing compliant with NFPA standard 1500 [NFPA 1997], leather boots, leather gloves, etc.) and that it is maintained in good condition.
- Ensure that rubber gloves and dielectric overshoes and tools (insulated sticks and cable cutters) for handling energized equipment are used by properly trained and qualified personnel.

Fire fighters should do the following:

- Assume all power lines are energized and call the power provider to de-energize the line(s).
- Wear appropriate personal protective equipment for the task at hand—Nomex® clothing compliant with NFPA standard 1500, rubber gloves, and dielectric overshoes and tools (insulated sticks and cable cutters).
- Do not stand or work in areas of dense smoke. Dense smoke can obscure energized electrical lines or equipment and can become charged and conduct electrical current.

3.5.11 Specialized Rescue Procedures

There are many common types of rescues such as building search, victim removal, and extrication from motor vehicles. On the other side there are certain specialized rescues such as water rescues, ice rescues, structural collapse rescues, and elevator/escalator rescues. These specialized rescues are generally low volume calls, depending on coverage area, and thus should be trained for fire fighters to better asses the situation.

Structural collapse, while not a common incident, may occur for any number of reasons: weakening from age or fire, environmental causes (earthquake, tornado, hurricane, flooding, rain, or snow buildup on roofs), or an explosion (accidental or intentional). Structural collapses can create numerous voids where victims could be trapped.

When arriving on scene the number of potential victims should be acquired. Certain hazards to look for in a structural collapse are a secondary collapse, live electrical wires, and gas leaks. The electricity and gas should be shut-off. Structural collapses come in three different types: pancake collapse, lean-to collapse, and v-type collapse.

- A pancake collapse is characterized by both supporting walls failing or from the anchoring system failing and the supported roof or upper floor falling parallel to the floor below. Small voids where victims can be found are created by debris.
- A lean-to collapse occurs when one only one side of the supporting walls or floor anchoring system fails. One side of the collapsed roof is attached to the remaining wall or anchoring system. The lean-to collapse creates a significant void near the remaining wall.
- A v-type collapse can happen when there is a large load in the center of the floor or roof above. The roof may be overloaded from a buildup of snow and/or have been weakened by fire, rot, termites, and improper removal of support beams. Both sides of the supporting wall are still standing but the center of the floor or roof above is compromised and thus collapsed in the center. The v-type collapse usually leaves a void on each side of the supporting walls.

Knowing where the voids are can help a fire fighter locate survivors of the structural collapse. After the firefighter has identified the type of collapse and where the voids may be they need to make a safe entranceway. The firefighter needs to have a basic knowledge of cribbing, shoring, and tunneling, all of which are useful in collapse operations. Cribbing is the use of various sizes of lumber arranged in systematic stacks to support an unstable load. Shoring is the use of wood to support and/or strengthen weakened structures such as roofs, floors, and walls. This avoids a secondary collapse during rescue. Tunneling could be necessary to reach survivors if there is no way other way to reach the victim. When tunneling there should be a set destination and a good indication that there is a victim in that void.

3.5.12 First Responder to Electrical Fire Incidents

This information is essential for anyone who might encounter emergency situations where energized electrical equipment creates a hazard. The language is intentionally non-technical in an effort to provide easily understandable information for those without an in-depth knowledge of electricity. Because of the wide variety of emergency situations first responders (i.e., law enforcement officers, fire fighters, ambulance attendants, etc.) might encounter, it is not possible to cover every situation.

- You must always maintain proper respect for downed wires even though some may appear harmless. Electrical equipment requires the same respect, awareness and caution you would accord a firearm – always "consider it loaded." In situations where no emergency exists and human life is not in any immediate danger, wait for the local utility personnel to secure the area; they have the knowledge and equipment to complete the job safely.
- Law enforcement officers, fire fighters or ambulance attendants are usually first on the scene when overhead wires are down – usually as a result of storms, damaged utility poles or fallen branches. They need to be aware of the hazards and procedures involved in dealing with emergencies resulting from fallen energized wires.
- Electricity seeks the easiest path to ground itself and does not "care" how it gets there. If you or your equipment creates that path, you will be placing yourself (and possibly others) in a life threatening situation. In some situations fallen wires snap and twist, sending out lethal sparks as they strike the ground. At other times the wires lie quietly; producing no sparks or warning signals – as quiet as a rattlesnake and potentially as dangerous. The first rule is to consider any fallen or broken wire extremely dangerous and not to approach within eight feet of it.
- Next, notify the local utility and have trained personnel sent to the scene. Have an ambulance or rescue unit dispatched if necessary. Remember, do not attempt to handle wires yourself unless you are properly trained and equipped.
- Set out flares and halt or reroute traffic. Keep all spectators a safe distance (at least 30.5 meters) from the

scene. Electric power emergencies often occur when it is raining; wet ground increases the hazard. After dark, light the scene as well as you can. Direct your spotlight on the broken or fallen wires. Remember that metal or cable guard-rails, steel wire fences and telephone lines may be energized by a fallen wire and may carry the current a mile or more from the point of contact.

- An ice storm, windstorm, tornado, forest fire or flood may bring down power lines by the hundreds. Under those circumstances electric companies customarily borrow skilled professionals from one another to augment their own work forces. First Responders have their own jobs to perform at such times – usually as part of a task force – which lessens the need for individual decision making. But every first responder should be prepared for when he or she faces an electric power emergency alone and must make decisions about people, power and the hazards involved. Remember that electricity from a power line (like lightening from a thundercloud) seeks to reach the ground, so it is imperative when working with fallen wires not to let yourself or others create a circuit between a wire and the ground.
- In a typical power emergency a car strikes a utility pole and a snapped power line falls on it. Advise the car's occupants they should stay in the car. Call the local power company, but remember: Do not come in contact with either the car or its occupants. If the car catches fire, instruct the occupants to leap, not step, from the car. To step out would put them in the circuit between the wire, the energized car and the ground – with deadly results. If fire fighters are on the scene, they may be best able to handle the situation; most full-time fire fighters are trained to deal with electric power emergencies and will have the proper equipment to do so. If you must extinguish a car fire without the aid of fire fighters, use only dry chemical or CO_2 extinguishers. If the car's occupants are injured and cannot leap to safety, you may be able to use your vehicle to push them out of contact with the wire. If you do this, it is critical to look around the vehicle

before leaving your car – there may be another fallen wire behind you or a wire hooked to your bumper. If there is (or you suspect there is), leap from your vehicle.

- Once a victim has been removed from the electric hazard, immediately check vital signs. If the victim has no pulse and is not breathing, begin cardio-pulmonary resuscitation immediately (and any other appropriate first aid treatment) until he or she is placed in the ambulance.
- In any rescue attempt it is essential that you protect yourself – it is a truism that dead heroes rescue no one. Do not, under any circumstance, rely on rubber boots, raincoats, rubber gloves or ordinary wire cutters for protection. Above all, do not touch (or allow your clothing to touch) a wire, a victim, or a vehicle that is possibly energized.

3.5.13 Evacuation Planning

Each facility should have a written plan for the orderly evacuation of each building at a facility. The plan should establish the necessary procedures for fire emergencies, bomb threats, etc. Each employee should be familiar with the plan. The following are recommended essential elements to include in the plan.

3.5.13.1 Designated Roles and Responsibilities

There should be a building coordinator for each building who is the designated responsible individual. The building coordinator is responsible for seeing that the plan is implemented and will appoint an adequate number of floor marshals, assure everyone is familiar with this plan and act as a liaison with the Health and Safety Officer (HSO) and first responders such as the fire department.

For the floor marshals, at least one individual per floor should be designated and there should be designated back-ups. Floor marshals will assist in the implementation of the plan by knowing and communicating evacuation routes to occupants during emergency evacuation and report the status of the evacuation to the building coordinator.

3.5.13.2 *Preparation & Planning for Emergencies*

Pre-planning for emergencies is a crucial element of the plan. The following steps should be taken in planning for emergency evacuation of each building:

1. All exits are labeled and operable.
2. Evacuation route diagrams have been approved by the HSO or safety division and are posted on all floors and at all exits, elevator lobbies, training/conference rooms and major building junctions.
3. Occupants do not block exits, hoses, extinguishers, corridors or stairs by storage or rearrangement of furniture or equipment. Good housekeeping is everyone's responsibility.
4. All floor marshals have been trained in their specific duties and all building occupants have been instructed in what to do in case of an emergency evacuation.
5. Fire evacuation drills are held at least annually in this building and are critiqued and documented. Prior to holding a fire evacuation drill where the alarm is to be triggered, the electric shop and the fire marshal are notified.

3.5.13.3 *Evacuation Procedure*

When a fire is discovered:

1. Anyone who receives information or observes an emergency situation should immediately call a central designated number. The Plan should identify the Emergency Number to call, which for most offices can be the Fire Department, Police Department, or Civil Defense.
2. In the incident building, occupants will be notified of emergencies by a notification system such as: fire alarm, paging system, and/or word of mouth.

Occupants will:

1. Know at least two exits from the building.

2. Be familiar with the evacuation routes posted on the diagram on your floor.
3. To report a fire or emergency, call the Emergency Number. Give your name, room number and the floor that locates the fire. State exactly what is burning, or what is smoking or what smells like a fire to you. Then notify the Building coordinator (or other designated person) and activate the building notification system.
4. When notified to evacuate, do so in a calm and orderly fashion:
   a. Walk, don't run.
   b. Keep conversation level down.
   c. Take your valuables and outer garments.
   d. Close all doors behind you.
   e. Use the stairs, not the elevators.
   f. Assist others in need of assistance.
5. Go to the designated assembly area or as instructed during the notification. Upon exiting the building, move at least 45.7 meters from the building to allow others to also safely exit the building (specify designated assembly areas or indicate areas on evacuation diagram).

3.5.13.4 General

During evacuation floor marshals should assure that every person on his/her floor has been notified and that evacuation routes are clear. If possible, the floor marshal - will check that all doors are closed and be the last one out. Upon leaving the floor, the floor marshal will report the status of the floor evacuation to the building coordinator.

Persons with disabilities (mobility, hearing, sight): Each individual who requires assistance to-evacuate is responsible for prearranging with someone else in their immediate work area to assist them. Anyone knowing of a person with a disability or injury who was not able to evacuate will report this immediately to a Floor Marshal, the Building coordinator or the Incident Commander.

The following are general evacuation instructions to include in Plan:

1. Know at least two exits from the building.

2. Be familiar with the evacuation routes posted on the diagram on your floor.
3. To report a fire or emergency, call a central emergency number designated in the plan. Give your name, room number and the floor that locates the fire. State exactly what is burning, or what is smoking or what smells like a fire to you. Then notify the building manager or other designated person and activate the building notification system.
4. When notified to evacuate, do so in a calm and orderly fashion:
   a. Walk, don't run.
   b. Keep conversation level down.
   c. Take your valuables and outer garments.
   d. Close all doors behind you.
   e. Use the stairs, not the elevators.
   f. Assist others in need of assistance.
5. Go to the designated assembly area or as instructed during the notification. If exiting the building, move at least 45.7 meters from the building to allow others to also safely exit the building.
6. Persons with disabilities that may impair mobility should establish a buddy system to help ensure that any needed assistance will be available to them in an emergency.

3.5.13.5 Template for Emergency Evacuation Plan

The following are general guidelines for implementation:

1. **Assignment of Responsibility** - Administrative responsibility for evacuation of each building must be clearly defined. The HSO shall designate an individual with a thorough understanding and appropriate authority as the building coordinator under the plan. The building coordinator shall appoint floor marshals as appropriate and ensure that they are adequately instructed in their duties and responsibilities. There must be adequate alternates to assume responsibilities in the absence of the building coordi-

nator or floor marshals. These designations should also be made in the pre-planning stage.

2. **Coordination -** In buildings that are under the control of more than one manager or HSO, there should be one evacuation plan for that building that has been coordinated with all key persons.
3. **Notification -** The building coordinator must assure that building occupants know who to call in case of an emergency and the proper sequence of notification. If it is a fire emergency, call a designated three digit number (in the United States 911 for example), then notify the Building coordinator and activate the building notification system. Similar emergency incidents may be covered under the same plan. If it is a tornado, the building coordinator or other designated person will notify occupants through the floor marshal of safe places of haven. If it is a bomb threat, calls a centralized emergency notification number identified in the plan, then notify the building coordinator, but do not activate the building notification system. Responsible parties will review the specific circumstances and mandate evacuation when deemed necessary or will otherwise advise management when evacuation is not necessary. Occupants must be aware of who to notify in the event the building coordinator is absent. The building coordinator must assure that there is an effective method to notify occupants of an emergency. Notification may be by means of an alarm system, public address system, telephone fan out system or oral communication, although this last method is not advisable for work areas with ten or more persons.
4. **Preparation and Planning -** Proper preparation and planning for emergencies is essential in order for evacuation to be effective and efficient. Emergency route diagrams can be produced using the building "key plans". These should be made accessible to all employees. Building/room diagrams should also be made available through the Facilities Management Information System (FMIS). Use the key plan to

highlight the main corridor exit ways and stairways and to identify the exits. You can also use this diagram to indicate where fire extinguisher and fire fighting equipment are located by using color coding or a symbol. The diagram can also be used to indicate designated assembly areas. Post the diagrams at: elevator lobbies, junctions in buildings where directions are routinely posted, information bulletin boards, training/conference rooms, or other strategic locations throughout the building. All emergency route diagrams must be reviewed the fire marshal and the HSO. Make sure all exits are labeled, operable and not blocked and there is proper illumination of pathways and exit signs. Floor marshals should be fully familiar with the building evacuation plan and their role. Make sure there are alternates to cover during absences. Decide on a designated meeting place outside the building where floor marshals can quickly report to the Building coordinator and where occupants can be accounted for. Communicate the program to all staff in each building.

5. **Fire Evacuation Drills -** All building occupants must be familiar with what they should do during an evacuation. The most effective method of familiarizing them is to hold a fire drill at least annually. Holding a fire drill has other advantages as well; it will provide you with an opportunity to evaluate your notification and evacuation procedures and it will give you an opportunity to test your fire alarm system and make occupants aware of the sound.

   Steps In Conducting a Fire Drill:

   a. If you have an alarm system you must first contact the Electric Shop to make the appropriate arrangements.
   b. Contact the Fire Marshal before holding your drill so that a representative can be on site to assist in critiquing the evacuation.
   c. Although the fire drills should be unannounced, you may need to give advanced notice to key personnel in your building.

3.5.14 Bibliography

Cheremisinoff, N. P., Handbook of Industrial Toxicology and Hazardous Materials, Marcel Dekker Publishers, New York, New York, 1999.

3.6 Excavations and Trenching

3.6.1 Introduction

This section will assist site supervisors to establish procedures for excavation work to help assure excavations are performed in a safe and proper manner. The following definitions are applied:

- Excavations are any man-made cut, cavity, trench or depression in the earth surface formed by earth removal. See examples in Figure 3.4 below. No distinction on the depth of excavation is made in this definition.
- Trenches are narrow excavations made below the surface of the ground. In general, the depth is greater than the width. However, the width of a trench is not greater than 4.6 meters. An excavation is also considered to be a trench.

Figure 3.4 Departmental supervisors and site supervisors must ensure their employees follow the procedures explained in this section.

3.6.2 Recommended Practices

The following recommended best practices should be followed prior to excavation or trenching:

1. **Identification of Buried Utilities** - All underground utilities must be clearly marked to identify potentially hazardous situations prior to starting work requiring excavation. The person responsible for oversight must contact the responsible parties to identify public utilities.
2. **Surface Encumbrances** - Remove, support or safeguard all surface encumbrances located at the site that may create a hazard to employees.
3. **Competent Person** - At each excavation site there must be a designated "competent" person whenever employees are digging or in the excavation. A "competent" person means someone capable of identifying existing and predictable hazards in the surroundings, or working conditions that unsanitary, hazardous, or dangerous to employees. The competent person has authorization to take prompt corrective measures to eliminate any unsafe condition. A "competent" person is someone who attends a four hour excavation/trenching course and successfully passes the course examination. There must be documentation of the training.

During excavation the following recommended best practices should be followed:

1. **Materials and Equipment** - Trench shoring and trench jacks shall be on the site before any excavation begins when the expected depth will exceed 1.22 meters or where unstable soil is expected. In addition, a radio equipped truck or portable radio shall be on the site at all times when work is in progress.
2. **Protection of Employees** - No one shall enter excavations of four feet or more in depth without having shoring in place or the banks cut back to a safe angle. (Refer to OSHA Standard 1926.652) No one shall enter an excavation of less than 1.22 meters without shoring

or proper angle of repose unless authorized by a competent person. Note that excavations less than 1.22 meters in depth may not require this degree of protection if examination of the ground by a competent person provides no indication of a potential cove in. Store excavated material at least 0.61 meters from the excavation edge. Workers must wear hard hats at the site when work is in progress. Provide a stairway, ladder, ramp or other safe means of egress in trench excavations that are 1.22 meters or more in depth. The travel distance to the ladder must be no more than 7.62 meters of lateral travel for employees. The ladder must extend 0.91 meters above the edge of the trench.

3. **Inspections** - A competent person must make daily inspections of excavations, the adjacent areas, and protective systems for evidence of possible cave-ins, indications of protective systems failure, hazardous atmospheres, or other hazardous conditions. The competent person must conduct an inspection:
   a. Prior to the start of work and as needed throughout the shift;
   b. After every rainstorm or other hazard increasing occurrence;
   c. As dictated by the activity taking place in the trench;
   d. When fissures, tension cracks, sloughing, under cutting, water seepage, bulging at the bottom or other similar circumstances occur;
   e. When there is any change in the size, location or placement of the soil pile; and
   f. When there is any indication of change or movement in adjacent structures.

Where the competent person finds evidence of a situation that could result in a possible cave-in, indications of failure of protective systems, hazardous atmosphere, or other hazardous conditions, remove exposed employees from the hazardous area until safety measures are in place.

1. **Barricades and Warnings** - Take special measures to protect the public because the danger excavation

work presents. Simple barricades are not enough protection in high traffic areas near sidewalks, and road ways subject to pedestrian traffic. In high traffic areas, place snow fencing around the excavation to provide maximum protection. An additional precaution is to use barricades with flashing lights whenever possible on pedestrian walkways and roadways.

Provide employees exposed to public vehicular traffic with warning vests or other suitable garments marked with reflector or high-visibility material.

2. **Competent Person Training** - To become a competent person, a person needs specialized training. The course should provide information on:
   a. Soil analysis
   b. Use of protective systems

3.6.3 Safety Guidelines

The following safety guidelines are recommended:

- Each side of an excavation or trench that is 1.52 meters or deeper must be protected by shoring/bracing and sheeting or be sloped – unless it is cut from rock.
- Each open side of an excavation or trench shall have a guardrail or a solid enclosure.
- Each excavation or trench shall have a way out, such as a ladder or ramp.
- Regularly check the walls of an excavation or trench for cracks, bulges and spalling and check the shoring for signs of distress -- especially after a rainstorm.
- Do not work in an excavation or trench filled with running or standing water.
- Do not work in an excavation or trench that is not properly protected.
- Do not store spoil, materials or equipment along the edge of an excavation or trench.
- Do not drive or park vehicles along the edge of an excavation or trench.

The following diagram (Figure 3.5) provides an illustrative guide to protecting trenches.

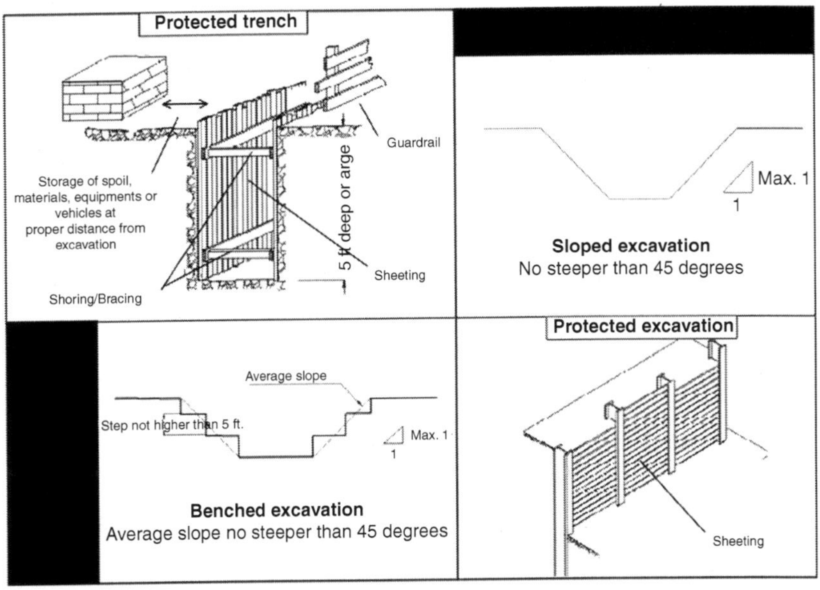

Figure 3.5 Guidelines for preparing protected trenches.

3.6.4 Excavation and Trenching Safety Program

3.6.4.1 Introductory Notes

The following provides the basis for the preparation of a formal excavation and trenching safety program. The purpose of the excavation and trenching safety program is to put in place work practices and procedures that will protect employees from hazards that may be found in or around excavations or trenches. This objective is met by:

1. Requiring each department that performs work in excavations to name one or more persons to serve as the departmental project manager.
2. Training project managers so they understand their duties and responsibilities.
3. Requiring that each project manager assure that all persons working in excavations have been provided adequate training as required by this program.

This program has been modeled after the U.S. Occupational Safety and Health Administration (OSHA) Standard 29 CFR 1926 Subpart P.

All departments that work in or around excavations must comply with the requirements of this program. Other key participants should include:

- Workers who work in or around excavations;
- Contractor's personnel, since the work that a contractor performs can directly affect the safety of persons working in or around excavations.

3.6.4.2 *Safety Coordinator*

The excavation and trenching safety program should:

- Monitor the overall effectiveness of the program;
- Assist with atmospheric testing and equipment selection as needed;
- Provide training for designated project managers;
- Assist the departmental project manager with training of other departmental employees;
- Provide technical assistance to the departments as needed; and
- Review and update the program on an annual basis as necessary.

Program elements are defined below.

3.6.4.3 *Training and Duties of Program Participants*

All personnel involved in trenching or excavation work shall be trained in the requirements of this program. Departmental personnel shall be trained by the departmental project manager with assistance from the health & safety coordinator (HSC).

1. Training shall be performed **before** the employee is assigned duties in excavations.
2. Retraining will be performed whenever work site inspections conducted by the project manager or HSC indicate that an employee does not have the necessary knowledge or skills to safely work in or around excavations.

3. Training records will be maintained by the project manager and will be copied to the HSC. These records shall include the date(s) of the training program, the instructor(s) of the training program, a copy of the written material presented, and the names of the employee(s) to whom the training was given.

3.6.4.4 *Training and Duties of Workers*

All personnel that perform work in excavations shall comply with the requirements of this program. These personnel shall receive appropriate training that shall include, at a minimum:

- The work practices that must be followed during excavating or working in excavations;
- The use of personal protective equipment that will typically be required during work in excavations, including but not limited to safety shoes, hard-hats, and fall protective devices;
- Procedures to be followed if a hazardous atmosphere exists or could reasonably be expected to develop during work in an excavation; and
- Emergency and non-entry rescue methods, and procedure for calling rescue services.

3.6.4.5 *Training and Duties of the Project Manager*

The project manager shall receive the training detailed above and shall, in addition, receive training on the requirements detailed in all recommended practices described below. The project manager shall:

- Coordinate and actively participate in the training of departmental employees. A copy of the training records shall be maintained by the project manager, and shall be copied to the HSC;
- Ensure on a daily basis, or more often as detailed in this program, that work site conditions are safe for employees to work in excavations;
- Determine the means of protection (sloping back the sides of the excavation, use of trench shields,

or shoring) that will be used for each excavation project; and

- Ensure, if required, that the design of a protective system has been completed and approved by a registered professional engineer before work is begun in the excavation.

3.6.4.6 *Contractor Awareness, Duties and Responsibilities*

- A Contractor that is performing excavation work shall coordinate trenching and excavation work with the project manager to assure the coordination of the work and shutdown of utilities if necessary.

3.6.4.7 *Specific Excavation Requirements*

Utilities and Pre-work Site Inspection - Prior to excavation the site shall be thoroughly inspected by the project manager to determine if special safety measures must be taken.

Surface encumbrances - All equipment, materials, supplies, permanent installations (for example, buildings or roadways), trees, brush, boulders and other objects at the surface that could present a hazard to employees working in the excavation shall be removed or supported as necessary to protect employees.

Underground installations - The location of sewers, telephone, fuel, electric, water lines, or any other underground installations that may be encountered during excavation work shall be determined and marked prior to opening an excavation. Arrangements shall be made as necessary by the project manager with the appropriate utility agency for the protection, removal, shutdown, or relocation of underground installations.

If it is not possible to establish the exact location of these installations, the work may proceed with caution if detection equipment or other safe and acceptable means are used to locate the utility.

Excavation shall be done in a manner that does not endanger the underground installations or the employees engaged in the work. Utilities left in place shall be protected by barricades, shoring, suspension or other means as necessary to protect employees.

3.6.4.8 *Protection of the Public*

Barricades, walkways, lighting and posting shall be provided as necessary for the protection of the public prior to the start of excavation operations.

Guardrails, fences, or barricades shall be provided on excavations adjacent to walkways, driveways and other pedestrian or vehicle thoroughfares. Warning lights or other illumination shall be maintained as necessary for the safety of the public and employees from sunset to sunrise.

Wells, holes, pits, shafts and all similar hazardous excavations shall be effectively barricaded or covered and posted as necessary to prevent unauthorized access. All temporary excavations of this type shall be backfilled as soon as possible.

Walkways or bridges protected by standard guardrails shall be provided where employees and the general public are permitted to cross over excavations. Where workers in the excavation may pass under these walkways or bridges, a standard guardrail and toeboard shall be used.

3.6.4.9 Protection of Workers in Excavations

Access and means of egress - Stairs, ladders or ramps shall be provided where employees are required to enter trench excavations over 1.22 meter deep. The maximum distance of lateral travel (e.g., along the length of the trench) required to reach the means of egress shall not exceed 7.62 meters

Structural ramps - Structural ramps used solely by employees as a means of access or egress from excavations shall be designed by a competent person. Structural ramps used for access or egress of equipment shall be designed by a person qualified in structural design, and shall be constructed in accordance with the design.

Ramps and runways constructed of two or more structural members shall have the structural members connected together to prevent movement or displacement.

Structural members used for ramps and runways shall be of uniform thickness.

Cleats or other appropriate means used to connect runway structural members shall be attached to the bottom of the runway or shall be attached in a manner to prevent tripping.

Structural ramps used in place of steps shall be provided with cleats or other surface treatments on the top surface to prevent slipping.

Ladders - When portable ladders are used, the ladder side rails shall extend a minimum of 91.44 centimeters above the upper surface of the excavation (see Figure 3.6). Ladders shall have nonconductive side rails if work will be performed near exposed energized equipment or systems. Two or more ladders, or a double-cleated ladder, will be provided where 25 or more employees will be conducting work in an excavation where ladders serve as the primary means of egress, or where ladders serve two-way traffic. Ladders will be inspected prior to use for signs of damage or defects. Damaged ladders will be removed from service and marked with "Do Not Use" until repaired.

Ladders shall be used only on stable and level surfaces unless secured. Ladders placed in any location where they can be displaced by workplace activities or traffic shall be secured, or barricades shall be used to keep these activities away from the ladder.

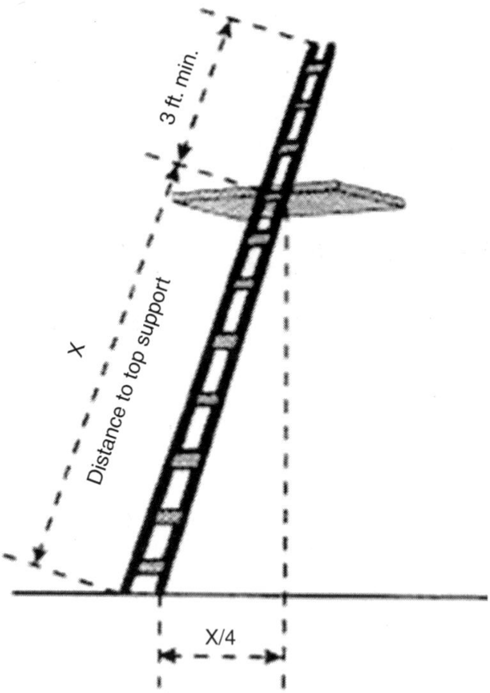

Figure 3.6 Proper positioning of non-self-supporting ladders.

Non-self-supporting ladders shall be positioned so that the foot of the ladder is one-quarter of the working length away from the support (see Figure 18.).

Employees shall not be allowed to carry any object or load while on the ladder that could cause them to lose their balance and fall.

Exposure to vehicular traffic - Employees exposed to vehicular traffic shall be provided with, and shall wear, warning vests or other suitable garments marked with or made of reflectorized or high-visibility material. Warning vests worn by flagmen shall be red or orange, and shall be of reflectorized material if worn during night work.

Employee exposure to falling loads - No employee shall be permitted underneath loads handled by lifting or digging equipment. Employees shall be required to stand away from any vehicle being loaded or unloaded to avoid being struck by any spillage or falling materials. Operators may remain in the cabs of vehicles being loaded or unloaded when the vehicles provide adequate protection for the operator during loading and unloading operations.

Warning system for mobile equipment - A warning system shall be used when mobile equipment is operated adjacent to the edge of an excavation if the operator does not have a clear and direct view of the edge of the excavation. The warning system shall consist of barricades, hand or mechanical signals, or stop logs. If possible, the grade should be away from the excavation.

Hazardous atmospheres - The project manager will test the atmosphere in excavations over 1.22 meters deep if a hazardous atmosphere exists or could reasonably be expected to exist. A hazardous atmosphere could be expected, for example, in excavations in landfill areas, in excavations in areas where hazardous substances are stored nearby, or in excavations near or containing gas pipelines.

Adequate precautions shall be taken to prevent employee exposure to atmospheres containing less than 19.5 percent oxygen and other hazardous atmospheres. These precautions include providing proper respiratory protection or forced ventilation of the workspace. Technical assistance on these precautions may be obtained by contacting EHSS.

Forced ventilation or other effective means shall be used to prevent employee exposure to an atmosphere containing a flammable gas in excess of 10 percent of the lower flammability limit of the gas.

When controls are used that are intended to reduce the level of atmospheric contaminants to acceptable levels, continuous air monitoring will be performed by the project manager. The device used for atmospheric monitoring shall be equipped with an audible and visual alarm.

Atmospheric testing will be performed using a properly calibrated direct reading gas monitor. Direct reading gas detector tubes or other acceptable means may also be used to test potentially toxic atmospheres.

Each atmospheric testing instrument shall be calibrated on a schedule and in the manner recommended by the manufacturer except:

- Any atmospheric testing instrument that has not been used within thirty (30) days shall be recalibrated by the department prior to use.
- Each atmospheric testing instrument shall be calibrated at least every six (6) months by the department.
- Copies of calibration records will be forwarded to HCS
- Each atmospheric testing instrument will be field checked immediately prior to use to ensure that it is operating properly.

3.6.4.10 *Personal Protective Equipment*

All employees working in trenches or excavations shall wear approved hard-hats and steel toed shoes or boots.

Employees exposed to flying fragments, dust, or other materials produced by drilling, sawing, sanding, grinding and similar operations shall wear approved safety glasses with side shields.

Employees exposed to hazards produced by, or performing, welding, cutting, or brazing operations shall wear, as determined by the project manager, approved spectacles or a welding faceshield or helmet.

Employees entering bell-bottom pier holes or other similar deep and confined footing excavations shall wear a harness with a lifeline securely attached to it. The lifeline shall be separate from any line used to handle materials and shall be individually attended at all times while the employee wearing the lifeline is in the excavation.

Employees shall wear, as determined by the project manager, approved gloves or other suitable hand protection.

Employees using, or working in the immediate vicinity of, hammer drills, masonry saws, jackhammers or similar high noise producing equipment shall wear suitable hearing protection.

Each employee at the edge of an excavation 1.83 meters or more deep shall be protected from falling. Fall protection shall be provided by guardrail systems, fences, barricades, covers, or a tie-back system.

Emergency rescue equipment, such as breathing apparatus, a safety harness and line, and a basket stretcher shall be readily available where hazardous atmospheric conditions exist or may develop during work in an excavation. This equipment shall be attended when in use. Only personnel that have received approved training and have appropriate equipment shall attempt retrieval that would require entry into a hazardous atmosphere. If entry into a known hazardous atmosphere must be performed, then the HSC shall be given advance notice so that the hazards can be evaluated and rescue personnel placed on standby if necessary.

3.6.4.11 Walkways and Guardrails

Walkways shall be provided where employees or equipment are permitted to cross over excavations. Guardrails shall be provided where walkways, accessible only to on-site project personnel, are 6 feet or more above lower levels.

3.6.4.12 Protection from Hazards Associated with Water Accumulation

Employees shall not work in excavations that contain or are accumulating water unless precautions have been taken to protect employees against the hazards posed by water accumulation. The precautions taken could include, for example, special support or shield systems to protect from cave-ins, water removal to control the level of accumulating water, or use of safety harnesses and lifelines.

If water is controlled or prevented from accumulating by the use of water removal equipment, the water removal equipment and operation shall be monitored by a person trained in the use of the equipment.

If excavation work interrupts the natural drainage of surface water (such as streams), diversion ditches, dikes, or other suitable means shall be used to prevent surface water from entering the excavation. Precautions shall also be taken to provide adequate drainage of the area adjacent to the excavation. Excavations subject to runoff from heavy rains shall be reinspected by the project manager to determine if special precautions should be taken.

The project manager shall inform workers of the precautions or procedures that are to be followed if water accumulates or is accumulating in an excavation.

3.6.4.13 *Stability of Adjacent Structures*

The project manager will determine if the excavation work could affect the stability of adjoining buildings, walls, sidewalks or other structures.

Support systems (such as shoring, bracing, or underpinning) shall be used to assure the stability of structures and the protection of employees where excavation operations could affect the stability of adjoining buildings, walls, or other structures.

Excavation below the level of the base or footing of any foundation or retaining wall that could be reasonably expected to pose a hazard to employees shall not be permitted except when:

- A support system, such as underpinning, is provided to ensure the safety of employees and the stability of the structure; or
- The excavation is in stable rock; or
- A registered professional engineer has approved the determination that the structure is sufficiently removed from the excavation so as to be unaffected by the excavation activity; or
- A registered professional engineer has approved the determination that such excavation work will not pose a hazard to employees.

Sidewalks, pavements and appurtenant structure shall not be undermined unless a support system or other method of protection is provided to protect employees from the possible collapse of such structures.

Where review or approval of a support system by a registered professional engineer is required, the Department shall secure this review and approval in writing before the work is begun. A copy of this approval shall be provided to HSC.

3.6.4.14 *Protection of Employees from Falling Objects and Loose Rocks or Soil*

Adequate protection shall be provided to protect employees from loose rock or soil that could pose a hazard by falling or rolling from an excavation face. Such protection shall consist of:

- Scaling to remove loose material;
- Installation of protective barricades, such as wire mesh or timber, at appropriate intervals on the face of the slope to stop and contain falling material; or
- Benching sufficient to contain falling material.

Excavation personnel shall not be permitted to work above one another where the danger of falling rock or earth exists.

Employees shall be protected from excavated materials, equipment or other materials that could pose a hazard by falling or rolling into excavations.

Protection shall be provided by keeping such materials or equipment at least .61 meters from the edge of excavations, by the use of restraining devices that are sufficient to prevent materials or equipment from falling or rolling into excavations, or by a combination of both if necessary.

Materials and equipment may, as determined by the project manager, need to be stored further than .61 meters from the edge of the excavation if a hazardous loading condition is created on the face of the excavation.

Materials piled, grouped or stacked near the edge of an excavation must be stable and self-supporting.

3.6.4.15 *Inspection By The Project Manager*

The project manager shall conduct daily inspections of excavations, adjacent areas, and protective systems for evidence of a situation that could result in possible cave-ins, failure of protective systems, hazardous atmospheres, or other hazardous conditions.

An inspection shall be conducted by the project manager prior to the start of work and as needed throughout the shift. Inspections shall also be made after every rainstorm or other hazard increasing occurrence. These inspections are only required when the trench will be or is occupied by employees.

Where the competent person finds evidence of a situation that could result in a possible cave-in, failure of protective systems, hazardous atmosphere, or other hazardous conditions, exposed employees shall be removed from the hazardous area until precautions have been taken to assure their safety.

The project manager shall maintain a written log of all inspections conducted. This log shall include the date, work site location, results of the inspection, and a summary of any action taken to correct existing hazards.

3.6.4.16 *Pre-Entry Checklist*

Prior to worker entry to an excavation certain factors should routinely be considered. The following is a checklist to facilitate a pre-entry safety inspection. The inspector should check Yes or No to the following questions:

YES	NO	
☐	☐	Is there a competent person on site? Note that a competent person is defined as one who is capable of identifying existing and predictable hazards in the surroundings, or working conditions which are unsanitary, hazardous, or dangerous to employees, and who has authorization to take prompt corrective measures to eliminate them.

1. Surface Encumbrances:

☐	☐	Have you checked for structures, trees, machines and equipment, or excavated spoils located near the excavation which might exert force on the excavation walls?
☐	☐	Have you checked for any evidence of cracking or subsidence along the excavation surface or within the excavation?

2. *Underground Installations:*

☐ ☐ Are all utilities, such as water, sewer, electrical, and telephone, located within the excavation properly supported, removed, or otherwise protected?

3. *Access and Egress:*

☐ ☐ Are structural ramps that are used solely by employees for access and egress designed by a competent person?

☐ ☐ Are ramps secured against displacement?

☐ ☐ Are ramps sound, trip hazard free, and slip resistant?

☐ ☐ Are trenches provided with ramps or ladders so that they are within a minimum of 25 feet of lateral travel?

4. *Traffic Safety:*

☐ ☐ Are signs and barricades placed appropriately and in sufficient quantity?

☐ ☐ Are personnel, where exposed to traffic, wearing appropriate warning vests or reflective high visibility clothing?

☐ ☐ Are personnel prohibited from working under lifting or digging equipment, and required to stand away from vehicles being loaded or unloaded?

☐ ☐ Where mobile equipment is operated adjacent to an excavation, or approaches an excavation; does the operator have a clear view of the edge?

☐ ☐ Where an operator's view of the edge of an excavation is obstructed, is there a warning system, such as hand signals, mechanical signals, or stop logs in place?

(If possible, the grade should be away from the excavation.)

5. *Hazardous Atmospheres:*

☐ ☐ Where hazardous conditions could reasonably be expected had the atmosphere been tested prior to entry?

☐ ☐ Where the atmosphere contains less than 19.5% oxygen, has appropriate respiratory protection and/or ventilation been provided?

☐ ☐ Where flammable atmospheres exist is there ventilation provided that is sufficient to reduce the flammable concentration to less than 20% of the lower flammable limit of the gas?

☐ ☐ Is frequent testing of the atmosphere conducted to ensure that the atmosphere remains safe?

6. Emergency Rescue Equipment:

☐ ☐ Is emergency rescue equipment, such as breathing apparatus, safety harness and lanyard, or a basket stretcher, readily available where hazardous atmosphere conditions exist or may be expected to develop?

☐ ☐ Are employees entering deep and confined excavations equipped with harness, an attached lifeline, and attended at all times while in the excavation?

7. Water Hazard Protection:

☐ ☐ Are adequate precautions taken to protect employees in excavations where water has accumulated or is accumulating (i.e., support or shield systems, water removal systems, or safety harness and lifelines.)?

☐ ☐ Are water removal equipment and operations monitored by a competent person?

☐ ☐ Are natural runoff and surface water adequately diverted to prevent entry to the excavation?

☐ ☐ Are excavations subjected to runoff from heavy rains inspected by a competent person prior to entry?

8. Adjacent Structures:

☐ ☐ Are adjoining buildings, walls or other structures adequately stabilized with approved shoring, bracing, and underpinning to assure protection of employees?

☐ ☐ Where the excavation is below the level of the base or footing of a foundation or retaining wall, have the excavation and support systems in use been determined appropriate by a registered professional engineer?

☐ ☐ Are sidewalks, pavement and other such structures adequately supported to prevent collapse?

9. Loose Rock and Soils:

☐ ☐ Are adequate measures taken to protect employees from loose and falling rock, soil, or other debris (i.e., scaling, protective barricades, etc.)?

☐ ☐ Are excavated spoils and other materials maintained at least 60.96 centimeters away from the edge of the excavation, and are retaining devices used where the 60.96 centimeter clear area is not possible?

10. Protective Systems:

☐ ☐ Has a competent person tested the soils to determine the proper classification as a basis for determining the required type of protective systems necessary for the excavation?

☐ ☐ Do protective systems have the capacity to resist without failure, all loads intended or expected to be applied to them?

☐ ☐ Are the designs of support systems, shield systems, or other protective systems in accordance with specifications, recommendations, and limitations of the manufacturer?

☐ ☐ Where there is deviation from specifications, recommendations and limitations has written manufacturer approval been obtained, with a copy on site?

☐ ☐ Where systems design is from other tabulated data or by a registered professional engineer, are such date and copies of the design maintained at the job site?

☐ ☐ Are materials and equipment used for protective systems free from defects that might impair their function?

☐ ☐ Are manufactured materials and equipment used for protective systems used and maintained in accordance with manufacturer recommendations?

☐ ☐ Are materials and equipment used for protective systems that have been damaged, inspected by a competent person to evaluate suitability for use?

☐ ☐ Are repaired materials and equipment evaluated and approved for use by a registered professional engineer before being returned to service?

☐ ☐ Are members of support systems securely connected to prevent falling, sliding, kickouts, or other predictable failure?

☐ ☐ Are support systems installed and removed in a manner that protects employees from cave-ins, structural collapses, or from being struck by members of the support system?

☐ ☐ Does system removal begin at, and progress from the bottom of the excavation?

☐ ☐ Is there excavation no greater than 60.96 centimeters below the bottom of the members of the support system where the system provides protection for the full depth of the trench and there is no potential for loss of soil from behind or below the bottom of the system?

☐ ☐ Are employees prohibited from working on the faces of sloped or benched excavations above other employees at lower levels unless there is adequate protection from falling, rolling, or sliding material or equipment?

☐ ☐ Are employees protected from the hazard of cave-ins when entering or exiting areas protected by shields?

☐ ☐ Are employees prohibited from being in shields during installation, removal, or vertical movement of them?

☐ ☐ Where the competent person finds evidence of a situation that could result in a possible cave-in, where there are indications of failure of protective systems, where there are hazardous atmospheres, or where there are other hazardous conditions, are exposed employees removed from the hazardous area until all necessary precautions are taken to ensure their safety?

3.6.5 Requirements for Protective Systems

3.6.5.1 Protection of Employees in Excavations

Employees in an excavation shall be protected from cave-ins by using either an adequate sloping and benching system or an adequate support or protective system (Section 5.3). The only exceptions are:

- Excavations made entirely in stable rock; or
- Excavations less than 5 feet in depth where examination of the ground by the project manager provides no indication of a potential cave-in.

Protective systems shall be capable of resisting all loads that could reasonably be expected to be applied to the system.

3.6.5.2 *Materials and Equipment*

Materials and equipment used for protective systems shall be free from damage or defects that might affect their proper function.

Manufactured materials and equipment used for protective systems shall be used and maintained in accordance with the recommendations of the manufacturer, and in a manner that will prevent employee exposure to hazards.

When material or equipment used for protective systems are damaged, the project manager shall ensure that these systems are examined by a competent person to evaluate its suitability for continued use. If the competent person can not assure the material or equipment is able to support the intended loads or is otherwise suitable for safe use, then such material or equipment shall be removed from service. These materials or equipment shall be evaluated and approved by a registered professional engineer before being returned to service.

3.6.5.3 *Installation and Removal of Support*

Members of support systems shall be securely connected together to prevent sliding, falling, kickouts, or other potential hazards.

Support systems shall be installed and removed in a manner that protects employees from cave-ins, structural collapses, or from being struck by members of the support system.

Individual members of support systems shall not be subjected to loads exceeding those which those members were designed to support.

Before temporary removal of individual support members begins, additional precautions shall be taken as directed by the project manager to ensure the safety of employees. These precautions could include, for example, the installation other structural members to carry the loads imposed on the support system.

Removal of support systems shall begin at, and progress from, the bottom of the excavation. Members shall be released slowly. If there is any indication of possible failure of the remaining members of the structure or possible cave-in of the sides of the excavation the work shall be halted until it can be examined by the project manager.

Backfilling shall progress together with the removal of support systems from excavations.

3.6.5.4 *Additional Requirements for Support Systems for Trench Excavations*

Excavation of material to a level no greater than 60.96 centimeters below the bottom of the members of a support system is allowed, but only if the system is designed to resist the forces calculated for the full depth of the trench. There shall be no indications while the trench is open of a possible loss of soil from behind or below the bottom of the support system.

Installation of a support system shall be closely coordinated with the excavation of trenches.

3.6.5.5 *Sloping and Benching Systems*

Employees shall not be permitted to work above other employees on the faces of sloped or benched systems except when employees at the lower levels are protected from the hazard of falling, rolling, or sliding material or equipment.

3.6.5.6 *Shield Systems*

Shield systems shall not be subjected to loads that are greater than those they were designed to withstand.

Shields shall be installed in a manner that will restrict lateral or other hazardous movement of the shield that could occur during cave-in or unexpected soil movement.

Employees shall be protected from the hazard of cave-ins when entering or exiting the areas protected by shields.

Employees shall not be allowed in shields when shields are being installed, removed, or moved vertically.

Excavation of material to a level no greater than 60.96 centimeters below the bottom of the shield system is allowed, but only if the system is designed to resist the forces calculated for the full depth of the trench. There shall be no indications while the trench is open of a possible loss of soil from behind or below the bottom of the shield system.

3.6.6 Definitions

Accepted engineering practices means the standards of practice required by a registered professional engineer.

Aluminum Hydraulic Shoring means a manufactured shoring system consisting of aluminum hydraulic cylinders (crossbraces) used with vertical rails (uprights) or horizontal rails (wales). Such system is designed to support the sidewalls of an excavation and prevent cave-ins.

Bell-bottom pier hole means a type of shaft or footing excavation, the bottom of which is made larger than the cross section above to form a belled shape.

Benching (Benching system) is a method of protecting employees from cave-ins by excavating the sides of an excavation to form one or more horizontal steps, usually with vertical or near-vertical surfaces between levels.

Cave-in means the movement of soil or rock into an excavation, or the loss of soil from under a trench shield or support system, in amounts large enough to trap, bury, or injure and immobilize a person.

Competent person means one who has been trained to identify hazards in the workplace, or working conditions that are unsafe for employees, and who has the authority to have these hazards corrected. The departmental project manager serves as the departmental competent person for the purposes of this program. The project manager will conduct all required tests and inspections as detailed in this program, and ensure that employees working in excavations have been trained and are following the requirements of this program.

Cross braces mean the horizontal members of a shoring system installed from side to side of the excavation. The cross braces bear against uprights.

Excavation means any man-made cut, cavity, trench, or depression in an earth surface formed by earth removal.

Faces or sides mean the vertical or inclined earth surfaces formed as a result of excavation work.

Failure means the movement or damage of a structural member or connection that makes it unable to support loads.

Hazardous atmosphere means an atmosphere that is explosive, flammable, poisonous, corrosive, oxidizing, irritating, oxygen deficient, toxic, or otherwise harmful that may cause death, illness, or injury.

Kickout means the accidental movement or failure of a cross brace.

Project Manager is the individual within the department that will oversee excavation work and that is responsible for assuring compliance with this program.

Protective system means a method of protecting employees from cave-ins, from material that could fall or roll from an excavation face into an excavation, or from the collapse of adjacent structures. Protective systems include support systems, sloping and benching systems, shield systems, and other systems that provide the necessary protection.

Ramp means an inclined walking or working surface that is used to gain access to one point from another. A ramp may be constructed from earth or from structural materials such as steel or wood.

Registered Professional Engineer means a person who is registered as a professional engineer.

Sheeting means the members of a shoring system that retain the earth in position and in turn are supported by other members of the shoring system.

Shield (Shield system) means a structure used in an excavation to withstand cave-ins and which will protect employees working within the shield system. Shields can be permanent structures or portable units moved along as work progresses. Shields used in trenches are usually referred to as "trench boxes" or "trench shields."

Shoring (Shoring system) means a structure that is built or put in place to support the sides of an excavation to prevent cave-ins.

Sides. See "Faces."

Sloping (Sloping system) means sloping the sides of the excavation away from the excavation to protect employees from cave-ins. The required slope will vary with soil type, weather, and surface or near surface loads that may affect the soil in the area of the trench (such as adjacent buildings, vehicles near the edge of the trench and so forth).

Stable rock means natural solid mineral material that can be excavated with vertical sides that will remain intact while exposed.

Structural ramp means a ramp built of steel or wood, usually used for vehicle access. Ramps made of soil or rock are not considered structural ramps.

Support system means a structure such as underpinning, bracing, or shoring, which provides support to an adjacent structure, underground installation, or the sides of an excavation.

Tabulated data means tables and charts approved by a registered professional engineer and used to design and construct a protective system.

Trench (Trench excavation) means a narrow excavation (in relation to its length) made below the surface of the ground.

Trench box or shield. See "Shield."

Uprights mean the vertical members of a trench shoring system placed in contact with the earth and usually positioned so that individual members do not contact each other. Uprights placed so that individual members are closely spaced, in contact with or interconnected to each other, are often called "sheeting."

Wales are horizontal members of a shoring system placed in the direction of the excavation face whose sides bear against the vertical members of the shoring system or earth (the uprights or sheeting).

3.6.7 Bibliography

The Occupational Safety and Health Administration (OSHA), Excavation and Trenching Standard, 1971

3.7 Confined Spaces

3.7.1 Identifying Confined Spaces

Recognition is an important aspect of making a safe entry into a confined space. Not all confined spaces will be considered permit-required confined spaces and being able to identify the difference between the two is important. To clarify what constitutes a Confined Space, the following definition will be used.

A Confined Space is any space that has the following characteristics:

1. It is large enough or so configured that an employee can bodily enter and perform assigned work.
2. It has limited or restricted means for entry or exit.

Confined-space openings are limited primarily by size and location. Openings may be small in size and may be difficult to move through easily. However, in some cases openings may be very large; for example, open-topped spaces such as pits or excavations. Entrance and exit may be required from top, bottom, or side. In some cases, having to access the work area by a fixed ladder may constitute limited or restricted entry or exit. Size or location may make rescue efforts difficult.

1. Is not designed for continuous employee occupancy.

Most confined spaces are not designed for employees to enter and work on a routine basis. They may be designed to store a product, enclose materials and processes, or transport products or substances. Because they are not designed for continuous occupancy, frequently they will not have good ventilation or lighting. Therefore, occasional employee entry for inspection, maintenance, repair, cleanup, or similar tasks can be difficult and dangerous. The danger associated with entry may come from chemical or physical hazards within the space.

A non-permit confined space is a confined space that does not contain, nor has the potential to contain, any hazard capable of causing death or serious physical harm (with respect to atmospheric hazards). Examples of non-permit required confined spaces might include the interiors of HVAC units, certain air plenums and pipe chases, attics, walk-in freezers or refrigerators, and some building crawl spaces.

A permit-required confined space (permit space) is a confined space that is potentially hazardous. A permit-required confined space has one or more of the following characteristics:

1. Contains or has a potential to contain a hazardous atmosphere.
2. Contains a material that has the potential for engulfing an entrant.
3. Has an internal configuration such that an entrant could be trapped or asphyxiated by inwardly-converging walls or by a floor that slopes downward and tapers to a smaller cross-section; or
4. Contains any other recognized serious safety or health hazard. Examples of serious safety or health hazards might include:
   a. Fall hazards;
   b. Unguarded machinery;
   c. Extreme heat or cold;
   d. Steam pipes or chemical lines;
   e. Hazardous noise levels;
   f. Electrical hazards;
   g. Presence of asbestos; and
   h. Potentially hazardous levels of dust (such as might occur at the Feed Mill).

Because of the lack of ventilation in most confined spaces, they will have the potential for a hazardous atmosphere. Therefore, they must be designated "permit-required," and the procedures for making entry into a permit-required space must be followed.

Any space that is accessed by lifting a manhole cover shall be considered a permit-required confined space. Additionally, some roofs, dam access tunnels, certain grain storage facilities, and equipment access areas may be designed permit-required confined spaces even though they don't technically meet the definition (i.e., they may not really have limited or restricted means of entry or exit). These areas shall be clearly marked as permit-required spaces.

Supervisors are directly responsible for ensuring the safety of their employees in regards to confined spaces. It is their responsibility to evaluate potentially hazardous spaces within their facilities and areas to ensure that the proper precautions are taken for safety. This includes clearly marking permit-required confined spaces, training employees, and ensuring proper entry procedures are followed. These responsibilities may be delegated to another competent person provided he/she is qualified.

Physical plant supervisors are responsible for ensuring their employees are properly trained to do the jobs they are sent to do. This includes recognition of confined spaces and proper procedures for making entry into permit-required confined spaces whenever necessary. No physical plant employee shall be sent on a job that potentially involves work in a confined space unless they have been properly trained in confined space entry procedures.

It may be determined that a space presents no real danger for employees. However, it is recommended that all spaces be considered potentially dangerous until they have been evaluated and tested. Once a space has been evaluated, the environmental health & safety department shall determine if the confined space requires a permit and will apply appropriate labeling.

3.7.2 Identifying Confined Space Hazards

Once a space has been identified as confined, the hazards that may be present within the confined space must be identified. Confined-space hazards can be grouped into the following categories: 1) Oxygen-deficient atmospheres, 2) flammable atmospheres, 3) toxic atmospheres, and 4) mechanical and physical hazards.

Every confined space must be evaluated for these four types of hazards. The three types of atmospheric hazards are often the most difficult to identify since they might not be detected without the assistance of a gas monitor.

3.7.2.1 *Oxygen-Deficient Atmospheres*

The normal atmosphere is composed of approximately 21% oxygen and 79% nitrogen. An atmosphere containing less than 19.5% oxygen shall be considered oxygen-deficient. The oxygen level inside a confined space may be decreased as the result of either consumption or displacement.

There are a number of processes that consume oxygen in a confined space. Oxygen is consumed during combustion of flammable materials, as in welding, cutting, or brazing. A more subtle consumption of oxygen occurs during bacterial action, as in the fermentation process. Oxygen can also be consumed during chemical reactions such as in the formation of rust on the exposed surfaces of a confined space. The number of people working in a confined space and the amount of physical activity can also influence oxygen consumption. Oxygen levels can also be reduced as the result of oxygen displacement by other gases.

3.7.2.2 *Flammable Atmospheres*

Flammable atmospheres are generally the result of flammable gases, vapors, dust mixed in certain concentrations with air, or an oxygen-enriched atmosphere.

Oxygen-enriched atmospheres are those atmospheres that contain an oxygen concentration greater than 22%. An oxygen-enriched atmosphere will cause flammable materials such as clothing and hair to burn violently when ignited.

Combustible gases or vapors can accumulate within a confined space when there is inadequate ventilation. Gases that are heavier than air will accumulate in the lower levels of a confined space. Therefore, it is especially important that atmospheric tests be conducted near the bottom of all confined spaces.

The work being conducted in a confined space can generate a flammable atmosphere. Work such as spray-painting, coating, or the use of flammable solvents for cleaning can result in the formation of an explosive atmosphere. Welding or cutting with oxyacetylene equipment can also be the cause of an explosion in a confined space and shall not be allowed without a hot work permit. Oxygen and acetylene hoses may have small leaks in them that could generate an explosive atmosphere and, therefore, should be removed when not in use. The atmosphere shall be tested continuously while <u>any</u> hot work is being conducted within the confined space.

3.7.2.3 Toxic Atmospheres

Toxic atmospheres may be present within a confined space as the result of one or more of the following:

1. The product stored in the confined space - When a product is stored in a confined space, the product can be absorbed by the walls and give off toxic vapors when removed or when cleaning the residual material. The product can also produce toxic vapors that will remain in the atmosphere due to poor ventilation.
2. The work being conducted in the confined space - Toxic atmospheres can be generated as the result of work being conducted inside the confined space. Examples of such work include: Welding or brazing with metals capable of producing toxic vapors, painting, scraping, sanding, etc. Many of the solvents used for cleaning and/or degreasing produce highly toxic vapors.
3. Areas adjacent to the confined space - Toxic fumes produced by processes near the confined space may enter and accumulate in the confined space. For example, if the confined space is lower than the adjacent area and the toxic fume is heavier than air, the toxic fume may "settle" into the confined space.

3.7.2.4 Mechanical and Physical Hazards

Problems such as rotating or moving mechanical parts or energy sources can create hazards within a confined space. All rotating or moving equipment such as pumps, process lines, electrical sources, etc., within a confined space must be identified.

Physical factors such as heat, cold, noise, vibration, and fatigue can contribute to accidents. These factors must be evaluated for all confined spaces.

Excavations could present the possibility of engulfment. Employees shall be protected from cave-ins by sloping, benching, or shoring systems when the depth of the excavation is more than four feet. In some circumstances, air monitoring may also be required.

3.7.3 Model Confined Space Entry Program

The recommended model for a confined space entry program is outlined below.

3.7.3.1 *Identifying All Confined Spaces*

1. All confined spaces located within a facility or under the facility's control should be identified. Once the space has been identified as confined, the HSC (Health & Safety Coordinator) shall determine if a permit is required.
2. All employees shall be made aware of these confined spaces through training or instruction provided by supervisors or their designated representatives. Assistance in this training shall be provided by HSC.

3.7.3.2 *Preventing Unauthorized Entry*

1. All employees shall be instructed by supervisors or their designated representatives that entry into a confined space is prohibited without an authorized permit.
2. Supervisors or their designated representatives shall instruct all employees to list their names on the authorized permit before they will be allowed to enter a confined space.

3.7.3.3 *The Permit System*

1. When a confined space must be entered, a permit shall be completed and authorized by department heads, supervisors, or their designated representatives prior to entry of the confined space. This permit shall serve as certification that the space is safe for entry. The permit shall contain the date, the location of the space, and the signature of the person providing the certification.
2. A permit shall not be authorized until all conditions of the permit have been met.

3.7.3.4 Planning the Entry

The first step towards conducting a safe confined-space entry is to plan the entry. This will allow for the identification of all hazards, and for the determination of all equipment necessary, to complete the project.

1. Gathering General Data
   a. Identify the confined space. Give the name or location of the confined space.
   b. Give the reason for entering the confined space. Be specific. Also, identify if hot work will be done.
   c. Identify the contents of the confined space. This refers to any chemicals or other materials and energy that are usually present in the confined space.
2. Identifying the Hazards
   NOTE: Atmospheric testing shall be conducted prior to entering permit-required confined spaces. It is recommended that the entry supervisor conduct these tests; however, any competent person certified in confined space entry may do so.
   a. The entry supervisor will determine the oxygen content and record this on the entry permit.
   b. The entry supervisor will determine flammable gas content and record this on the entry permit.
   c. The entry supervisor will determine levels of H_2S and Carbon Monoxide and record this on the entry permit.
   d. If a toxic substance is determined to be in the confined space during testing by the entry supervisor, environmental health & safety shall be contacted to assist in obtaining a material safety data sheet or other chemical information to determine what type of personal protective equipment is required, the potential health effects, the permissible exposure limits, and any other information needed to safely conduct the work.
   e. Entry supervisors will determine mechanical and physical hazards. They should list all items and energy that will require lockout/tagout, blanking and bleeding, disconnecting, or securing. Physical hazards should also be listed.

3. Ventilation of the Confined Space
    a. Indicate whether mechanical or natural ventilation will be used. Describe the procedures to be used. NOTE: If mechanical ventilation is to be used, the exhaust must be pointed away from personnel or ignition sources. Also, mechanical ventilators should be bonded to the confined space.
4. Isolating the Confined Space
    a. Describe the procedures for disconnecting equipment or lockout and tagout. All mechanical, electrical, or heat-producing equipment should be disconnected or locked and tagged out. This would also include any pumps that pull fluid from, or pump fluid into, the confined space.
5. Purging/Cleaning the Confined Space
    a. Indicate if the confined space will be purged. Purging with inert gas is not recommended. If the space must be purged, describe the procedures.
    b. Indicate the type of cleaning methods to be used. If chemical cleaners are to be used, name the type and describe the procedures. The MSDS for the chemical should be consulted prior to use. NOTE: When introducing a chemical into a confined space, the compatibility of that chemical with the contents of the confined space must be checked. If in doubt, consult environmental health & safety. NOTE: If steam is to be used, the hose should be bonded to the confined space.
6. Placement of Warning Signs
    a. Indicate if warning signs or barriers will be needed to prevent unauthorized entry or to protect workers from external hazards. If the confined space will be left open and unattended for any length of time, warning signs and barriers such as barricades and/or caution tape will be required.
7. Identifying All Personnel
    a. List all employees that will be required to prepare the confined space and complete the work inside the space.

8. Identifying Necessary Equipment
   a. List all equipment that will be necessary to complete the project.

3.7.3.5 *Conducting Pre-Entry Training*

Once the entry has been planned, department heads or their designated representatives must train all employees who will be involved in the entry. The training should be conducted no earlier than one day before entry is to be made.

The following outline should be used for the training:

1. Identify the confined space and the reason(s) for entry.
2. Identify work detail
   a. Assign each employee the job(s) he/she is to perform in the entry project (entrant, standby person, etc.).
   b. If an employee is required to use a piece of equipment, be sure that he/she is capable of using the equipment properly.
   c. Inform all personnel that no one is to enter the confined space unless the attendant is present at the work site.
3. Inform entrants of all known or suspected hazards
   a. Inform personnel of any access or exit problems.
   b. Inform personnel of all equipment that must be locked out or tagged out.
   c. Inform personnel of the contents of the confined space.
   d. Inform personnel of all atmospheric levels that must be maintained before entering and while working in the confined space.

If a toxic atmosphere or substance is present or could become present, the following additional training must be completed:

   e. If respiratory protection is not going to be used, inform personnel of the maximum permissible exposure level (PEL) that can exist within the confined space, and the method used to monitor PEL.

f. Inform personnel of the potential health effects of exposure to the toxic atmosphere or substance.
g. Inform personnel of the signs and symptoms of exposure to the toxic fume.
h. Inform personnel of the personal protective equipment (PPE) that they will be required to wear.
i. If entrants are unaware of the proper use of the PPE, they must be trained in the proper use of this equipment.

NOTE: Supervisors may request assistance from environmental health & safety in providing the above-mentioned training.

j. Persons should not be assigned to tasks requiring use of respirators unless it has been determined that they are physically able to perform the work and use the equipment. A local physician shall determine what health and physical conditions are pertinent. The respirator user's medical status should be reviewed periodically (annually).

1. Identify isolation procedures
   a. Inform the personnel responsible for the lockout/tagout of all equipment that must be isolated.
   b. Inform the personnel responsible for performing this function of the methods to be used.
2. Identify purging and/or ventilation procedures
   a. Inform all personnel responsible for performing this function of the methods to be used.
3. Identify all equipment needed
   a. Inform personnel involved in the project of all equipment that will be necessary to complete the project.
   b. Make sure that all employees are capable of using their assigned equipment properly.
4. Determine necessary personal protective equipment
   a. Inform personnel of all PPE that must be used to ensure their safety.
   b. Make sure that all personnel required to use PPE are trained in the proper use of the equipment.

5. Establish communication
   a. Inform all entrants that they are required to maintain communication with the attendant.
   b. Inform attendant that he/she must maintain constant contact with all entrants.
   c. Inform personnel of the type of communication they are to use.
6. Protect from external hazards
   a. Inform personnel where signs and barriers will be placed to prevent unauthorized entry and protect entrants from external hazards.
7. Pre-plan rescue procedures
   a. The designated attendant(s) should be informed of the rescue procedures to be followed. Rescue procedures to be used are listed in Item 10 of this section.
   b. The attendant should be informed that he/she can have no other duty but to maintain contact with personnel inside the confined space.
   c. Inform the attendant(s) that they must not enter the confined space under any circumstances.
8. Place the confined space back into service
   a. Inform personnel of the steps to be taken to place the confined space back into service.

3.7.3.6 Preparing the Confined Space for Entry

Once the entry has been planned and personnel have been trained, the next step is to prepare the confined space for entry.

The following steps are to be followed when preparing the confined space for entry:

1. Place warning signs or barriers around the confined space to prevent unauthorized entry as necessary.
2. Place all tools, safety equipment, monitoring equipment, etc., near the confined space.
3. Isolate all mechanical and/or electrical hazards as necessary.
4. Purge/ventilate the confined space as necessary.

5. Test the atmosphere using an appropriate gas monitor.
   a. If oxygen content is less than 19.5% or greater than 21.5%, perform additional ventilation. Then shut off ventilation equipment and re-test the oxygen content.
   b. If oxygen content is between 19.5% and 21.5%, continue entry preparation.
6. Test for flammable gases.
   a. If the meter reading is less than 10% of the lower explosive limit (LEL), continue entry preparations.
   b. If the meter reading is above 10% of the LEL, continue ventilation of the confined space. Then shut off the ventilation and have the atmosphere re-tested.
   c. If the meter reading is still above 10% of the LEL, the confined space must be cleaned before entry is permitted. If the confined space must be entered for cleaning purposes, the procedures outlined in Item 9 of this section must be followed.
7. Test for toxics (If a toxic atmosphere is present, no person should be permitted to enter the confined space at a level exceeding the permissible exposure limit without proper personal protective equipment). Environmental health & safety should be called to assist in identifying proper precautions and the protective measures to be taken.
8. Assemble all personnel involved and review rescue procedures. The entry supervisor will then add any needed information, then complete and sign the permit.
9. Notify the department head or supervisor that entry is commencing. If department head or supervisor is unavailable, notify EHS department.

3.7.3.7 *Utilizing Safety Equipment*

Where practical, all personnel entering a confined space should be equipped with a retrieval line secured at one end to the entrant by a full-body harness with its other end secured to a tripod lifting hoist.

3.7.3.8 Atmospheric Testing Procedures

1. All of the manufacturer's operating instructions must be followed.
2. The test equipment should be tested in a known atmosphere to insure its accuracy.
3. Ventilation equipment must be shut off before conducting any atmospheric tests.
4. The atmosphere must be tested at the bottom, top, and middle of all confined spaces.
5. The atmosphere must be continuously monitored while work is being conducted in the confined space.
6. If the confined space is left for any reason, the atmosphere must be re-tested before re-entering the space.

3.7.3.9 Confined Space Cleaning Procedures

If cleaning must be conducted in a confined space to achieve acceptable atmospheric conditions, the following procedures must be followed:

1. All entrants must be equipped with the safety equipment designated under 3(g).3(vii).
2. All entrants must be equipped with an SCBA.
3. No spark-producing tools will be allowed for use.

3.7.3.10 Rescue Procedures

In the event of an emergency, the attendant should:

1. Immediately summon the fire department by radio or telephone.
2. Attempt to remove the victim by use of the retrieval line from outside the confined space if this can be accomplished without creating further hazard for the entrant or the attendant.

3. If the attendant is able to remove the victim with the retrieval line, he/she should administer aid within the limits of his/her training until emergency medical services (EMS) arrive.
4. If the attendant is unable to remove the victim by using the retrieval line, he or she must wait for help to arrive. The attendant(s) is not to enter the confined space for any reason.
5. Give EMS personnel any information they request.

3.7.4 Personnel Responsibilities and Training

Everyone involved in a confined-space entry project has certain responsibilities and requires a certain amount of training. It is very important that every individual is familiar with his/her responsibilities. This section outlines the responsibilities and training requirements of each individual involved in a project.

3.7.4.1 Responsibilities of the Director of Environmental Health & Safety

The director of environmental health & safety or his/her designated representative shall be responsible for the following:

1. Reviewing and updating the Oklahoma State University confined space entry program to conform to current CFR standards.
2. Ensuring compliance with standards set forth in the program by periodic inspection of entry sites and canceling permits where unsafe conditions are present.
3. Assisting Supervisors with:
   a. providing training as set forth in the program;
   b. identification of confined spaces;
   c. identifying spaces that require a permit for entry; and
   d. labeling permit-required confined spaces.
4. Performing a single annual review covering all entries performed during a 12-month period to ensure employees participating in entry operations are protected from permit space hazards.

3.7.4.2 *Responsibilities and Training Requirements of Supervisors or Their Designated Representatives*

Supervisors or their designated representatives shall be responsible for the following:

1. Identifying confined spaces within facilities or areas under their control.
2. Identifying hazards within a confined space under their control.
3. Documenting that all training requirements for a specific confined space entry have been met by signing the pre-entry authorization space on the entry permit.

3.7.4.3 *Responsibilities and Training Requirements of Entry Supervisors*

Entry supervisors shall be responsible for the following:

1. Ensuring that the required atmospheric tests are performed at the confined space and results recorded on the permit prior to entry authorization.
2. Obtaining and maintaining all equipment necessary to complete the confined-space entry project.
3. Authorize entry by signing the entry authorization space on the entry permit after all conditions for a safe entry have been met.
4. Terminating the entry and canceling the permit when:
   a. Entry operations covered by the entry permit have been completed.
   b. A condition that is not allowed under the entry permit arises in or near the permit space.
5. Determining, whenever responsibility for a permit space entry operation is transferred, and at intervals dictated by the hazards and operations performed within the space, that entry operations remain consistent with terms of the entry permit and that acceptable entry conditions are maintained.

3.7.4.4 *Responsibilities and Training Requirements of Authorized Entrants*

The person(s) authorized to enter a confined space shall be responsible for and receive training in the following:

1. The knowledge of hazards that may be faced during entry, including the mode, signs or symptoms, and consequences of the exposure.
2. Proper use of equipment, which includes:
   a. Atmospheric testing and monitoring equipment.
   b. Ventilating equipment needed to obtain acceptable entry conditions.
   c. Communication equipment necessary to maintain contact with the attendant.
   d. Personal protective equipment as needed.
   e. Lighting equipment as needed.
   f. Barriers and shields as needed.
   g. Equipment, such as ladders, needed for safe ingress and egress.
   h. Rescue and emergency equipment as needed.
   i. Any other equipment necessary for safe entry into and rescue from permit spaces.
3. Communication with the attendant as necessary to enable the attendant to monitor entrant status and to enable the attendant to alert entrants of the need to evacuate the space if required.
4. Alert the attendant (standby person) whenever:
   a. The entrant recognizes any warning sign or symptom of exposure to a dangerous situation; or
   b. The entrant detects a prohibited condition.
5. Exiting the permit space as quickly as possible whenever:
   a. An order to evacuate has been given by the attendant or the entry supervisor;
   b. The entrant recognizes any warning sign or symptom of exposure to a dangerous situation;
   c. The entrant detects a prohibited condition; or
   d. An evacuation alarm is activated.

3.7.4.5 Responsibilities and Training Requirements of Attendants

Persons authorized to perform duties as attendant shall be responsible for and receive training in the following:

1. Knowing the hazards that may be faced during entry, including information on the mode, signs or symptoms, and consequences of exposure.
2. Awareness of possible behavioral effects of hazard exposure in authorized entrants.
3. Continuously maintaining an accurate count of authorized entrants in the permit space and ensuring that the means used to identify authorized entrants accurately identifies who is in the permit space.
4. Remains outside the permit space during entry operations until relieved by another attendant.
5. Attempting non-entry rescue if proper equipment is in place and the rescue attempt will not present further hazards to the entrant or attendant.
6. Communicating with authorized entrants as necessary to monitor entrant status and to alert entrants of the need to evacuate the space when conditions warrant.
7. Monitoring activities inside and outside the space to determine if it is safe for entrants to remain in the space and ordering the authorized entrants to evacuate the permit space immediately under any of the following conditions:
    a. If the attendant detects a prohibited condition.
    b. If the attendant detects the behavioral effects of hazard exposure in an authorized entrant.
    c. If the attendant detects a situation outside the space that could endanger the authorized entrants.
    d. If the attendant cannot effectively and safely perform all the duties required by this program.
8. Summoning rescue and other emergency services as soon as the attendant determines that authorized entrants may need assistance to escape from permit space hazards.

9. Taking the following actions when unauthorized persons approach or enter a permit space while entry is underway:
   a. Warning the unauthorized persons that they must stay away from the permit space.
   b. Advising the unauthorized persons that they must exit immediately if they have entered the permit space.
   c. Informing the authorized entrants and the entry supervisor if unauthorized persons have entered the permit space.
10. Performing no duties that might interfere with the attendant's primary duty to monitor and protect the authorized entrants.

3.7.5 Definitions

Attendant - A person designated by the department head in charge of entry to remain outside the confined space and to be in constant communication with the personnel working inside the confined space.

Authorized Entrant - A person who is approved or assigned by the department head in charge of the entry to perform a specific type of duty or duties or to be at a specific location at the job site.

Bonding - The joining of two or more items with an electrical conductor so that all ends joined have the same electrical charge or potential.

Confined Space - A Confined Space is any space that has the following characteristics: 1) It is large enough or so configured that an employee can bodily enter and perform assigned work; 2) It has limited or restricted means for entry or exit.

Entry - The action by which a person passes through an opening into a permit-required confined space. Entry includes ensuing work activities in that space and is considered to have occurred as soon as any part of the entrant's body breaks the plane of an opening into the space.

Entry Permit - The written or printed document that is provided by the employer to allow and control entry into a permit space and that contains the information specified in this program.

Entry Supervisor - Department Head or the designated representative (such as the foreman or crew chief) responsible for determining if acceptable entry conditions are present at a permit space where entry is planned, for authorizing entry and overseeing entry operations, and for terminating entry as required by this program. NOTE: An entry supervisor also may serve as an attendant or as an authorized entrant, as long as that person is trained and equipped as required by this program for each role he or she fills. Also, the duties of entry supervisor may be passed from one individual to another during the course of entry operation.

Hazardous Atmosphere - An atmosphere that may expose employees to the risk of death, incapacitation, impairment of ability to self-rescue (that is, escape unaided from a permit space), injury, or acute illness from one or more of the following causes:

- Flammable gas, vapor, or mist in excess of 10% of its lower flammable limit (LFL).
- Airborne combustible dust at a concentration that meets or exceeds its LFL. NOTE: This concentration may be approximated as a condition in which the dust obscures vision at a distance of 1.52 meters or less.
- Atmospheric oxygen concentration below 19.5% or above 23.5%.
- Atmospheric concentration of any substance for which a dose or a permissible exposure limit is published in Subpart G, Occupational health and Environmental Control, or in Subpart Z, Toxic and Hazardous Substances, of 29 CFR 1910 and that could result in employee exposure in excess of its dose or permissible exposure limit. NOTE: An atmospheric concentration of any substance that is not capable of causing death, incapacitation, impairment of ability to self-rescue, injury, or acute illness due to its health effects is not covered by this provision.
- Any other atmospheric condition that is immediately dangerous to life or health. NOTE: For air contaminants

for which OSHA has not determined a dose or permissible exposure limit, other sources of information, such as Material Safety Data Sheets that comply with the Hazard Communication Standard, section 1910.1200, published information, and internal documents can provide guidance in establishing acceptable atmospheric conditions.

Hot Work - Any work involving burning, welding or similar fire-producing operations. Also, any work that produces a source of ignition, such as grinding, drilling, or heating.

Hot Work Permit - The employer's written authorization to perform operations (for example, riveting, welding, cutting, burning, and heating) capable of providing a source of ignition.

Immediately Dangerous to Life or Health - An atmosphere that poses an immediate threat of loss of life: May result in irreversible or immediate severe health effects; may result in eye damage/irritation; or other condition that could impair escape from a confined space.

Lower Explosive Limit (LEL) - The minimum concentration of a combustible gas or vapor in air that will ignite if an ignition source is introduced.

Oxygen-Deficient Atmosphere - An atmosphere that contains an oxygen concentration of less than 19.5% by volume.

Oxygen-Enriched Atmosphere - An atmosphere that contains an oxygen concentration greater than 22% by volume.

PPE - Personal Protective Equipment - Any devices or clothing worn by the worker to protect against hazards in the environment. Examples are respirators, gloves, and chemical splash goggles.

PEL - Permissible Exposure Level - Concentration of a substance to which an individual may be exposed repeatedly without adverse effect.

Purging - The removal of gases or vapors from a confined space by the process of displacement.

Standby Person - A person designated by the department head in charge of entry to remain outside the confined space and to be in constant communication with the personnel working inside the confined space.

3.7.6 Bibliography

The following references were consulted.

Title 29 of the Code of Federal Regulations Part 1910.146 - Permit-Required Confined Spaces. U.S. Government Printing Office.

N.I.O.S.H. Training and Resource Manual - Safety and Health in Confined Workspaces for the Construction Industry.

N.I.O.S.H. 87-113 - A Guide to Safety in Confined Spaces.

City of Stillwater, Oklahoma - Confined Space Entry Manual 1990.

Title 29 of the Code of Federal Regulations Part 1926.652 - Requirements for Protective Systems.

Title 29 of the Code of Federal Regulations Part 1910.150 - The Control of Hazardous Energy.

Title 29 of the Code of Federal Regulations Part 1910.134 - Respiratory Protection.

4

Safe Use of Equipment

4.1 Hand Tools and Workshop Machines

4.1.1 Introduction

Employees should be trained in the proper use of all tools. Workers should be able to recognize the hazards associated with the different types of tools and the safety precautions necessary.

Five basic safety rules can help prevent hazards associated with the use of hand and power tools:

- Keep all tools in good condition with regular maintenance;
- Use the right tool for the job;
- Examine each tool for damage before use and do not use damaged tools;
- Operate tools according to the manufacturers' instructions; and
- Provide and use properly the right personal protective equipment.

Employees and employers should work together to establish safe working procedures. If a hazardous situation is encountered, it should be brought immediately to the attention of the proper individual for hazard abatement.

4.1.2 What are the Hazards of Hand Tools?[1]

Employees should be trained in the proper use and handling of tools and equipment. The greatest hazards posed by hand tools result from misuse and improper maintenance.

Some examples include:

- Using a chisel as a screwdriver, the tip of the chisel may break and fly off, hitting the user or other employees.
- If a wooden handle on a tool is loose, splintered, or cracked, the head of the tool may fly off and strike the user or other employees.
- If the jaws of a wrench are sprung, the wrench might slip.
- If impact tools such as chisels, wedges, or drift pins have mushroomed heads, the heads might shatter on impact, sending sharp fragments flying toward the user or other employees.
- When working in close proximity, employees should target saw blades, knives, or other tools away from away aisle areas and other employees. Knives and scissors must be sharp; dull tools can cause more hazards than sharp ones. Cracked saw blades must be removed from service.

Wrenches must not be used when jaws are sprung to the point that slippage occurs. Impact tools such as drift pins, wedges, and chisels must be kept free of mushroomed heads.

Iron or steel hand tools may produce sparks that can be an ignition source around flammable substances. Where this hazard exists, spark-resistant tools made of non-ferrous materials should be used where flammable gases, highly volatile liquids, and other explosive substances are stored or used.

1. Hand and Power Tools, U.S. Department of Labor, Occupational Safety & Health Administration, http://www.osha.gov/Publications/osha3080.html

4.1.3 What are the Dangers of Power Tools?[2]

Power tools are determined by their power source: electric, pneumatic, liquid fuel, hydraulic, and powder-actuated. Power tools should be equipped with guards and safety switches. Personal protective equipment such as safety goggles and gloves should be worn to protect against hazards that may be encountered while using power tools.

To prevent hazards associated with the use of power tools, workers should observe the following general precautions:

- Never carry a tool by the cord or hose.
- Never yank the cord or the hose to disconnect it from the receptacle.
- Keep cords and hoses away from heat, oil, and sharp edges.
- Disconnect tools when not using them, before servicing and cleaning them, and when changing accessories such as blades, bits, and cutters.
- Keep all people not involved with the work at a safe distance from the work area.
- Secure work with clamps or a vise, freeing both hands to operate the tool.
- Avoid accidental starting. Do not hold fingers on the switch button while carrying a plugged-in tool.
- Maintain tools with care; keep them sharp and clean for best performance.
- Follow instructions in the user's manual for lubricating and changing accessories.
- Be sure to keep good footing and maintain good balance when operating power tools.
- Wear proper apparel for the task. Loose clothing, ties, or jewelry can become caught in moving parts.

Remove all damaged portable electric tools from use and tag them: "Do Not Use."

---

2. Hand and Power Tools, U.S. Department of Labor, Occupational Safety & Health Administration, http://www.osha.gov/Publications/osha3080.html

4.1.4 Guards[3]

The exposed moving parts of power tools need to be safeguarded. Belts, gears, shafts, pulleys, sprockets, spindles, drums, flywheels, chains, or other reciprocating, rotating, or moving parts of equipment must be guarded.

Appropriate machine guards must be provided to protect the operator and others from the following:

- Point of operation;
- In-running nip points;
- Rotating parts; and
- Flying chips and sparks.

Safety guards must never be removed when a tool is being used. Portable circular saws having a blade greater than 5.08 centimeters in diameter must be equipped at all times with guards. An upper guard must cover the entire blade of the saw. A retractable lower guard must cover the teeth of the saw, except where it makes contact with the work material. The lower guard must automatically return to the covering position when the tool is withdrawn from the work material.

4.1.5 Operating Controls and Switches[4]

The following hand-held power tools must be equipped with a constant-pressure switch or control that shuts off the power when pressure is released:

- drills;
- tappers;
- fastener drivers;
- horizontal, vertical, and angle grinders with wheels more than 5.08 centimeters in diameter;
- disc sanders with discs greater than 5.08 centimeters; belt sanders; reciprocating saws;
- saber saws, scroll saws, and jigsaws with blade shanks greater than 0.63 centimeters wide; and
- other similar tools.

3. Ibid.
4. Hand and Power Tools, U.S. Department of Labor, Occupational Safety & Health Administration, http://www.osha.gov/Publications/osha3080.html

These tools also may be equipped with a "lock-on" control, if it allows the worker to also shut off the control in a single motion using the same finger or fingers.

The following hand-held power tools must be equipped with either a positive "on-off" control switch, a constant pressure switch, or a "lock-on" control:

- disc sanders with discs 5.08 centimeters or less in diameter;
- grinders with wheels 5.08 centimeters or less in diameter;
- platen sanders, routers, planers, laminate trimmers, nibblers, shears, and scroll saws; and
- jigsaws, saber and scroll saws with blade shanks a nominal 6.35 millimeters or less in diameter.

Circular saws having a blade diameter greater than 5.08 centimeters, chain saws, and percussion tools with no means of holding accessories securely must be equipped with a constant-pressure switch.

4.1.6 Electric Tools[5]

Electrical shocks, which can lead to injuries such as heart failure and burns, are among the major hazards associated with electric-powered tools. In certain instances, even a small amount of electric current can result in fibrillation of the heart and death.

Electric tools must have a three-wire cord with a ground and be plugged into a grounded receptacle, be double insulated, or be powered by a low-voltage isolation transformer. Three-wire cords contain two current-carrying conductors and a grounding conductor. Any time an adapter is used to accommodate a two-hole receptacle, the adapter wire must be attached to a known ground. Never remove the third prong from the plug.

Double-insulated tools provide protection against electrical shock without third-wire grounding. On double-insulated tools, an internal layer of protective insulation completely isolates the external housing of the tool.

---

5. Ibid.

The following general practices should be followed when using electric tools:

- Operate electric tools within their design limitations;
- Use gloves and appropriate safety footwear when using electric tools;
- Store electric tools in a dry place when not in use;
- Do not use electric tools in damp or wet locations unless they are approved for that purpose;
- Keep work areas well lighted when operating electric tools; and
- Ensure that cords from electric tools do not present a tripping hazard.

In the construction industry, employees who use electric tools must be protected by ground-fault circuit interrupters or an assured equipment-grounding conductor program.

4.1.7 Portable Abrasive Wheel Tools[6]

Abrasive wheel tools must be equipped with guards that: (1) cover the spindle end, nut, and flange projections; (2) maintain proper alignment with the wheel; and (3) do not exceed the strength of the fastenings.

An abrasive wheel should be:

- Inspected for damage and should be sound- or ring-tested to ensure that it is free from cracks or defects; and
- Tested by tapping gently with a light, non-metallic instrument. A stable and undamaged wheel, when tapped, will give a clear metallic tone or "ring."

To prevent an abrasive wheel from cracking, it must fit freely on the spindle. The spindle nut must be tightened enough to hold the wheel in place without distorting the flange. Always follow the manufacturer's recommendations. Take care to ensure that the spindle speed of the machine will not exceed the maximum operating speed marked on the wheel.

6. Hand and Power Tools, U.S. Department of Labor, Occupational Safety & Health Administration, http://www.osha.gov/Publications/osha3080.html

An abrasive wheel may disintegrate or explode during start-up. Allow the tool to come up to operating speed prior to grinding or cutting. Never stand in the plane of rotation of the wheel as it accelerates to full operating speed. Portable grinding tools need to be equipped with safety guards to protect workers not only from the moving wheel surface, but also from flying fragments in case of wheel breakage.

When using a powered grinder:

- Always use eye or face protection;
- Turn off the power when not in use; and
- Never clamp a hand-held grinder in a vise.

4.1.8 Pneumatic Tools[7]

Pneumatic tools are powered by compressed air and include chippers, drills, hammers, and sanders.

There are several dangers associated with the use of pneumatic tools. First and foremost is the danger of getting hit by one of the tool's attachments or by some kind of fastener the worker is using with the tool.

Pneumatic tools must be checked to see that the tools are fastened securely to the air hose to prevent them from becoming disconnected. A short wire or positive locking device attaching the air hose to the tool must also be used and will serve as an added safeguard.

If an air hose is more than 12.7 millimeters in diameter, a safety excess flow valve must be installed at the source of the air supply to reduce pressure in case of hose failure.

In general, the same precautions should be taken with an air hose that are recommended for electric cords, because the hose is subject to the same kind of damage or accidental striking, and because it also presents tripping hazards.

When using pneumatic tools, a safety clip or retainer must be installed to prevent attachments such as chisels on a chipping hammer from being ejected during tool operation.

Pneumatic tools that shoot nails, rivets, staples, or similar fasteners and operate at pressures more than 6,890 kPa, must be equipped

7. Hand and Power Tools, U.S. Department of Labor, Occupational Safety & Health Administration, http://www.osha.gov/Publications/osha3080.html

with a special device to keep fasteners from being ejected, unless the muzzle is pressed against the work surface.

Airless spray guns that atomize paints and fluids at pressures of 6,890 kPa or more must be equipped with automatic or visible manual safety devices that will prevent pulling the trigger until the safety device is manually released.

Eye protection is required, and head and face protection is recommended for employees working with pneumatic tools.

Screens must also be set up to protect nearby workers from being struck by flying fragments around chippers, riveting guns, staplers, or air drills.

Compressed air guns should never be pointed toward anyone. Workers should never "dead-end" them against themselves or anyone else. A chip guard must be used when compressed air is used for cleaning.

Use of heavy jackhammers can cause fatigue and strains. Heavy rubber grips reduce these effects by providing a secure handhold. Workers operating a jackhammer must wear safety glasses and safety shoes that protect them against injury if the jackhammer slips or falls. A face shield also should be used.

Noise is another hazard associated with pneumatic tools. Working with noisy tools such as jackhammers requires proper, effective use of appropriate hearing protection.

4.1.9 Liquid Fuel Tools[8]

The most serious hazard associated with the use of fuel-powered tools comes from fuel vapors that can burn or explode and also give off dangerous exhaust fumes. The worker must be careful to handle, transport, and store gas or fuel only in approved flammable liquid containers.

Before refilling a fuel-powered tool tank, shut down the engine and allow it to cool to prevent accidental ignition of hazardous vapors. When a fuel-powered tool is used inside a closed area, effective ventilation and/or proper respirators such as atmosphere-supplying respirators must be utilized to avoid breathing carbon monoxide. Fire extinguishers must also be available in the area.

---

8. Hand and Power Tools, U.S. Department of Labor, Occupational Safety & Health Administration, http://www.osha.gov/Publications/osha3080.html

4.1.10 Powder-Actuated Tools[9]

Powder-actuated tools operate like a loaded gun and must be treated with extreme caution.

When using powder-actuated tools, an employee must wear suitable ear, eye, and face protection. Select a powder level that is suitable for the powder-actuated tool and necessary to do the work.

The muzzle end of the tool must have a protective shield or guard centered perpendicular to and concentric with the barrel to confine any fragments or particles that are projected when the tool is fired. A tool containing a high-velocity load must be designed not to fire unless it has this kind of safety device.

To prevent the tool from firing accidentally, two separate motions are required for firing. The first motion is to bring the tool into the firing position, and the second motion is to pull the trigger. The tool must not be able to operate until it is pressed against the work surface with a force of at least 2.2 kg greater than the total weight of the tool.

If a powder-actuated tool misfires, hold the tool in the operating position for at least 30 seconds before trying to fire it again. If it still will not fire, hold the tool in the operating position for another 30 seconds and then carefully remove the load in accordance with the manufacturer's instructions. This procedure will make the faulty cartridge less likely to explode. The bad cartridge should then be put in water immediately after removal. If the tool develops a defect during use, it should be tagged and must be immediately removed from service.

Safety precautions that must be followed when using powder-actuated tools include the following:

- Do not use a tool in an explosive or flammable atmosphere.
- Inspect the tool before using it to determine that it is clean, that all moving parts operate freely, and that the barrel is free from obstructions and has the proper shield, guard, and attachments recommended by the manufacturer.
- Do not load the tool unless it is to be used immediately.
- Do not leave a loaded tool unattended, especially where it would be available to unauthorized persons.
- Keep hands clear of the barrel end.
- Never point the tool at anyone.

---

9. Ibid.

When using powder-actuated tools to apply fasteners, several additional procedures must be followed:

- Do not fire fasteners into material that would allow the fasteners to pass through to the other side.
- Do not drive fasteners into very hard or brittle material that might chip or splatter or make the fasteners ricochet.
- Always use an alignment guide when shooting fasteners into existing holes.
- When using a high-velocity tool, do not drive fasteners more than 7.62 centimeters from an unsupported edge or corner of material such as brick or concrete.
- When using a high velocity tool, do not place fasteners in steel any closer than 1.27 centimeters from an unsupported corner edge unless a special guard, fixture, or jig is used.

4.1.11 Hydraulic Power Tools[10]

The fluid used in hydraulic power tools must be an approved fire-resistant fluid and must retain its operating characteristics at the most extreme temperatures to which it will be exposed. The exception to fire-resistant fluid involves all hydraulic fluids used for the insulated sections of derrick trucks, aerial lifts, and hydraulic tools that are used on or around energized lines.

Do not exceeded the manufacturer's recommended safe operating pressure for hoses, valves, pipes, filters, and other fittings.

All jacks – including lever and ratchet jacks, screw jacks, and hydraulic jacks – must have a stop indicator, and the stop limit must not be exceeded. Also, the manufacturer's load limit must be permanently marked in a prominent place on the jack, and the load limit must not be exceeded.

Never use a jack to support a lifted load. Once the load has been lifted, it must immediately be blocked up. Put a block under the base of the jack when the foundation is not firm, and place a block between the jack cap and load if the cap might slip.

---

10. Hand and Power Tools, U.S. Department of Labor, Occupational Safety & Health Administration, http://www.osha.gov/Publications/osha3080.html

To set up a jack, make certain of the following:

- The base of the jack rests on a firm, level surface;
- The jack is correctly centered;
- The jack head bears against a level surface; and
- The lift force is applied evenly.

All jacks must be lubricated regularly. Additionally, each jack must be inspected according to the following schedule:

- for jacks used continuously or intermittently at one site – inspected at least once every 6 months;
- for jacks sent out of the shop for special work – inspected when sent out and inspected when returned; and
- for jacks subjected to abnormal loads or shock – inspected before use and immediately thereafter.

4.1.12 General Requirements of Safety in Workshops Policy[11]

The following rules apply to all workshop personnel, whether they are permanently employed in the workshop or just occasional users:

- Keep the workshop clean and tidy at all times.
- Always seek instruction before using an unfamiliar piece of equipment.
- Only use tools and machines for their intended purpose.
- Report all damaged equipment and do not use it until it has been repaired by a qualified person.
- Where machine guards are provide they must be kept in place.
- Never distract the attention of another staff member when that person is operating equipment and never indulge in horseplay.

11. The University of Western Australia, *Safety in Workshops*, Australian Standard, Safety and Health, http://www.safety.uwa.edu.au/policies/safety_in_workshops

- Always use the appropriate personal protective devices and check that they are clean and in good repair before and after use.
- Long hair needs to be restrained by either a tie or hat.
- Never use compressed air for cleaning clothing and machinery.
- Report all hazards and unsafe conditions and work practices.

4.1.13 Machinery Installation[12]

Inspect machinery, plant and equipment on delivery to ensure its safety features comply with the requirements of the department. Inspect each machine prior to commencement of work to ensure that all guards are correctly fitted.

Install machinery, plant and equipment so as to ensure that sufficient space and safe footholds are provided around an individual machine to allow for normal operation, group instruction, adjustment and ordinary repairs.

4.1.14 Machine Controls[13]

Machine controls should be in accordance with the following requirements:

- Start-stop controls of the push button type easily visible, readily accessible and incorporating both no-volt and overload release.
- Start buttons should be shrouded or recessed, colored green and the word "START" shall be indicated on or near the button. Starting levers and handles should have a provision for automatic retention in the "off" position.
- Stop buttons shall be long, easy to locate, colored red and clearly marked with an identifying symbol or the

12. The University of Western Australia, *Safety in Workshops*, Australian Standard, Safety and Health, http://www.safety.uwa.edu.au/policies/safety_in_workshops
13. Ibid.

word "STOP". Each machine shall have a stop control for disconnecting power and the control should be readily and safely accessible to the operator from the normal operating position.

- Emergency stop buttons of the mushroom-head type, prominently and suitably labeled, should be installed at selected positions so that pressing any one of the buttons will immediately operate the circuit breaker and disconnect the supply from the machines.

4.1.15 Machine Guards[14]

Use of any power machinery introduces the danger of personal injury due to pinching, cutting, tearing or crushing. Minimize this danger by the wearing of suitable clothing and fitting suitable guards to protect both the operator and passing traffic. Guards should be made of unperforated material, but designed to allow access for inspection and maintenance. They should not make the operation of the machine more difficult.

4.1.16 Service Installations[15]

Design and construct electrical equipment and apparatus to prevent danger from shock and fire and should always be maintained in a safe and good condition.

Any electrical equipment should be checked by a qualified electrician at least quarterly and a tagged to identify the type of item concerned, the name of the owner of the item, the license number of the electrical worker who carried out the inspection and test and the date on which the inspection and test was carried out. Unqualified persons must not interfere with or alter any electrical installation. All electrical power distribution circuitry should be protected by core-balance earth-leakage protection of 30mA sensitivity.

The gas supply must be automatically interrupted at the occurrence of system failure.

14. Ibid.
15. Ibid.

4.1.17 Grinding and Polishing Machines[16]

Provide every grinding or polishing machine which generates dust with an efficient exhaust system or dust abatement system. The exhaust system should consist of a hood ducted to an exhaust fan in such a manner as to carry away the dust to a device whereby the dust is separated from the air and is prevented from entering the workroom.

All personnel engaged in grinding or polishing operations must wear suitable eye protection.

Properly mount grinding wheels, and where necessary, fit with a bush of suitable material between the wheel and the spindle. A guard of sufficient mechanical strength should enclose the grinding wheel.

It is necessary to prevent vibration, which can be caused by incorrect wheel balance, lack of rigidity in the machine, loose bearings or incorrect use of the work rest. Additionally, incorrect fitting of the belt fasteners for a belt-driven wheel may cause the vibration.

Provide an eye screen for hand-held work when using pedestal or bench-type grinding machines. The area of the screen should be large enough to discourage the operator from looking around it. The screen should always be in place and maintained at an adequate transparency.

Every grinding wheel should be positioned so that when in use the plane of rotation is not in line with any door, passageway, entrance or a place where someone regularly works.

Finishing machines should be guarded with only the working face of the belt exposed and the belt should be mounted such that it rotates away from the operator wherever practicable. Before use the condition of abrasive belt should be examined and replaced if worn and the correctness of the tracking of the belt should be checked by rotating the belt by hand. If necessary the belt should be adjusted and finally checked with a trial run. Where possible suitable jigs or fixtures should be used to hold or locate the work piece. The work piece should never be held in a cloth or any form of pliers and gloves must not be worn when using a finishing machine.

---

16. The University of Western Australia, *Safety in Workshops*, Australian Standard, Safety and Health, http://www.safety.uwa.edu.au/policies/safety_in_workshops

4.1.18 Milling Machines[17]

Operators of milling machines should observe the following:

- Exercise care when using fast traverse levers in order to avoid running the job into the cutter and never attempt to remove the arbor nut by applying power to the machine.
- Clamp the job or vice firmly on the table before starting the machine, and, where necessary, to provide steady supports to prevent vibration.
- Utilize the correct type of handling equipment when heavy cutters are involved and the use of a chip guard when a fly cutter is used.
- Keep hands and fingers away from the cutters when an unguarded cutter is in motion.

4.1.19 Metal-Cutting Guillotines[18]

The following requirements apply to the safe use of metal-cutting guillotines:

- Guards must be provided to prevent the operator's fingers from contacting the knife or clamp from either the front or rear of the machine.
- Only one person should be allowed to operate the machine at the one time and where long material is being cut and cannot be adequately supported by the work table, additional supports should be provided.
- The shear edges of the blades should be maintained in good condition and blade clearance must be adjusted in accordance with the manufacturer's recommendation appropriate to the thickness of the material being cut.
- Waste scrap metal provides a hazard for the hands and protective gloves should be worn when the metal is handled. A container should be provided for waste material from the guillotine.

---

17. Ibid.
18. The University of Western Australia, *Safety in Workshops*, Australian Standard, Safety and Health, http://www.safety.uwa.edu.au/policies/safety_in_workshops

4.1.20 General Considerations[19]

There are several points in relation to chemical safety which are particularly relevant to workshops including:

- Harmful or potentially harmful processes should be carried out using properly designed and well maintained equipment and where practicable in separate areas restricted to a minimum of persons.
- If harmful concentrations of fumes or gases develop in certain processes, specific provision should be made for their extraction using local exhaust ventilation in addition to the general ventilation of the workshop.
- Provision should be made to afford protection against chemical agencies such as harmful dusts, mists, vapors.
- Chemicals bearing trade names should not be used unless the supplier or manufacturer provides a material safety data sheet giving full information on the precautions which need to be taken when handling the chemical.
- The possibility of toxic or flammable gases existing or being generated should be indicated by prominently displayed notices.

4.1.21 Solvent Degreasing[20]

The following solvents are permitted for use in workshops:

- 1,1,1, trichloroethane - should always be used in a fume cupboard and only for small scale operations.
  - Trichloroethylene & perchloroethylene
  - Should only be used in equipment specifically designed and in a well ventilated area free from draught. These solvents have anesthetic properties

19. Ibid.
20. The University of Western Australia, *Safety in Workshops*, Australian Standard, Safety and Health, http://www.safety.uwa.edu.au/policies/safety_in_workshops

and are harmful when inhaled or on contact with the skin producing:
- headaches, nausea, vomiting, mental confusion, visual disturbances, even unconsciousness, and dermatitis.

Do no use caustic alkalis used with trichloroethylene or perchloroethylene as they produce an explosive mixture.

The following solvents are prohibited for use in workshops:

- petrol, kerosene, alcohol, ketones, esters; and
- carbon tetrachloride.

Solvent degreasing processes should not be carried out near open flames or electric heaters.

Mop up spillages with rags or by absorbing in sawdust, dry sand or earth and removing to an open space. Do no use incinerators.

4.1.22 Bibliography

U.S. Department of Labor, Occupational Safety and Health, Hand and Power Tools, http://www.osha.gov/Publications/osha3080.html.

The University of Western Australia, Safety in Workshops, Australian Standard, Safety and Health, http://www.safety.uwa.edu.au/policies/safety_in_workshops.

4.2 Ladder Safety

4.2.1 Introduction

Ladders are major sources of injuries and fatalities among construction workers for example, and many of the injuries are serious enough to require time off the job. Although ladders are uncomplicated, planning and care are still required to use them safely.

4.2.2 General Requirements[21]

- Provide a ladder at all worker points of access if there is a break in elevation of 48 cm or more and

21. U.S. Occupational Safety and Health Administration, 29 CFR 3124-12R, 2003. http://www.osha.gov/Publications/osha3124.pdf

no ramp, runway, embankment or personnel hoist is available.
- Keep the access point clear of obstacles to permit free passage by workers if there is only one point of access between levels. Employers must provide a second point of access if free passage becomes restricted.
- Ensure that at least one point of access remains clear when there are more than two points of access between levels.

Install all ladder fall protection systems required by these rules and ensure that their worksite meets all requirements of the ladder rules before employees use ladders.

4.2.2.1 Rules for All Ladder Types[22]

Apply the following rules when using all types of ladders:

- Maintain ladders free of oil, grease and other slipping hazards.
- Do not load ladders beyond their maximum intended load nor beyond their manufacturer's rated capacity.
- Only use ladders for their designed purpose.
- Only use ladders on stable and level surfaces unless secured to prevent accidental movement.
- Do not use ladders on slippery surfaces unless secured or provided with slip-resistant feet to prevent accidental movement.
- Secure ladders placed in areas such as passageways, doorways or driveways, or where they can be displaced by workplace activities or traffic to prevent accidental movement.
- Use a barricade to keep traffic or activity away from the ladder.
- Keep areas clear around the top and bottom of ladders.
- Do not move, shift or extend ladders while in use.

22. Ibid.

- Use ladders equipped with nonconductive side rails if the worker or the ladder could contact exposed energized electrical equipment.
- Face the ladder when moving up or down.
- Use at least one hand to grasp the ladder when climbing.
- Do not carry objects or loads that could cause loss of balance and falling.
- Apply the additional general requirements to all ladders, including ladders built at the jobsite:
  - Double-cleated ladders or two or more ladders must be provided when ladders are the only way to enter or exit a work area where 25 or more employees work or when a ladder serves simultaneous two-way traffic.
  - Ladder rungs, cleats and steps must be parallel, level and uniformly spaced when the ladder is in position for use.
  - Rungs, cleats and steps of portable and fixed ladders (except as provided below) must not be spaced less than 25 cm apart, nor more than 36 cm apart, along the ladder's side rails.
  - Rungs, cleats and steps of step stools must not be less than 20 cm apart, nor more than 31 cm apart, between center lines of the rungs, cleats and steps.
  - Rungs, cleats and steps at the base section of extension trestle ladders must not be less than 20 cm nor more than 46 cm apart, between center lines of the rungs, cleats and steps. The rung spacing on the extension section must not be less 15 cm nor more than 31 cm.
  - Ladders must not be tied or fastened together to create longer sections unless they are specifically designed for such use.
  - When splicing side rails, the resulting side rail must be equivalent in strength to a one-piece side rail made of the same material.
  - Two or more separate ladders used to reach an elevated work area must be offset with a platform or landing between the ladders, except when portable ladders are used to gain access to fixed ladders.

  - Ladder components must be surfaced to prevent snagging of clothing and injury from punctures or lacerations.
  - Wood ladders must not be coated with any opaque covering except for identification or warning labels, which may be placed only on one face of a side rail.

4.2.2.2 Rules for Specific Types of Ladder[23]

- Do not use single-rail ladders.
- Use non-self-supporting ladders at an angle where the horizontal distance from the top support to the foot of the ladder is approximately one-quarter of the working length of the ladder.
- Use wooden ladders built at the jobsite with spliced side rails at an angle where the horizontal distance is one-eighth of the working length of the ladder.

Stepladders[24]

- Do not use the top or top step of a stepladder as a step.
- Do not use cross bracing on the rear section of stepladders for climbing unless the ladders are designed and provided with steps for climbing on both front and rear sections.
- Provide metal spreader or locking devices must be on stepladders to hold the front and back sections in an open position when ladders are being used.

Portable Ladders[25]

- Minimum clear distance between side rails for all portable ladders must be 29 cm.

---

23. U.S. Occupational Safety and Health Administration, 29 CFR 3124-12R, 2003. http://www.osha.gov/Publications/osha3124.pdf
24. Ibid.
25. U.S. Occupational Safety and Health Administration, 29 CFR 3124-12R, 2003. http://www.osha.gov/Publications/osha3124.pdf

- Rungs and steps of portable metal ladders must be corrugated, knurled, dimpled, coated with skid-resistant material or treated to minimize slipping.
- Non-self-supporting and self-supporting portable ladders must support at least four times the maximum intended load; extra heavy-duty type 1A metal or plastic ladders must sustain 3.3 times the maximum intended load.
- Extend side rails at least .9 m above the upper landing surface when portable ladders are used for access to an upper landing surface. If such an extension is not possible, secure the ladder and use a grasping device such as a grab rail to assist workers in mounting and dismounting the ladder.

Fixed Ladders[26]

If the total length of the climb on a fixed ladder equals or exceeds 7.3 m, the ladder must be equipped with:

- self-retracting lifelines and rest platforms at intervals not to exceed 45.7 m; or
- a cage or well and multiple ladder sections with each ladder section not to exceed 15.2 m in length.

Sections ladders must be offset from adjacent sections and landing platforms must be provided at maximum intervals of 15.2 m.

Fixed ladders should also meet the following requirements and standards:

- Able to support at least two loads of 114 kg each, concentrated between any two consecutive attachments. Must also support added anticipated loads caused by ice buildup, winds, rigging and impact loads resulting from using ladder safety devices.
- Extend individual rung/step ladders at least 1.1 m above an access level or landing platform either by the continuation of the rung spacings as horizontal grab bars or by providing vertical grab bars that must have

26. Ibid.

the same lateral spacing as the vertical legs of the ladder rails.

- Each step or rung of a fixed ladder must be able to support a load of at least 114 kg applied in the middle of the step or rung.
- Minimum clear distance between the sides of individual rung/step ladders and between the side rails of other fixed ladders must be 41 cm.
- Rungs of individual rung/step ladders must be shaped to prevent slipping off the end of the rungs.
- Minimum perpendicular clearance between fixed ladder rungs, cleats, and steps and any obstruction behind the ladder must be 18 cm, except that the clearance for an elevator pit ladder must be 11 cm.
- Minimum perpendicular clearance between the centerline of fixed ladder rungs, cleats and steps, and any obstruction on the climbing side of the ladder must be 76 cm. If obstructions are unavoidable, clearance may be reduced to 61 cm, provided a deflection device is installed to guide workers around the obstruction.
- Step-across distance between the center of the steps or rungs of fixed ladders and the nearest edge of a landing area must be no less than 18 cm and no more than 30 cm.
- Provide a landing platform if the step-across distance exceeds 30 cm.
- Fixed ladders without cages or wells must have at least a 38 cm clearance width to the nearest permanent object on each side of the centerline of the ladder.
- Provide fixed ladders with cages, wells, ladder safety devices or self-retracting lifelines where the length of climb is less than 7.3 m but the top of the ladder is at a distance greater than 7.3 m above lower levels.
- Side rails of through or side-step fixed ladders must extend 1.1 m above the top level or landing platform served by the ladder.
- Parapet ladders must have an access level at the roof if the parapet is cut to permit passage through it. If the parapet is continuous, the access level is the top of the parapet.

- Steps or rungs for through-fixed-ladder extensions must be omitted from the extension; and the extension of side rails must be flared to provide between 61 cm and 76 cm clearance between side rails.
- Maximum clearance distance between side rail extensions must not exceed 91 cm when safety devices are provided.
- Use fixed ladders at a pitch no greater than 90 degrees from the horizontal, measured from the back side of the ladder.

Cages for Fixed Ladders[27]

The requirements for cages for fixed ladders are as follows:

- Fastened horizontal bands to the side rails of rail ladders or directly to the structure, building or equipment for individual-rung ladders.
- Position vertical bars on the inside of the horizontal bands.
- Fasten vertical bars to the horizontal bands.
- Cages must not extend less than 68 cm, or more than 76 cm from the centerline of the step or rung and must not be less than 68 cm wide.
- Clear insides of cages of projections.
- Space horizontal bands at intervals not more than 1.2 m apart measured from centerline to centerline.
- Space vertical bars at intervals not more than 24 cm, measured centerline to centerline.
- Bottoms of cages must be between 2.1 m and 2.4 m above the point of access to the bottom of the ladder.
- Flare bottom of the cage not less than 10 cm between the bottom horizontal band and the next higher band.
- Tops of cages must be a minimum of 1.1 m above the top of the platform or the point of access at the top of the ladder. There must be a way to access the platform or other point of access.

---

27. U.S. Occupational Safety and Health Administration, 29 CFR 3124-12R, 2003. http://www.osha.gov/Publications/osha3124.pdf

Wells for Fixed Ladders[28]

- Wells must completely encircle the ladder.
- Wells must be free of projections.
- Inside faces of wells on the climbing side of the ladder must extend between 68 cm and 76 cm from the centerline of the step or rung.
- Inside widths of wells must be at least 76 cm.
- Bottoms of wells above the point of access to the bottom of the ladder must be between 2.1 m and 2.4 m.

Ladder Safety Devices and Related Support Systems for Fixed Ladders[29]

- Connection between the carrier or lifeline and the point of attachment to the body belt or harness must not exceed 23 cm in length.
- All safety devices must be able to withstand, without failure, a drop test consisting of a 226 kg weight dropping 41 cm.
- All safety devices must permit the worker to ascend or descend without continually having to hold, push or pull any part of the device, leaving both hands free for climbing.
- All safety devices must be activated within .61 m after a fall occurs and limit the descending velocity of an employee to 2.1 m/sec or less.

Requirements for Mounting Ladder Safety Devices for Fixed Ladders[30]

- Attach mountings for rigid carriers at each end of the carrier, with intermediate mountings spaced along the entire length of the carrier.
- Attach mountings for flexible carriers at each end of the carrier. Installed cable guides for flexible carriers

---

28. U.S. Occupational Safety and Health Administration, 29 CFR 3124-12R, 2003. http://www.osha.gov/Publications/osha3124.pdf
29. Ibid.
30. Ibid.

with a spacing between 7.6 m and 12.2 m along the entire length of the carrier.
- Side rails and steps or rungs for side-step fixed ladders must be continuous in extension.

Defective Ladders[31]

Ladders needing repairs are subject to the following rules:

- Immediately label portable ladders with structural defects as defective or tagged with "Do Not Use" or similar language. Withdraw ladder from service until repaired.
- Withdraw fixed ladders with structural defects from service until repaired.
- Ladder must be restored to a condition meeting its original design criteria before the ladder is returned to use.

4.2.3 Ladder Hazards/Prevention Tips[32]

Ladder accidents usually are caused by improper selection, care or use, not by manufacturing defects. Some of the more common hazards involving ladders, such as instability, electrical shock, and falls, can be predicted and prevented. Prevention requires proper planning, correct ladder selection, good work procedures and adequate ladder maintenance.

Prevention tips:

- Do not hand-carry loads on a ladder.
- Move the ladder rather than to over-extend or overreach.
- Non-skid feet or spurs may prevent a ladder from slipping on a hard, smooth surface.
- Do not stand on top three rungs of the ladder.
- A damaged side rail may cause one side of a ladder to give way.

31. Ibid.
32. U.S. Occupational Safety and Health Administration, 29 CFR 3124-12R, 2003. http://www.osha.gov/Publications/osha3124.pdf

- The base should be spaced 30.5 centimeters away for every 121.9 centimeters it reaches up.
- Ladders used to reach a walking surface or roof must extend at least 91.4 centimeters beyond.
- Extension ladders need both locks holding to prevent overloading a rail.
- Step ladders should be securely spread open. Never use a folding step ladder in an unfolded position.
- Electrical shock can occur with metal or wet wooden ladders.

4.2.4 Ladder Selection[33]

Ladders are constructed under three general classes:

- Type I Industrial - Heavy-duty with a load capacity not more than 113.4 kg;
- Type II Commercial - Medium-duty with a load capacity not more than 102.7 kg (suited for painting and similar tasks); and
- Type III Household - Light-duty with a load capacity of 90.72 kg.

4.2.5 Ladder Maintenance[34]

Protect wooden ladders with a clear sealer varnish, shellac, linseed oil or wood preservative. Do not paint wooden ladders. Check and inspect for cracks, rot, splinters, broken rungs, loose joints and bolts and hardware in poor condition.

Inspect aluminum or steel ladders for rough burrs and sharp edges before use. Inspect closely for loose joints and bolts, faulty welds and cracks. Make sure the hooks and locks on extension ladders are in good condition. Replace worn or frayed ropes on extension ladders at once.

Fiberglass ladders should have a surface coat of lacquer maintained. If scratched beyond normal wear, lightly sand before applying a coat of lacquer.

---

33. Ibid.
34. Ibid.

4.2.5.1 *Using Extension Ladders*[35]

- Place ladders on a firm, level surface and ensure the footing is secure.
- Erect extension ladders so that the upper section rests on (e.g., in front of) the bottom section. This means the bottom section "faces" a wall or other supporting surface.
- Place the ladder feet so that the horizontal distance between the feet and the top support is 1/4 of the working length of the ladder. The ladder will be leaning at a 75° angle from the ground.
- Raise and lower ladders from the ground. Ensure that locking ladder hooks are secure before climbing.
- Erect ladders so that a minimum of 1 m extends above a landing platform. Tie the top at support points.
- Where a ladder cannot be tied off at the top, station a person at the foot to prevent it from slipping. This method is only effective for ladders up to 5 m long. The person at the foot of the ladder should face the ladder with a hand on each side rail and with one foot resting on the bottom rung.
- Leave all tie-off devices in place until they must be removed before taking the ladder down.
- Maintain the minimum overlap of sections as shown on a ladder label. Refer to safety regulations.
- Do not use ladders near electrical wire.
- Do not set up or take a ladder down when it is extended.
- Do not overextend. Maintain minimum overlap of sections.
- Do not climb higher than the fourth rung from the top of a ladder.
- Do not use ladders on ice, snow or other slippery surfaces without securing ladders' feet.
- Do not extend top section of a ladder from above or by "bouncing" on a ladder.
- Do not leave ladders unattended.

---

35. U.S. Occupational Safety and Health Administration, 29 CFR 3124-12R, 2003. http://www.osha.gov/Publications/osha3124.pdf.

4.2.5.2 Setting Up an Extension Ladder[36]

- Lay a ladder on the ground close to intended location.
- Brace ladder base using helpers' feet.
- Grasp the top rung with both hands, raise the top end over your head and walk toward the base of a ladder.
- Grasp the centre of the rungs to maintain stability.
- Move the erect ladder to the desired location. Lean it forward against the resting point.

To erect a shorter extension ladder:

Place the bottom of a ladder firmly against the base of a building or stationary object.

- Lift the top of ladder, and pull upwards to raise a ladder to a vertical position.
- Transfer a ladder to its required position when it is erect.
- Keep a ladder upright and close to the body with a firm grip.

4.2.5.3 Climbing a Fixed Ladder[37]

- Wait until the other person has exited before ascending or descending.
- Use the appropriate safety devices (e.g., restraint belt, traveling fixture).
- Maintain three-point contact by keeping two hands and one foot, or two feet and one hand on a ladder always.
- Face ladder and use both hands to grip the rungs firmly.
- Place feet firmly on each rung.
- Wear footwear with heels. Ensure that footwear is in good condition.
- Clean muddy or slippery boot soles before mounting a ladder.

---

36. Ibid.
37. U.S. Occupational Safety and Health Administration, 29 CFR 3124-12R, 2003. http://www.osha.gov/Publications/osha3124.pdf

- Rise or lower tools and materials using a hand-line.

4.2.6 Ladder Inspections[38]

Inspect any ladder for:

- missing or loose steps or rungs (they are loose if you can move them by hand);
- damaged or worn non-slip feet;
- loose nails, screws, bolts or nuts;
- loose or faulty spreaders, locks, and other metal parts in poor repair;
- rot, decay or warped rails in wooden ladders;
- cracks and exposed fiberglass in fiberglass ladders;
- cracked, split, worn or broken rails, braces, steps or rungs;
- sharp edges on rails and rungs;
- rough or splintered surfaces;
- corrosion, rust, oxidization and excessive wear, especially on treads twisted or distorted rails;
- distortion by sighting along the rails;
- wobble;
- loose or bent hinges and hinge spreaders; and
- broken stop on a hinge spreader.

Inspect extension ladders for:

- loose, broken or missing extension locks;
- defective locks that do not seat properly when ladder is extended;
- sufficient lubrication of working parts;
- defective cords, chains and ropes; and
- missing or defective pads or sleeves.

After inspecting any ladder:

- Tag any defective ladders and take them out of service;

38. Ibid.

- Clean fiberglass ladders every three months;
- Spray lightly with a clear lacquer or paste wax;
- Protect wooden ladders with a clear sealer or wood preservative;
- Replace worn or frayed ropes on extension ladders; and
- Lubricate pulleys on extension ladders regularly.

4.2.7 Bibliography

U.S. Occupational Safety and Health Administration, 29 CFR 3124-12R, 2003. http://www.osha.gov/Publications/osha3124.pdf.

4.3 Forklift Safety

4.3.1 Introduction

Appropriate procedures and caution are necessary to avoid personal injury during the movement of products and materials via a forklift or power industrial truck.

4.3.2 Pre-Qualifications for Forklift Operators[39]

Forklift training prerequisites:

- Driver's license and good driving record.
- No vision problems that cannot be rectified by glasses or contacts.
- No hearing loss that cannot be rectified by a hearing aid.
- No physical disabilities that would impair safe operation.
- No neurological disorders that may affect balance or consciousness.
- No medication can be taken that affects perception, vision, or physical abilities.

39. U.S. Department of the Interior, Occupational Health and Medical Services. http://medical.smis.doi.gov/FORKLIFT.html

4.3.3 Safe Operating Procedures[40]

- Only authorized and trained personnel should operate powered industrial trucks.
- Powered industrial trucks must contain a headache rack, fire extinguisher, rotating beacon, back-up alarm, and seat belts.
- Seatbelts must be worn by operators at all times.
- Inspections must be conducted by operators.
- Report safety defects immediately, and repair or remove the powered industrial truck removed from service.
- Proper recharging/refueling safety procedures must be followed.
- Operators must wear hard hats where overhead hazards exist.
- Sound the horn and use extreme caution when around pedestrians, making turns, and cornering.
- Prohibit passengers from riding on any portion of a powered industrial truck.
- Use a suitable man lift platform that is attached to the mast if powered industrial trucks are used to elevate a person.
- The path of the vehicle must be free from obstructions, aisles must be marked, and wide enough (minimum of 1.83 meters).
- Clearly mark the lift capacity on the industrial power truck, and the load should not exceed that weight limit.
- A multi-purpose dry chemical fire extinguisher must be present in all powered industrial trucks (with the exception of pallet jacks).
- Report all accidents to a supervisor.
- Follow proper docking/undocking procedures, and rail cars and trailers should parked "squarely" to the loading area and the wheels locked in place.

40. Southern Methodist University, Risk Management and Environmental Health. & Safety Procedure, March 2004.

- Changes must be approved by the manufacturer. The new rated load capacities established and posted on the truck, and written approval is necessary.

4.3.4 Changing and Charging Storage Batteries[41]

- Place battery charging installations in designated areas.
- Provide facilities for flushing and neutralizing spilled electrolyte, fire protection, protecting charging equipment from damage by trucks, and for ventilation for diffusion of fumes from gassing batteries.
- Use a provided conveyor, overhead hoist, or equivalent equipment when handling batteries.
- Reinstalled batteries must be correctly positioned and secured in the truck.
- Handled electrolytes by using a provided carbon filter or siphon.
- Pour acid into water when charging a battery.
- Do not pour water into acid.
- Before charging or changing the batteries, trucks must be correctly positioned and brakes applied.
- Ensure that the vent caps are functioning. The battery cover must be open to disperse heat.
- Smoking is prohibited in the battery charging area.
- Avoid open flames, sparks, or electric arcs in battery charging areas.
- Tools and other metallic objects can not be on the top of uncovered batteries.

4.3.5 Operations[42]

- Remove a powered industrial truck from service if it is discovered to be defective, or unsafe.
- Do not drive a truck or forklift up to an individual standing in front of a fixed object.

---

41. Southern Methodist University, Risk Management and Environmental Health & Safety Procedure, March 2004.

42. U.S. Department of the Interior, Occupational Health and Medical Services. http://medical.smis.doi.gov/FORKLIFT.html

- Individuals are prohibited from standing or passing under an elevated portion of a truck.
- Unauthorized personnel are prohibited from riding on an industrial truck. Passengers may only ride in a truck if there is an extra seat with a seatbelt.
- Individuals are prohibited from placing their arms or legs between the uprights of the mast or outside the running lines of the truck.
- When a powered industrial truck is left unattended, the load engaging means must be completely lowered, controls neutralized, power shut off, and brakes set. The wheels of the truck must be blocked if the truck is parked on an incline.
- Maintain a safe distance from the edge of the ramps or the platforms on any elevated dock, platform, or freight car. The trucks can not be used for opening or closing freight doors.
- There must be sufficient headroom under overhead installations, lights, pipes, sprinkler system, etc.
- Use an overhead guard for protection from falling objects. An overhead guard offers protection from impact of small packages, boxes, bagged material, etc.
- A load backrest extension should be used to decrease the chance of the load or part of it from falling backwards.

4.3.6 Traveling[43]

- Observe all traffic regulations, including authorized speed limits and yielding to pedestrians.
- Maintain a safe distance, approximately three truck lengths from the truck ahead, and keep the truck under control at all times.
- Yield the right of way to ambulances, fire trucks, or other vehicles in emergency situations.
- Refrain from passing other trucks traveling in the same direction at intersections, blind spots, or other dangerous locations.

43. U.S. Department of the Interior, Occupational Health and Medical Services. http://medical.smis.doi.gov/FORKLIFT.html

- Slow down and use the horn at cross aisles and areas where vision is obstructed.
- Travel with the load trailing if the load carried is obstructing forward view.
- The loads should be tilted back and raised no more than six inches above the ground.
- Cross railroad tracks diagonally. Do not park closer than eight feet from the center of the railroad tracks.
- Look in the direction of the path you are traveling and maintain a clear view ahead.
- Ascend and descend grades slowly, and position the load uphill in relation to the operator during ascend or descend of grades.
- Use dock plate when loading rail cars and trailers. Secure dock board or bridge plates before driving them over. Drive dock board or bridge plate slowly and carefully, never exceeding their rated capacity.
- Be careful not to run over loose objects on the road.
- During turns, slow down to a safe speed and turn the steering wheel in a smooth motion. When maneuvering at a low speed, turn the steering wheel at a moderate and even rate

4.3.7 Loading[44]

- When possible, handle stable or safely arranged loads that are centered – always use caution when handling loads. Adjust the long or high loads that may affect capacity.
- Only handle loads within the rated capacity of the truck.
- Trucks equipped with attachments must be operated as partially loaded trucks when not handling a load.
- A load engaging means must be placed under the load as far as possible. The mast must be carefully tilted backward to stabilize the load.
- Use extreme caution when tilting the load forward or backward.

---

44. Southern Methodist University, Risk Management and Environmental Health & Safety Procedure, March 2004.

- Do not tilt an elevated load forward unless when the load is in a depositing position. During stacking or tiering, only tilt enough to stabilize the load.

4.3.8 Fueling[45]

- Do not fill the fuel tanks while the engine is running.
- Do not spill.
- Ensure that if oil or fuel spills, it is either washed away or evaporated, and that the fuel tank cap is replaced before starting the engine.
- Do not operate truck with a leak in the fuel system.
- Do not use an open flame while checking the electrolyte level in storage batteries or gasoline levels in fuel tanks.

4.3.9 Maintenance[46]

- All power operated trucks that are unsafe must be removed from service and repaired by authorized personnel.
- Conduct repairs to the fuel and ignition systems of industrial trucks that involve fire hazards in designated area.
- Disconnect the battery in trucks when repairing the electric system.
- Use parts recommended by the manufacturer of the truck or forklift.
- Examine all industrial trucks prior to placing them in service. Remove the truck if it is deemed unsafe. Examine trucks and forklifts before each shift if it is used round-the-clock and correct necessary defects.
- Remove any truck if its temperature exceeds the normal operating temperature, therefore creating potentially harmful conditions. Only put truck or forklift back in service if the cause of temperature excess is rectified.

---

45. Ibid.
46. Southern Methodist University, Risk Management and Environmental Health. & Safety Procedure, March 2004.

- Keep industrial trucks and forklifts in a clean condition, free of lint, excess oil, and grease.
- Use noncombustible agents to clean the trucks and forklifts. Low flash point (below 37.8 degrees C) solvents must not be used. High flash point (at or above 37.8 degrees C) solvents may be used.

4.3.10 Training[47]

Choose an experienced operator to undergo forklift training. The training occurs under close supervision.

Trainees are permitted to operate a powered industrial truck under the following conditions:

- Under the direct supervision of qualified personnel, who have the knowledge, training, and experience to train operators and evaluate their competence; and
- Where such operation does not endanger the trainee or other employees.

Training should consist of a combination of formal instruction, practical training (demonstrations performed by the trainer and practical exercises performed by the trainee), and evaluation of the operator's performance in the workplace.

Initial training for truck/forklift-related and workplace-related topics[48]

Truck-related topics:

- Operating instructions, warnings, and precautions for the type of truck the operator will be authorized to operate,
- Differences between the truck and automobiles,
- Truck controls and instrumentation,
- Engine or motor operation,
- Steering and maneuvering,
- Visibility (including restrictions due to loading),

47. U.S. Department of the Interior, Occupational Health and Medical Services. http://medical.smis.doi.gov/FORKLIFT.html
48. Ibid.

- Fork and attachment adaptation, operation, and use limitations,
- Vehicle capacity,
- Vehicle stability,
- Vehicle inspection and maintenance that the operator will be required to perform,
- Refueling and/or charging and recharging of batteries,
- Operating limitations, and
- Operating instructions, warnings, or precautions listed in the operator's manual for the types of vehicle that the employee is being trained to operate.

Workplace-related topics:

- Surface conditions where the vehicle will be operated.
- Composition of loads to be carried and load stability.
- Load manipulation, stacking, and unstacking.
- Pedestrian traffic in areas where the vehicle will be operated.
- Narrow aisles and other restricted places where the vehicle will be operated.
- Hazardous (classified) locations where the vehicle will be operated.
- Ramps and other sloped surfaces that would affect the vehicles' stability.
- Closed environments and other areas where insufficient ventilation or poor vehicle maintenance could cause a buildup of carbon monoxide or diesel exhaust and
- Other unique or potentially hazardous environmental conditions in the workplace that could affect safe operation.

4.3.11 Refresher Training and Evaluation[49]

Refresher training, including an evaluation of the effectiveness of that training, must be conducted to ensure that the operator has

49. U.S. Department of the Interior, Occupational Health and Medical Services. http://medical.smis.doi.gov/FORKLIFT.html

the knowledge and skills needed to operate the powered industrial truck safely.

Refresher training in relevant topics must be provided to the operator in the following situations:

- The operator has been observed operating the vehicle in an unsafe manner.
- The operator has been involved in an accident or near-miss incident.
- The operator has received an evaluation that reveals that the operator is not operating the truck safely.
- The operator is assigned to drive a different type of truck.
- A condition in the workplace changes in a manner that could affect safe operation of the truck.
- Once every three years, an evaluation will be conducted of each powered industrial truck operator's performance.

4.3.12 Bibliography

ACT WorkCover, WorkSafe Victoria publications, http://www.workcover.act.gov.au/docs/publications.htm.

Southern Methodist University, Risk Management and Environmental Health. & Safety Procedure, March 2004.

U.S. Department of the Interior, Occupational Health and Medical Services. http://medical.smis.doi.gov/FORKLIFT.html.

4.4 Crane Operation Safety

4.4.1 Lifting Principles

There are four basic lifting principles that govern a crane's mobility and safety during lifting operations:

1. Center of Gravity - The center of gravity of any object is the point in the object where its weight can be assumed to be concentrated or, stated in another way, it is the point in the object around which its weight is evenly distributed. The location of the center of gravity of a

mobile crane depends primarily on the weight and location of its heaviest components (boom, carrier, upperworks and counterweight).

2. Leverage Cranes use the principle of leverage to lift loads. Rotation of the upperworks (cab, boom, counterweight, and load) changes the location of the crane's center of gravity, its leverage point or fulcrum.

As the upperworks rotates, the leverage of a mobile crane fluctuates. This rotation causes the crane's center of gravity to change and causes the distance between the crane's center of gravity and its tipping axis to also change. Stability can be affected by the fluctuating leverage the crane exerts on the load as it swings. The crane's rated capacity is therefore altered in the load chart to compensate for those changes in leverage.

Provided the ground is capable of supporting the load, a crane can be made more stable by moving the tipping axis further away from its center of gravity. The extra stability gained by moving the tipping axis can then be used to carry larger/heavier loads.

Increased Stability = More Load

1. Stability is the relationship of the load weight, angle of the boom and its radius (distance from the cranes center of rotation to the center of load) to the center of gravity of the load. The stability of a crane could also be affected by the support on which the crane is resting. A crane's load rating is generally developed for operations under ideal conditions, i.e., a level firm surface. Unlevel surfaces or soft ground therefore must be avoided. In areas where soft ground poses a support problem for stability, mats and or blocking should be used to distribute a crane's load and maintain a level stable condition.

In addition to overturning (stability failure), cranes can fail structurally if overloaded enough. Structural failure may occur before a stability failure. In other words, a mobile crane's structure may fail long before it tips. As loads are added beyond its rated capacity, a crane may fail structurally before there is any sign of tipping. Structural failure is not limited to total fracture; it includes all permanent damage such as overstressing, bending and twisting of any of the components. When a crane is overstressed, the damage may not be apparent. Nevertheless, a structural failure has occurred and overstressed components are then subject to catastrophic failure at some future time.

1. Structural Integrity - The crane's main frame, crawler track and/or outrigger supports, boom sections, and attachments are all considered part of the structural integrity of lifting. In addition, all wire ropes, including stationary supports or attachment points, help determine lifting capacity and are part of the overall structural integrity of a crane's lifting capacity. The following elements may also affect structural integrity:
    - The load chart capacity in relationship to stability;
    - The boom angle limitations which affect stability and capacity; and
    - The knowledge of the length of boom and radius in determining capacity.

Stability failures are foreseeable, but in structural failure it is almost impossible to predict what component will fail at any given time. No matter what the cause, if the crane is overloaded, structural failure can occur.

4.4.2 Operational Considerations

Cranes are carefully designed, tested, and manufactured for safe operation. When used properly they can provide safe reliable service to lift or move loads. Because cranes have the ability to lift heavy loads to great heights, they also have an increased potential for catastrophic accidents if safe operating practices are not followed.

Crane operators and personnel working with cranes need to be knowledgeable of basic crane capacities, limitations, and specific job site restrictions, such as location of overhead electric power lines, unstable soil, or high wind conditions. Personnel working around crane operations also need to be aware of hoisting activities or any job restrictions imposed by crane operations, and ensure job site coordination of cranes. Crane inspectors therefore should become aware of these issues and, prior to starting an inspection, take time to observe the overall crane operations with respect to load capacity, site coordination, and any job site restrictions in effect.

4.4.3 Construction Requirements

Some key requirements are as follows:

1. The employer shall comply with the manufacturer's specifications and limitations applicable to the operation of any and all cranes and derricks. Where manufacturer's specifications are not available, the limitations assigned to the equipment shall be based on the determinations of a qualified engineer competent in this field and such determinations will be appropriately documented and recorded. Attachments used with cranes shall not exceed the capacity, rating, or scope recommended by the manufacturer.
2. Rated load capacities, and recommended operating speeds, special hazard warnings, or instruction, shall be conspicuously posted on all equipment. Instructions or warnings shall be visible to the operator while he is at his control station.
3. The employer shall designate a competent person who shall inspect all machinery and equipment prior to each use, and during use, to make sure it is in safe operating condition. Any deficiencies shall be repaired, or defective parts replaced, before continued use.
4. A thorough, annual inspection of the hoisting machinery shall be made by an authorized and competent person, or by a government or private agency. The employer shall maintain a record of the dates and results of inspections for each hoisting machine and piece of equipment.
5. Except where electrical distribution and transmission lines have been de-energized and visibly grounded at

point of work or where insulating barriers, not a part of or an attachment to the equipment or machinery, have been erected to prevent physical contact with the lines, equipment or machines shall be operated approximate to power lines only in accordance with the following:

a. For lines rated 50 kV or below, minimum clearance between the lines and any part of the crane or load shall be 3.05 meters;
b. For lines rated over 50 kV, minimum clearance between the lines and any part of the crane or load shall be 3.05 meters plus 1.02 centimeters for each 1 kV over 50 kV, or twice the length of the line insulator, but never less than 3.05 meters;
c. A person shall be designated to observe clearance of the equipment and give timely warning for all operations, where it is difficult for the operator to maintain the desired clearance by visual means;
d. Any overhead wire shall be considered to be an energized line unless and until the person owning such line or the electrical utility authorities indicate that it is not an energized line and it has been visibly grounded.

6. No modifications or additions which affect the capacity or safe operation of the equipment shall be made by the employer without the manufacturer's written approval. In no case shall the original safety factor of the equipment be reduced.

4.4.4 Inspection Guidelines

Since cranes impact such a large segment of work going on at any job site, crane inspections must include a survey, or walk around, of the entire operation that questions how the crane will be operating and how other crafts will be effected by working with and around the crane.

Observation of crane operations prior to an inspection, or simply asking how cranes have or will be used, can indicate possible problem areas that may need a closer review during the inspection process.

4.4.4.1 *Preinspection*

Before the actual inspection, some general information about the crane operator's qualifications and the crane's certifications should be gathered, such as:

- Operator Qualifications - Observe the operator in action and when the opportunity permits ask a few question concerning the cranes capacity and restrictions imposed, either due to activity involved in or functional limitations.
- Crane Records - Ask for inspection and maintenance records and verify that the appropriate operator's manual and load charts are available for that particular crane in use.

4.4.4.2 *Crane Setup*

In your initial survey of crane operations, look for crane stability, physical obstructions to movement or operation, and proximity of electrical power lines, as well as the following:

1. Leveling - Has the crane operator set the crane up level and in a position for safe rotation and operation?
2. Outriggers - Are the outriggers, where applicable, extended and being used in accordance with manufacturer's recommendations?
3. Stability - The relationship of the load weight, angle of boom, and its radius (the distance from the cranes center of rotation to the center of load) to the center of gravity of the load. Also, the condition of crane loading where the load moment acting to overturn the crane is less than the moment of the crane available to resist overturning.
4. Structural Integrity - The crane's main frame, crawler, track and outrigger supports, boom sections, and attachments are all considered part of structural components of lifting. In addition, all wire ropes, including stationary supports, help determine lifting capacity and are part of the structural elements of crane operations.

4.4.4.3 Electrical Hazards

Working around or near electrical power lines is one of the most dangerous practices for crane operations. Limit crane operations to a minimum clearance of 3.05 meters.

Cranes should not be used to handle materials or loads stored under electric power lines. In addition, operation of mobile cranes near de-energized electric power lines is not recommended until the following steps have been taken:

- The power company or owner of the power line has deenergized the lines.
- The lines are visibly grounded and appropriately marked at jobsite.
- Durable warning signs are installed at the operator's station and on the outside of the crane identifying the clearance requirements between the crane/load and electrical power lines.
- Qualified representatives of the power company or owner of the electrical power line are on the job site to verify that the power lines have been de-energized or properly grounded.

4.4.4.4 Load Charts

Load charts are the principle set of instructions and requirements for boom configurations and parts of line which establish crane capacity for safe crane operations.

- Availability - The crane operator must have in his/her possession the appropriate load charts related to the crane in use and for the loads being lifted.
- Correct Use - The crane operator must show adequate understanding and proficient use of the load charts as related to the equipment in use and the loads being lifted.

4.4.4.5 Safe Operating Precautions

Because cranes have the ability to lift heavy loads to great heights, they also have an increased potential for catastrophic accidents if safe, operating practices are not followed.

Accidents can be avoided by careful job planning. The person in charge must have a clear understanding of the work to be performed and consider all potential dangers at the job site. A safety plan must be developed for the job and must be explained to all personnel involved in the lift.

Before operations begin for the day, a walkaround inspection needs to be conducted to ensure that the machine is in proper working condition. Only qualified and properly designated people shall operate the crane. Regular inspections are important; they provide a means of detecting potential hazards or conditions that could contribute to a sequence of events leading to an accident. Safe, reliable, and the economic operation of lifting equipment, cannot be ensured without regular safety inspections and thorough preventive maintenance programs. A thorough inspection program can forecast maintenance needs or potential equipment failures or malfunctions. The lack of such a program could result in serious deterioration of the equipment which might lead to excessive replacement, or repair charges, as well as an increased potential for accidents.

Due to the wide variation of conditions under which a crane may operate, it is impossible for the manufacturer to determine inspection intervals appropriate for every situation. Inspection intervals recommended in manufacturer's publications represent minimum intervals for average operating conditions. More frequent inspection intervals should be required if use and site conditions are severe and warrant it. Inspections are also designed as maintenance checks and/or as a verification that proper repairs or modifications of equipment have been completed which, if not checked could affect capacities as well as personnel safety. Since the initial load rating for cranes was determined and set under ideal conditions, inspections are required by manufacturers to guarantee optimal operating efficiency and capacity as determined by the load charts.

OSHA requires inspections to be divided into two categories: frequent and periodic. In addition to the performance of these regular inspections, equipment is required to be inspected and tested to ensure that it is capable of safe and reliable operation when initially set or placed in service and after any major repairs or any design modification.

4.4.4.6 *Inspection Types*

1. Frequent Inspections (daily to monthly intervals). Frequent inspections are usually performed at the

start of each shift by the operator who walks around the crane looking for defects or problem areas. Components that have a direct bearing on the safety of the crane and whose status can change from day to day with use must be inspected daily, and when possible, observed during operation for any defects that could affect safe operation. To help determine when the crane is safe to operate, daily inspections should be made at the start of each shift. Frequent inspections should include, but are not limited to the following:

- Check that all exposed moving parts are guarded. A removed guard may indicate that a mechanic is still working on part of the crane.
- Visually inspect each component of the crane used in lifting, swinging, or lowering the load or boom for any defects that might result in unsafe operation.
- Inspect all wire rope (including standing ropes), sheaves, drums rigging, hardware, and attachments. Remember, any hook that is deformed or cracked must be removed from service. Hooks with cracks, excessive throat openings of 15%, or hook twists of 10 degrees or more, must be removed from service.
- Check for freedom of rotation of all swivels.
- Visually inspect the boom and jib for straightness and any evidence of physical damage, such as cracking, bending, or any other deformation of the welds. Look for corrosion under any attachments that are connected to the chords and lacing. Watch carefully for cracking or flaking of paint. This may indicate fatigue of the metal which often precedes a failure. On lattice booms, look for bent lacing. If they are kinked or bent, the main chord can lose substantial support in that area. When lacing is bent, the ends also tend to draw together which pulls the main chords out of shape. This precaution is especially important on tubular booms where every component must be straight and free from any dents. Do not attempt to straighten these members by hammering or heating them and drawing them out. They must be cut out

and replaced with lacing to the manufacturer's specifications, procedures, and approval.
- Inspect tires for cuts, tears, breaks, and proper inflation.
- Visually inspect the crane for fluid leaks, both air and hydraulic.
- Visually check that the crane is properly lubricated. The fuel, lubricating oil, coolant and hydraulic oil reservoirs should be filled to proper levels.
- Check that the crane is equipped with a fully charged fire extinguisher and that the operator knows how to use it.
- Check all functional operating mechanisms such as: sheaves, drums, brakes, locking mechanisms, hooks, the boom, jib, hook rollers brackets, outrigger components, limit switches, safety devices, hydraulic cylinders, instruments, and lights.
- Check the turntable connections for weld cracks and loose or missing bolts. If they are loose, there is a good chance that they have been stretched.
- When checking the outriggers be sure that neither the beams nor the cylinders are distorted. Check that the welds are not cracked and that both the beams and cylinders extend and retract smoothly and hold the load. Check the condition of the floats, and check that they are securely attached.
- Inspect and test all brakes and clutches for proper adjustment and operation.
- Always inspect boom hoist lockout and other operator aids, such as anti-two-block devices (ATB) and load moment indicators (LMI), for proper operation and calibration.
- While the engine is running, check all gauges and warning lights for proper readings and operate all controls to see that they are functioning properly.
- Check for any broken or cracked glass that may affect the view of the operator.

2. Periodic Inspections (1 to 12 month intervals). The periodic inspection procedure is intended to determine the need for repair or replacement of components to keep the machine in proper operating

condition. It includes those items listed for daily inspections as well as, but not limited to, structural defects, excessive wear, and hydraulic or air leaks.

Inspection records of the inspected crane shall be maintained monthly on critical items in use, such as brakes, crane hooks, and ropes. These inspection records should include, the date of inspection, the signature of the person who performed the inspection, and the serial number, or other identifier. This inspection record should be kept readily available for review. The manufacturer's maintenance and inspection records, forms/checklist, or equivalent should be used.

1. Inspect the entire crane for structural damage. Be careful to check for distortion or cracks in main frame, outrigger assemblies, and structural attachments of the upperworks to the carrier.
2. Inspect all welded connections for cracks. Inspect the main chords and lacings and other structural items for paint flaking and cracking which may indicate potential failure, as well as for dents, bends, abrasions, and corrosion. Check hydraulic booms for bending, side sway, or droop.
3. Check for deformed, cracked, or corroded members in the load/stress bearing structure. Magnetic particle or other suitable crack detecting inspection should be performed at least once each year by an inspection agency retained by the owner. Inspection reports should be requested and retained in the crane file.
4. Inspect cracked or worn sheaves and drums.
5. Inspect for worn, cracked, or distorted parts such as: pins, bearings, shafts, gears, rollers, locking devices, hook roller brackets, removable outrigger attachments lugs, and welds.
6. Inspect for excessive wear on brake and clutch system parts, linings, pawls, and ratchets.
7. Inspect all indicators, including load and boom angle indicators, for proper operation and calibration.

8. Inspect all power plants for proper operation.
9. Inspect for excessive wear on drive sprockets and/or chain stretch.
10. Inspect for correct action of steering, braking, and locking devices.
11. Check that the counterweight is secure.
12. Check that the identification number is permanently and legibly marked on jibs, blocks, equalizer beams, and all other accessories.
13. Inspect all hydraulic and pneumatic hoses, fittings, and tubing. Any deterioration of any system component should cause the inspector to question whether further use would constitute a safety hazard. Conditions, such as the following, require replacement of the part in question:
    - Any evidence of oil or air leaks on the surfaces of flexible hoses or at the point at which the hose in question joins the metal end couplings.
    - Any abnormal deformation of the outer covering of hydraulic hose, including any enlargement, local or otherwise.
    - Any leakage at connections which cannot be eliminated by normal tightening.
    - Any evidence of abrasive wear that could have reduced the pressure retaining capabilities of the hose or tube effected. The cause of the rubbing or abrasion must be immediately eliminated.

4.4.4.7 *Starting the Inspection*

Since most crane inspections begin with a general walkaround and observation of the overall crane set up and operation, followed by a specific inspection of items or components, the following guidelines are presented in that order. The first section addresses the general items and operational considerations when inspecting any type of crane, followed by the specific inspection items for two specific types of cranes; Grove Rough Terrain 45 U.S. Ton (hydraulic) and Manitowoc 4100 150 U.S. Ton Crawler (lattice boom friction) cranes.

In general, the following should be considered when inspecting any crane:

1. Request for and review all inspection and maintenance documents for the crane being inspected, including the crane manufacturer's inspection and maintenance requirements.
2. Conduct a walkaround inspection, paying particular attention to mechanical systems leaks or damage (oil, hydraulic, air) and structural deficiencies.
3. Look at crane cab for properly marked controls, damaged instruments and for properly displayed and legible load charts.
4. Ask the operator, ground crew (riggers), and/or supervisor's appropriate questions on load charts, rigging and load weight determinations, and capacities.
5. Request the operator to raise and lower the boom/load line, where practical, and inspect, from the cab position, the running line or rope of the main hoist drum and secondary line or jib line. Check brake action and its ability to stop.
6. If practical, request the operator to lower boom to look at the condition of booms sections, lacing, lifting components, anti-two-block devices, jib back stops, and the condition of the hook.
7. Check crane set up and stability of outriggers on hydraulics and/or the effectiveness of cribbing on crawlers. if possible, request that the crane be rotated to check all clearances and overall stability.

4.4.4.8 *Specific Inspection Items and References*

Table 4.1 identifies the specific inspection items for cranes as well as a brief description and purpose to help the inspector to have a better understanding of what and why the item is being inspected.

Table 4.2 shows the items that need to be examined for the Grove Rough Terrain 45 U.S. Ton Hydraulic Crane and their corresponding applicable OSHA 29 CFR 1926 and ANSI B30.5 Standards.

Table 4.3 shows the items that need to be examined for the Manitowoc 4100 150 U.S. Ton Lattice Boom Crawler Crane and their corresponding applicable OSHA 1926 and ANSI B30.5 Standards.

Table 4.1 Inspection items and description.

Item	Description/Purpose
(1) Manufacturer's Operating and Maintenance Manuals	Manufacturer's operating and maintenance manuals shall accompany all mobile hoisting equipment. These manuals set forth specific inspection, operation and maintenance criteria for each mobile crane and lifting capacity.
(2) Guarding	All exposed moving parts such as gears, chains, reciprocating or rotating parts are guarded or isolated.
(3) Swing Clearance Protection	Materials for guarding rear swing area.
(4) High-Voltage Warning Sign	High-voltage warning signs displaying restrictions and requirements should be installed at the operator's station and at strategic locations on the crane.
(5) Boom Stops	Shock absorbing or hydraulic type boom stops are installed in a manner to resist boom overturning.
(6) Jib Boom Stops	Jib stops are restraints to resist overturning.
(7) Boom Angle Indicator	A boom angle indicator readable for the operator station is installed accurately to indicate boom angle.
(8) Boom Hoist Disconnect, Automatic Boom Hoist Shutoff	A boom hoist disconnect safety shutoff or hydraulic relief automatically stops the boom hoist when the boom reaches a predetermined high angle.
(9) Two-Blocking Device	Cranes with telescoping booms should be equipped with a two-blocking damage prevention feature that has been tested on-site in accordance with manufacturer's requirements. All cranes hydraulic and fixed boom used to hoist personnel must be equipped with two-blocking devices on all hoistlines intended to be used in the operation. The anti-two blocking device has automatic capabilities for controlling functions that may cause a two-blocking condition.

Table 4.1 (cont.) Inspection items and description.

Item	Description/Purpose
(10) Power Controlled Lowering	Cranes for use to hoist personnel must be equipped for power controlled lowering operation on all hoistlines. Check clutch, chains, and sprockets for wear.
(11) Leveling Indicating Device	A device or procedure for leveling the crane must be provided.
(12) Sheaves	Sheave grooves shall be smooth and free from surface defects, cracks, or worn places that could cause rope damage. Flanges must not be broken, cracked, or chipped. The bottom of the sheave groove must form a close fitting saddle for the rope being used. Lower load blocks must be equipped with close fitting guards. Almost every wire rope installation has one or more sheaves – ranging from traveling blocks with complicated reeving patterns to equalizing sheaves where only minimum rope movement is noticed.
(13) Main Hoist and Auxiliary Drums System	Drum crushing is a rope condition sometimes observed which indicates deterioration of the rope. Spooling is that characteristic of a rope which affects how it wraps onto and off a drum. Spoiling is affected by the care and skill with which the first larger of wraps is applied on the drum. Manufacturer's criteria during inspection usually specify: minimum number of wraps to remain on the drum; condition of drum grooves; condition of flanges at the end of drum; rope end attachment; spooling characteristics of rope; and rope condition.
(14) Main Boom, Jib Boom, Boom Extension	Boom jibs, or extensions, must not be cracked or corroded. Bolts and rivets must be tight. Certification that repaired boom members meet manufacturers original design standard shall be documented. Non-certified repaired members shall not be used until recertified.

Table 4.1 (cont.) Inspection items and description.

Item	Description/Purpose
(15) Load Hooks and Hook Blocks	Hooks and blocks must be permanently labeled with rated capacity. Hooks and blocks are counterweighted to the weight of the overhaul line from highest hook position. Hooks must not have cracks or throat openings more than 15% of normal or twisted off center more than 10° from the longitudinal axis. All hooks used to hoist personnel must be equipped with effective positive safety catches especially on hydraulic cranes.
(16) Hydraulic Hoses Fittings and Tubing	Flexible hoses must be sound and show no signs of leaking at the surface or its junction with the metal and couplings. Hoses must not show blistering or abnormal deformation to the outer covering and no leaks at threaded or clamped joints that cannot be eliminated by normal tightening or recommended procedures. There should be no evidence of excessive abrasion or scrubbing on the outer surfaces of hoses, rigid tubing, or hydraulic fittings.
(17) Outriggers	Outrigger number, locations, types and type of control are in accordance with manufacturer's specifications. Outriggers are designed and operated to relieve all weight from wheels or tracks within the boundaries of the outriggers. If not, the manufacturer's specifications and operating procedures must be clearly defined. Outriggers must be visible to the operator or a signal person during extension or setting.
(18) Load Rating Chart	A durable rating chart(s) with legible letters and figures must be attached to the crane in a location accessible to the operator while at the controls. The rating charts shall contain the following: • A full and complete range of manufacturer's crane loading ratings at all stated operating radii.

Table 4.1 (cont.) Inspection items and description.

Item	Description/Purpose
	• Optional equipment on the crane such as outriggers and extra counterweight which effect ratings. • A work area chart for which capacities are listed in the load rating chart, i.e. over side, over rear, over front. • Weights of auxiliary equipment, i.e. load block, jibs, boom extensions. • A clearly distinguishable list of ratings based on structural, hydraulic or other factors rather than stability. • A list of no-load work areas. • A description of hoistline reeving requirements on the chart or in operator's manual.
(19) Wire Rope	Main hoist and auxiliary wire rope inspection should include examining for broken wires, excess wear, and external damage from crushing, kinking, cutting or corrosion.
(20) Cab	Contains all crane function controls in addition to mechanical boom angle indicators, electric wipers, dash lights, warning lights and buzzers, fire extinguishers, seat belts, horn, and clear unbroken glass.
(21) Braking Systems	Truck cranes and self-propelled cranes mounted on rubber-tired chassis or frames must be equipped with a service brake system, secondary stopping emergency brake system and a parking brake system. Unless the owner/operator can show written evidence that such systems were not required by the standards or regulations in force at the date of manufacture and are not available from the manufacturer. The braking systems must have been inspected and tested and found to be in conformance with applicable requirements.

Table 4.1 (cont.) Inspection items and description.

Item	Description/Purpose
	Crawler cranes are provided with brakes or other locking devices that effectively hold the machine stationary on level grade during the working cycle. The braking system must be capable of stopping and holding the machine on the maximum grade recommended for travel. The brakes or locks are arranged to engage or remain engaged in the event of loss of operating pressure or power.
(22) Turntable/Crane Body	Make sure that the rotation point of a crane gears and rollers are free of damage, wear and properly adjusted and the components are securely locked and free of cracks or damage. The swing locking mechanism must be functional (pawl, pin) and operated in the cab.
(23) Counterweight	The counterweight must be approved and installed according to manufacturer's specifications with attachment points secured.

Table 4.2 Rough terrain 45 ton hydraulic crane.

Standard (29 CFR 1926.550)	**Inspection Items**	**ANSI B 30.5**
	Outriggers 1. Lubrication 2. Structural Condition 3. Pressure Hoses/Connections	 5-1.9.9 5-2.1.3 5-2.1.2
	Turntable/Crane Body 1. Ensure Level/Stability 2. Wear/Gear/Teeth/Rollers 3. Cracks 4. Bolts/Ensure Securely Attached	5-1.1 & 5-1.2
	Counter Weight 1. Proper Size 2. Attachment Connection/Bolts	5-3.4.2

Table 4.2 (cont.) Rough terrain 45 ton hydraulic crane.

Standard (29 CFR 1926.550)	Inspection Items	ANSI B 30.5
	Engine Housing	
1926.550(a)(8)	1. Cleanliness/No Rags/Trash	5-1.9.6
1926.550(a)(13) (ii) & (iii)	2. Gear/Machinery Guards	5-1.8.2 & 3
----------	3. Clear Access/Walkways	----------
1926.550(a)(4)	4. Brakes/Clutch Adjustments	5-2 (Fig.16)
----------	5. Hand Signal Illustration	5-1.4
----------	6. Swing Break	
	Cab	
1926.550(a)(12)	1. Glass/Visibility	----------
----------	2. Instruments and Controls	5-5.1.6.1.1
----------	3. Functioning Horn (Warning signal)	----------
1926.550(a)(14)(i)	4. Fire Extinguisher	5-3.4.9
1926.550(a)(2)	5. Appropriate Load Charts and Warning Signs	5-5.1.1.3
1926.550(a)(13) & 1926.550(a)(13)(iii)	6. Proper and Adequate Access (Steps/Walkway)	5-1.8.2 & 3
	Drum	
	1. Proper Size and Spoiling of Hoistlines	5-1.3.1 & 2
	2. Drum Sides/Shields for cracks	
	3. Dogs/Pawls/Locking Devices	
	4. Drum Rotation vs. Control Motion	
	Boom Sections (Boom sections correspond with crane model)	5-2.1.3
----------	1.	
----------	2.	
----------	3. Boom Stops	
----------	4.	
----------	5. Hoist Line Guides/Sheaves	
----------	6.	

Table 4.2 (cont.) Rough terrain 45 ton hydraulic crane.

Standard (29 CFR 1926.550)	**Inspection Items**	**ANSI B 30.5**
---------- 1926.550(b)	7. 8. Jib Attachment/Backstops/ Belly Slings	
	Sheave System 1. Ensure Hoist Line and Sheave Size Match 2. Worn 3. Lubrication/Move Freely	5-1.7.4
	Load/Auxiliary Hook and Block System 1. Sheaves Function Smoothly 2. Hook Rotates Freely/ Lubricated 3. Proper Becket 4. Properly Reeved	5-1-7.1-6
1926.550(a)(7) 1926.550(a)(7)(v) ----------- -----------	Wire Rope/Hoist Line 1. Overall Condition 2. End Connections 3. Lubrication 4. Clips	5-1.7.6 ---------- ----------
---------- ---------- 1926.550(a)(9) 1926.550(a)(15)	Safety Devices 1. Anti-Two Block Devices 2. Boom Backstop Devices 3. Swing Radius Warning Devices 4. Job or Site Specific Devices/ system (near electric power/ personnel hoisting platforms)	5-1.1.9 5-2 (Fig. 17)

Additional References:

1926.550(a)(1) ---- Crane used in accordance with manufactures specification.

1926.550(a)(5) ---- Inspection: Competent Person.

1926.550(a)(6) ---- Annual Inspection Record.

1926.550(a)(16) --- No modifications without written approval of manufacturer.

Table 4.3 (cont.) One-fifty ton lattice boom crawler crane.

Standard (29 CFR 1926.550)	**Inspection Items**	**ANSI B 30.5**
	Track Crawler System	
	1. Lubrication	5-1.9.9
	2. Connection Bolts	5-2.1.3
	3. Drive Chain (slack & wear)	5-2.1.2
	Turntable/Crane Body (Upper Works)	
	1. Assure Level/Stability	
	2. Wear/Gear/Teeth/Rollers	5-1.1 & 5-1.2
	3. Cracks	
	4. Bolts/Pins - Assure Securely Attached	
	Counter Weight	
	1. Proper Size	5-3.4.2
	2. Attachment Connection/Bolts	
	Engine Housing	
1926.550(a)(8)	1. Cleanliness/No Rags/Trash	5-1.9.6
1926.550(a)(13)(ii) & (iii)	2. Gear/Machinery Guards	5-1.8.2 & 3
	3. Clear Access/Walkways	----------
---------	4. Brakes/Clutch Adjustments	5-2 (Fig.16)
1926.550(a)(4)	5. Hand Signal Illustration	5-1.4
----------	6. Swing Break	

	Cab	
	1. Glass/Visibility	
1926.550(a)(12)	2. Instruments and Controls	----------
----------- ----------	3. Functioning Horn (warning signal)	5-5.1.6.1.1 ----------
1926.550(a)(14)(i)	4. Fire Extinguisher	
1926.550(a)(2) 1926.550(a)(13) &	5. Appropriate Load Charts and Warning Signs	5-3.4.9 5-5.1.1.3
1926.550(a)(13)(iii)	6. Proper and Adequate Access (steps/walkway)	5-1.8.2 & 3
	Hoist Drum System	
	1. Proper Size and Spoiling of Hoistlines	5-1.3.1 & 2

Table 4.3 (cont.) One-fifty ton lattice boom crawler crane.

Standard (29 CFR 1926.550)	**Inspection Items**	**ANSI B 30.5**
	2. Drum Sides/Shields for Cracks 3. Dogs/Pawls/Locking Devices 4. Drum Rotation vs. Control Motion 5. Clutch and Brakes	
	Boom Sections (Boom sections correspond with crane model)	
----------	1. Base Section Properly Attached	5-2.1.3
----------	2. Pin Clearance	
----------	3. Boom Lacing/Cord Damage	
----------	4. Boom Stops	
----------	5. Gantry System A-Frame	
----------	6. Hoist Line Guides/Sheaves	
----------	7. Boom Section Connection Pins/Keys	
1926.550(b)	8. Boom and Gantry Support System	
	9. Jib Attachment/Backstops/ Belly Slings (Jib Security Device)	
	Sheave System 1. Ensure Hoistline and Sheave Size Match 2. Worn 3. Lubrication/Move freely	5-1.7.4
	Load/Auxiliary Hook and Block System 1. Sheaves Function Smoothly 2. Hook Rotates Freely/ Lubricated 3. Proper Becket 4. Properly Reeved	5-1-7.1-6

Table 4.3 (cont.) One-fifty ton lattice boom crawler crane.

Standard (29 CFR 1926.550)	**Inspection Items**	**ANSI B 30.5**
	Wire Rope/Hoist Line	
1926.550(a)(7)	1. Overall Condition	5-1.7.6
1926.550(a)(7)(v)	2. End Connections	----------
-----------	3. Lubrication	----------
-----------	4. Clips	
	Safety Devices	
	1. Anti-Two Block Devices	
----------	2. Boom Backstop Devices	5-1.1.9
----------	3. Swing Radius Warning Devices	
1926.550(a)(9)		
1926.550(a)(15)	4. Job or Site Specific Devices/ System/Program for work (near electric power and use of personnel hoisting platforms)	5-2 (Fig. 17)

Additional references:

1926.550(a)(1) — Crane used in accordance with manufactures specification.

1926.550(a)(5) — Inspection: Competent Person.

1926.550(a)(6) — Annual Inspection Record.

1926.550(a)(16) — No modifications without written approval from manufacturers.

4.4.5 Definitions

Auxiliary Hoist - A supplemental hoisting unit, usually of lower load rating and higher speed than the main hoist.

Axis of Rotation - The vertical axis around which the crane's superstructure rotates.

Boom - In cranes and derricks usage, an inclined spar, strut, or other long member supporting the hoisting tackle. Also defined as a structural member attached to the revolving superstructure used for guiding and acting as a support for the load.

Boom Angle Indicator - An accessory device that measures the angle of the boom base section centerline to horizontal.

Boom Stops - A devise used to limit the angle of the boom at its highest position.

Brake – A device used for retarding or stopping motion by friction or power means.

Block - Sheaves or grooved pulleys in a frame provided with hook, eye, and strap.

Crane - A machine consisting of a rotating superstructure for lifting and lowering a load and moving it horizontally on either rubber tires or crawler treads.

Counterweight - Weights used for balancing loads and the weight of the crane in providing stability for lifting.

Deck - The revolving superstructure or turntable bed.

Drum - The spool or cylindrical member around which cables are wound for raising and lowering loads.

Gantry - A structural frame work (also known as an A Frame) mounted on the revolving superstructure of the crane to which the boom supporting cables are reeved.

Headache Ball - Heavy weight attached above the hook on a single line or whip line to provide sufficient weight to lower the hook when unloaded.

Holding Brake - A brake that automatically sets to prevent motion when power is off.

Jib - An extension attached to the boom point to provide added boom length for lifting specified loads.

Load - The weight of the object being lifted or lowered, including load block, ropes, slings, shackles, and any other ancillary attachment.

Load Block - The assembly of the hook or shackles, swivel, sheaves, pins, and frame suspended from the boom point.

Main Hoist - Hoist system or boom used for raising and lowering loads up to maximum rated capacity.

Mechanical Load Brake - An automatic type of friction brake used for controlling loads in the lowering direction. This device requires torque from the motor to lower a load but does not impose additional loads on the motor when lifting a load.

Outriggers - Support members attached to the crane's carrier frame which are used to the crane and may be blocked up to increase stability.

Pawl - Also known as "dog". It is a gear locking device for positively holding the gears against movement.

Pendants - Stationary cables used to support the boom.

Radius - The horizontal distance from the axis of rotation of the crane's superstructure to the center of the suspended load.

Reeving - The path that a rope takes in adapting itself to all sheaves and drums of a piece of equipment.

Running Sheave - Sheaves that rotate as the hook is raised or lowered.

Superstructure - The rotating frame, gantry and boom or other operating equipment.

Test Load - Any load or force, expressed in pounds, used for testing or certifying the limitations within acceptable tolerances of the anticipated load.

Two-Block - The condition in which the lower load lock or hook assembly comes in contact with the upper load block or boom point sheave assembly.

Quadrant of Operation - The area of operation that the lift is being made in. Usually divided into four quadrants i.e. front, rear and side(s) - left side and right side.

4.4.6 Bibliography

The American National Standards Institute, ANSI B30.5, 1968.

Occupational Safety and Health Administration, Regulations 29 CFR, 1926.

American National Standards Institute, Overhead and Gantry Cranes B30.5 Standards.

4.5 Scaffolds and Other Work Platforms

4.5.1 Introduction

When scaffolds, conductive tools, or other materials contact overhead power lines (see Figure 4.1), workers receive serious and often fatal injuries. Data from the U.S. Institute of Occupational Safety and Health (NIOSH), indicate that nearly 6,500 traumatic work-related deaths occur each year in the United States; an estimated 7% of these fatalities are electrocutions [NIOSH 1991]. The NTOF data base also shows that from 1980 through 1986, at least 25 deaths resulted when workers contacted overhead power lines while erecting or moving scaffolds or while using conductive tools on scaffolds.

Figure 4.1 Metal scaffold contacting an overhead power line.

A review of the NTOF data has revealed that many occupational groups (e.g., brickmasons, carpenters, painters, construction laborers, and plasterers) as well as the general public are at risk of electrocution because their jobs involve working from scaffolds near overhead power lines [NIOSH 1991].

4.5.2 Scaffold Safety

4.5.2.1 General Safety Guidelines

No scaffold shall be erected, moved, dismantled, or altered except under the supervision of competent persons. A competent person is defined as one who is capable of identifying existing and predictable hazards in the surroundings or working conditions, and who has the authorization to take prompt corrective action to eliminate them.

The following are recommended general safe practices:

- Provide safe access to get on and off scaffolds and work platforms safely. Use ladders safely.
- Keep scaffolds and work platforms free of debris. Keep tools and materials as neat as possible on scaffolds and platforms. These practices will help prevent materials from falling and workers from tripping.
- Erect scaffolds on firm and level foundations.
- Finished floors will normally support the load for a scaffold or work platform and provide a stable base.
- Place scaffold legs on firm footing and secure from movement or tipping, especially surfaces on dirt or similar surfaces.
- Erect and dismantle scaffolds only under the supervision of a competent person.
- Each scaffold must be capable of supporting its own weight and 4 times the maximum intended load.
- The competent person must inspect scaffolds before each use.
- Use manufactured base plates or mud sills made of hardwood or equivalent to level or stabilize the footings. Don't use blocks, bricks, or pieces of lumber.

For Planking:

- Fully plank a scaffold to provide a full work platform or use manufactured decking. The platform decking and/or scaffold planks must be scaffold grade and must not have any visible defects.
- Keep the front edge of the platform within 35.56 centimeters of the face of the work.
- Extend planks or decking material at least 15.24 centimeters over the edges or cleat them to prevent movement. The work platform or planks must not extend more than 30.48 centimeters beyond the end supports to prevent tipping when workers are stepping or working.
- Be sure that manufactured scaffold planks are the proper size and that the end hooks are attached to the scaffold frame.

4.5.2.2 *Safe Usage of Non-mobile Welded Frame Scaffolding*

The following are important characteristics of a safe scaffolding frame:

- The frame scaffold must be capable of supporting four times the maximum intended load.
- Using a horizontal frame separation distance of 2.13 meters, the maximum plank extension over the end support shall be no more than 30.48 centimeters if guardrails are not used to block access to the area.
- On a 10-section wide frame scaffold, the overlap for each plank on top of each other must be 30.48 centimeters.
- A scaffold five feet wide, 10.67 meters in length and 9.14 meters high with frame spacing of seven feet must have horizontal securement every 6.1 meters and vertical tie-in every 6.1 meters.
- A frame scaffold, three sections high, using a cantilevered outrigger platform, shall use ties, guys, braces, outriggers or scaffold manufacturer's stabilizer legs or equivalent.

The following are important characteristics of safe scaffolding rails:

- Guardrails are required on welded frame scaffolds that are 3.05 meters high or more.
- Top rail height shall be at least 91.44 centimeters to a maximum of 114.3 centimeters.
- Top rail strength must be at least 90.7 kg for welded frame scaffolds.
- Midrail strength on a welded frame scaffold must be at least 68 kg.
- Scaffold legs must bear on base plates and these shall rest on either firm foundation or mudsills.

The following are important characteristics of safe scaffolding planks:

- The space between the scaffold planks must be no more than one inch.
- The maximum distance permitted between a plank and a frame upright is nine inches.
- An uncleated scaffold plank must extend over the end support a minimum of six inches.

The following are important characteristics of safe scaffolding cross bracing and pins:

- Cross bracing is acceptable in place of midrails when the "X" is between 50.8 centimeters and 76.2 centimeters above the work surface.
- Cross bracing is acceptable in place of a top rail when the "X" is between 76.2 centimeters and 121.9 centimeters above the work surface.
- The purpose of using pins to lock a scaffold vertically together is to prevent uplift. This can occur with a rough terrain forklift grabbing scaffold. Wind, climbing a scaffold or the use of a hoist can also cause it to lift up.

4.5.2.3 Maximum Weight

The maximum intended load of a 25.4 centimeters wide, two-inch nominal thickness wood scaffold plank platform with the scaffold frames 2.14 meters apart horizontally is approximately 79.38 kg.

4.5.2.4 *Other Specs*

- Scaffold platforms above one section high must be provided with safe access.
- When hook-on-ladders are used, they must be provided with a rest platform every 10.67 meters.
- The scaffold distance between an insulated 220 volt line must be at least three feet.
- Work shall not be permitted during high winds. High winds are classified as 30 mph (48.28 km/h) gusts or higher. Ultimately, the decision is to be made by the competent person.

These recommendations were developed using generally accepted safety standards that are reported by OSHA and NIOSH. Compliance with these recommendations is not a guarantee that you will be in conformance with any building code, federal, state or local regulation regarding safety or fire. Compliance with these recommendations does not ensure the absolute safety of your occupation or place of business. It is the property owner's duty to warn any tenants or occupants of the property of any safety hazards that may exist.

4.5.2.5 *Don'ts for Safe Scaffold Use*

- DO NOT use damaged parts that affect the strength of the scaffold.
- DO NOT allow employees to work on scaffolds when they are feeling weak, sick, or dizzy.
- DO NOT work from any part of the scaffold other than the platform.
- DO NOT alter the scaffold.
- DO NOT move a scaffold horizontally while workers are on it, unless it is a mobile scaffold and the proper procedures are followed.
- DO NOT allow employees to work on scaffolds covered with snow, ice, or other slippery materials.
- DO NOT erect, use, alter, or move scaffolds within 3.05 meters of overhead power lines.
- DO NOT use shore or lean-to scaffolds.
- DO NOT swing loads near or on scaffolds unless you use a tag line.

- DO NOT work on scaffolds in bad weather or high winds unless the competent person decides that doing so is safe.
- DO NOT use ladders, boxes, barrels, or other makeshift contraptions to raise your work height.
- DO NOT let extra material build up on the platforms.
- DO NOT put more weight on a scaffold than it is designed to hold.

4.5.2.6 *Scaffold Guardrails*

- Guard scaffold platforms that are more than 3.05 meters above the ground or floor surface with a standard guardrail. If guardrails are not practical, use other fall protection devices such as safety harnesses and lanyards.
- Place the top-rail approximately 106.68 centimeters above the work platform or planking with a mid-rail about half that high at 53.34 centimeters.
- Install toe boards when other workers are below the scaffold.

4.5.2.7 *Scaffold Safety Checklist*

For set up and use, ask the following questions:

- Is the scaffold being erected under the direction of a competent person?
- Are all employees involved with (or near) the scaffold wearing hard hats?
- Are footings sound and rigid - not set on soft ground, frozen ground (that could melt), or resting on blocks?
- Is the scaffold level?
- Are wheels/castors locked?
- Is the scaffold able to hold four times its maximum intended load?
- Is the platform complete front to back and side to side (fully planked or decked, with no gaps greater than 2.54 centimeters)?
- Are guardrails and toeboards in place on all open sides?

- Are all sections pinned or appropriately secured?
- Is there a safe way to get on and off the scaffold, such as a ladder (without climbing on crossbraces)?
- Is the front face within 35.56 centimeters of the work (or within 91.44 centimeters for outrigger scaffolds)?
- Does the scaffold meet electrical safety clearance distances?
- Is the scaffold inspected by a competent person before being put in use?
- If the scaffold is over 3.05 meters, is personal fall protection provided, or are guardrails over 96.52 centimeters high?
- Are hardhats worn by workers on and around the scaffold?
- Are scaffold loads (including tools and other equipment) kept to a minimum and removed when the scaffold is not in use (like at the end of a day)?
- Are employees removed from scaffolds during high winds, rain, snow, or bad weather?
- Are materials secured before moving a scaffold?
- Are employees removed from the scaffold before they are moved?
- Are heavy tools, equipment, and supplies hoisted up (rather than carried up by hand)?

It is critical that workers understand that serious injury or death can result from failure to familiarize oneself, and comply with all safety guidelines before erecting, using or dismantling this scaffold. The following checklist has been compiled to assist site supervisors:

Prior to erection-all scaffold assemblies:

- Job site should be inspected to determine ground conditions or strength of supporting structure, and for proximity of electric power lines, overhead obstructions, wind conditions, and the need for overhead protection or weather protection coverings. These conditions must be evaluated and adequately provided for.
- Frame spacing and mud sill size can only be determined after the total loads to be imposed on the scaffold and the strength of the supporting soil or structure

are calculated and considered. This analysis must be done by a qualified person. Load carrying information on components is available from the manufacturer.

- Stationary scaffolds over 38.1 meters in height and rolling scaffolds over 18.29 meters in height must be designed by a professional engineer.
- All equipment must be inspected to see that it is in good condition and is serviceable. Damaged or deteriorated equipment should not be used.
- Wood plank should be inspected to see that it is graded for scaffold use, is sound and in good condition, straight grained, free from saw cuts, splits and holes. (Not all species and grades of lumber can be used as scaffold plank. Wood planks used for scaffolding must be specifically graded for scaffold use by an approved grading agency).
- The scaffold assembly must be designed to comply with local, state and federal safety requirements.

For erection of fixed scaffold the site supervisor shall use the following safety checklist:

- Scaffold must be erected, moved, or disassembled only under the supervision of qualified persons. Hard hats must be worn by all persons erecting, moving, dismantling or using scaffolding.
- Mud sills must be adequate size to distribute the loads on the scaffolding to the soil or supporting structure. Special care is needed when scaffolding is to be erected on fill or other soft ground or on frozen ground. Sills should be level and in full contact with the supporting surface.
- Base plates or screwjacks with base plates must be in firm contact with both the sills and the legs of the

scaffolding. Compensate for uneven ground with screwjacks with base plates. DO NOT USE unstable objects such as blocks, loose bricks, etc.

- Plumb and level scaffold until connections can be made with ease. Do not force members to fit. Be sure scaffold stays level and plumb as erection progresses.
- Ties, guys, bracing and/or outriggers may be needed to assure a safe stable scaffold assembly. The height of the scaffold in relation to the minimum base width, wind loads, the use of brackets or cantilevered platforms and imposed scaffold loads determines the need for stability bracing. The following general guides are minimum requirements.
- Scaffolding must always be secure when the height of the scaffold exceeds four (4) times the minimum base width.
- The bottom tie must be placed no higher than four (4) times the minimum base width and every 7.92 meters vertically thereafter. Ties should be placed as close to the top of the scaffold as possible and, in no case, less than four (4) times the minimum base width of the scaffold from the top.
- Vertical ties should be placed at the ends of scaffold runs and at no more than 9.14 meters horizontal intervals in between.
- Ties should be installed as the erection progresses and not removed until the scaffold is dismantled to that height.
- Side brackets, cantilevered platforms, pulleys or hoist arms and wind conditions introduce overturning and uplift forces that must be considered and compensated for. These assemblies may require additional bracing, tieing or guying.
- Circular scaffolds erected completely around or within a structure may be restrained from tipping by the use of "stand off" bracing members.
- Each leg of a free standing tower must be guyed at the intervals outlined above or otherwise restrained to prevent tipping or overturning.
- Work platforms must be fully planked either with scaffold graded solid sawn or laminated plank, in good

sound condition, or with fabricated platforms in good condition.

- Each plank must overlap the support by a minimum of 15.24 centimeters or be cleated, i.e. 2.44 meters planks on 2.13 meter spans must be cleated.
- Plank should not extend beyond the support by more than 45.72 centimeters. Such overhangs should be separated from the work platform by guard-railing so that they cannot be walked on.
- Plank on continuous runs must extend over the supports and overlap each other by at least 30.48 centimeters.
- Spans of full thickness, 5.08 centimeters by 25.4 centimeters scaffold grade planks, should never exceed 3.05 meters. Loads on plank should be evenly distributed and not exceed the allowable loads for the type of plank being used. No more than one person should stand on an individual plank at one time.
- Planks and/or platforms should be secured to scaffolding when necessary to prevent uplift of displacement because of high winds or other job conditions.
- Guardrails must be used on all open sides and ends of scaffold platforms. Both top and midrails are required. Local codes specify the minimum heights where guardrails are required, however, use at lower heights if falls can cause injury.
- Toeboards are required whenever people are required to work or pass under or around the scaffold platform.
- Access must be provided to all work platforms. If it is not available from the structure, access ladders, frames with built-in ladders, or stairways must be provided. When frames with built-in ladders are used, cleated plank or fabricated plank must be used at platform levels to minimize or eliminate platform overhang. Access ladders must extend at least 91.44 centimeters above platforms.
- Side and end brackets are designed to support people only. Materials should never be placed on cantilevered platforms unless the assembly has been designed to support material loads by a qualified person. (These types of platforms cause overturning and uplift forces which

must be compensated for. All frames should be fastened together to prevent uplift an overturning moment compensated for with counterweights or adequate ties).

- Putlogs must never be used for the storage of materials. They are designed for personnel use only. Special care should be taken when putlogs are used.
- Putlogs should overhang the support points by at least 15.24 centimeters. Use putlogs hangers with bolts fastened to support putlogs on frames.
- Putlog spans of greater than 3.66 meters require knee-bracing and lateral support.
- Putlogs used as side or end brackets need special bracing.
- Bridging between towers should not be done with plank or stages unless the assembly is designed by a qualified person and overturning moments have been compensated for.
- Scaffold should not be used as material hoist towers or for mounting derricks unless the assembly is designed by a qualified person.
- Check the erected assembly before use. A qualified person should thoroughly inspect the completed assembly to see that is complies with all safety codes, that nuts and bolts are tightened, that it is level and plumb, that work platforms are fully planked, that guardrails are in place and safe access is provided.

For erection of rolling scaffolds the site supervisor shall use the following safety checklist:

- Height of the tower must not exceed four (4) times the minimum base dimension (three (3) times in California). Outrigger frames or outrigger units on both sides of the tower may be used to increase base width dimension when necessary.
- All casters must be secured to frame legs or screwjacks with a nut and bolt or other secure means. Total weight of tower should not exceed the capacity of the casters.
- Screwjacks must not be extended more than 30.48 centimeters above caster base. Tower must be kept level and plumb at all times.

- Horizontal/diagonal bracing must be used at the bottom and top of tower and at intermediate levels of 6.1 meters. Fabricated planks with hooks may replace the top diagonal brace.

- All frames must be fully cross-braced.
- Only prefabricated plank or cleated plank should be used.
- Casters must be locked at all times the scaffold is not being moved.

In the use of all scaffolds the site supervisor shall use the following safety checklist:

- Inspect the scaffold assembly before each use to see that it is assembled correctly, that it is level and plumb, base plates are in firm contact with sills, bracing is in place and connected, platforms are fully planked, guardrails in place, safe access is provided, that it is properly tied and/or guyed and that there are no overhead obstructions or electric lines within 3.66 meters of the scaffold assembly.
- Use only the safe means of access that is provided. Do not climb bracing or frames not specifically designed for climbing. If such access is not provided, insist that it be provided.
- Climb safely.
- Face the rungs as you climb up or down.
- Use both hands.
- Do not try to carry materials while you climb.
- Be sure of your footing and balance before you let go with your hands. Keep one hand firmly on frame or ladder at all times.

- Do not work on slippery rungs to avoid slipping.
- Do not overload platforms with materials.
- Working heights should not be extended by planking guardrails or by use of boxes or ladders on scaffold platforms.
- Do not remove any component of a completed scaffold assembly except under the supervision of a qualified person. Any component that has been removed should be immediately replaced.

For rolling towers all of the above precautions should be checked plus:

- Do not ride manually propelled rolling scaffold. No personnel should be on the tower while it is being moved.
- Lock all casters before getting on the tower.
- Work only within the platform area: do not try to extend overhead work area by reaching out over guardrailing.
- Do not bridge between two rolling towers with plank or stages.
- Secure all materials before moving scaffolds.
- Be sure floor surface is clear of obstructions or holes before moving scaffold.
- Be sure there are no overhead obstructions or electric power lines in the path of rolling scaffold.
- Rolling towers must only be used on level surfaces.
- Move rolling towers by pushing at the base level only. Do not pull from the top.

4.5.2.8 Fall Protection

When dealing with floor and wall openings:

- Install guardrails around openings in floors and across openings in walls when the fall distance is 1.22 meters or more. Be sure the top rails can withstand a 90.72 kg load.
- Construct guardrails with a top rail approximately 1.07 meters high with a mid-rail about half that high at 53.34 centimeters.

- Install toe boards when other workers are to be below the work area.
- Cover floor openings and holes or protect the area with guardrails.

When working on roofs:

- Inspect for and remove frost and other slipping hazards before getting onto roof surfaces.
- Cover and secure all skylights and openings, or install guardrails to keep workers from falling through the openings.
- Fall protection is required when working at heights at or above 3.05 meters and when working on walking working surfaces at or above 1.22 meters.
- Stop roofing operations when storms, high winds or other adverse weather conditions create unsafe conditions.
- Remove or properly guard any impalement hazards.
- Wear shoes with slip-resistant soles.

4.5.2.9 Additional Guidelines

- Wear proper slip-resistant shoes or footwear to lessen slipping hazards.
- Train workers in safe work practices before performing work on foundation walls, roofs, trusses, or before they perform exterior wall erections and floor installations.

4.5.3 Bibliography

U.S. National Institute of Occupational Safety and Health (NIOSH), National Traumatic Occupational Fatalities (NTOF) database, 1991.

4.6 Compressed Gas Cylinder Safety

4.6.1 Introduction

Compressed gases present a unique hazard. Depending on the particular gas, there is a potential for simultaneous exposure to both mechanical and chemical hazards. Gases may be:

- Flammable or combustible;

- Explosive;
- Corrosive;
- Poisonous;
- Inert; or
- Combination of hazards.

If the gas is flammable, flash points lower than room temperature compounded by high rates of diffusion present a danger of fire or explosion. Additional hazards of reactivity and toxicity of the gas, as well as asphyxiation, can be caused by high concentrations of even "harmless" gases such as nitrogen. Since the gases are contained in heavy, highly pressurized metal containers, the large amount of potential energy resulting from compression of the gas makes the cylinder a potential rocket or fragmentation bomb.

Careful procedures are necessary for handling the various compressed gases, the cylinders containing the compressed gases, regulators or valves used to control gas glow, and the piping used to confine gases during flow.

4.6.2 Identification

The contents of any compressed gas cylinder must be clearly identified. Such identification should be stenciled or stamped on the cylinder or a label. Commercially available, three-part tag systems may also be used for identification and inventory.

No compressed gas cylinder should be accepted for use that does not legibly identify its contents by name. If the labeling on a cylinder becomes unclear or an attached tag is defaced to the point the contents cannot be identified, the cylinder should be marked "contents unknown" and returned directly to the manufacturer.

Never rely on the color of the cylinder for identification. Color coding is not reliable because cylinder colors may vary with the supplier. Additionally, labels on caps have little value because caps are interchangeable.

All gas lines leading from a compressed gas supply should be clearly labeled to identify the gas, the laboratory or area served, and the relevant emergency telephone numbers. The labels should be color coded to distinguish hazardous gases (such as flammable, toxic, or corrosive substances) (e.g., a yellow background and black letters). Always read the label.

Signs should be clearly posted in areas where flammable compressed gases are stored, identifying the substances and appropriate precautions (e.g., HYDROGEN - FLAMMABLE GAS - NO SMOKING - NO OPEN FLAMES).

4.6.3 Handling & Use

Cylinders may be attached to a bench top, individually to the wall, placed in a holding cage, or have a non-tip base attached. Chains or sturdy straps may be used to secure them.

Gas cylinders must be secured at all times to prevent tipping.

If a leaking cylinder is discovered, move it to a safe place (if it is safe to do so) and inform the environmental health & safety department. You should also call the vendor as soon as possible. Under no circumstances should any attempt be made to repair a cylinder or valve.

Standard cylinder-valve outlet connections have been devised by the Compressed Gas Association (CGA) to prevent mixing of incompatible gases. The outlet threads used vary in diameter; some are internal, some are external; some are right-handed, some are left-handed. In general, right-handed threads are used for non-fuel and water-pumped gases, while left-handed threads are used for fuel and oil-pump gases.

To minimize undesirable connections, only CGA standard combinations of valves and fittings should be used in compressed gas installations; the assembly of miscellaneous parts should be avoided. The threads on cylinder valves, regulators and other fittings should be examined to ensure they correspond and are undamaged.

Cylinders should be placed with the valve accessible at all times. The main cylinder valve should be closed as soon as it is no longer necessary that it be open (i.e., it should never be left open when the equipment is unattended or not operating). This is necessary not only for safety when the cylinder is under pressure, but also to prevent the corrosion and contamination resulting from diffusion of air and moisture into the cylinder after it has been emptied.

Cylinders are equipped with either a hand wheel or stem valve. For cylinders equipped with a stem valve, the valve spindle key should remain on the stem while the cylinder is in service. Only wrenches or tools provided by the cylinder supplier should be used to open or close a valve. At no time should pliers be used to open a cylinder valve. Some valves may require washers; this should be checked before the regulator is fitted.

Cylinder valves should be opened slowly. Oxygen cylinder valves should be opened all the way. Open up the oxygen cylinder valve stem just a crack. Once the needle on the high pressure gauge has stopped, open up the valve all the way. This back-seats the valve. Oxygen cylinders must have the valve opened up all the way because of the high pressure in the cylinder. There is a back-seating valve on the oxygen cylinder. This prevents the high-pressure gas from leaking out through the threaded stem.

When opening the valve on a cylinder containing an irritating or toxic gas, the user should position the cylinder with the valve pointing away from them and warn those working nearby.

Cylinders containing flammable gases such as hydrogen or acetylene must not be stored in close proximity to open flames, areas where electrical sparks are generated, or where other sources of ignition may be present.

Cylinders containing acetylene shall never be stored on their side.

An open flame shall never be used to detect leaks of flammable gases. Hydrogen flame is invisible, so "feel" for heat. One common practice is to use a natural bristle broom to "sweep" the air in front of you. All cylinders containing flammable gases should be stored in a well-ventilated area.

Oxygen cylinders, full or empty, shall not be stored in the same vicinity as flammable gases. The proper storage for oxygen cylinders

requires that a minimum of 6.1 meters be maintained between flammable gas cylinders and oxygen cylinders or the storage areas be separated, at a minimum, by a fire wall 1.52 meters high with a fire rating of 0.5 hours. Greasy and oily materials shall never be stored around oxygen; nor should oil or grease be applied to fittings.

Regulators are gas specific and not necessarily interchangeable.

Always make sure that the regulator and valve fittings are compatible.

If there is any question as to the suitability of a regulator for a particular gas, check with environmental health & safety services or call your vendor for advice. After the regulator is attached, the cylinder valve should be opened just enough to indicate pressure on the regulator gauge (no more than one full turn) and all the connections checked with a soap solution for leaks. Never use oil or grease on the regulator of a cylinder valve.

The following rules should always be followed in regards to piping:

- Copper piping shall not be used for acetylene.
- Plastic piping shall not be used for any portion of a high-pressure system.
- Do not use cast iron pipe for chlorine.
- Do not conceal distribution lines where a high concentration of a leaking hazardous gas can build up and cause an accident.
- Distribution lines and their outlets should be clearly labeled as to the type of gas contained.
- Piping systems should be inspected for leaks on a regular basis.
- Special attention should be given to fittings as well as possible cracks that may have developed.

A cylinder should never be emptied to a pressure lower than 172 kPa (the residual contents may become contaminated if the valve is left open). When work involving a compressed gas is completed, the cylinder must be turned off, and if possible, the lines bled.

When the cylinder needs to be removed or is empty, all valves shall be closed, the system bled, and the regulator removed. The valve cap shall be replaced, the cylinder clearly marked as "empty," and returned to a storage area for pickup by the supplier.

Empty and full cylinders should be stored in separate areas.

Where the possibility of flow reversal exists, the cylinder discharge lines should be equipped with approved check valves to prevent inadvertent contamination of cylinders connected to a closed system. "Sucking back" is particularly troublesome where gases are used as reactants in a closed system. A cylinder in such a system should be shut off and removed from the system when the pressure remaining in the cylinder is at least 172 kPa. If there is a possibility that the container has been contaminated, it should be so labeled and returned to the supplier.

Liquid bulk cylinders may be used in laboratories where a high volume of gas is needed. These cylinders usually have a number of valves on the top of the cylinder. All valves should be clearly marked as to their function. These cylinders will also vent their contents when a preset internal pressure is reached, therefore, they should be stored or placed in service where there is adequate ventilation.

Always use safety glasses (preferably with a face shield) when handling and using compressed gases, especially when connecting and disconnecting compressed gas regulators and lines.

Always use safety glasses (preferably with a face shield) when handling and using compressed gases, especially when connecting and disconnecting compressed gas regulators and lines.

All compressed gas cylinders, including lecture-size cylinders, must be returned to the supplier when empty or no longer in use.

4.6.4 Transportation of Cylinders

The cylinders that contain compressed gases are primarily shipping containers and should not be subjected to rough handling or abuse.

Such misuse can seriously weaken the cylinder and render it unfit for further use or transform it into a rocket having sufficient thrust to drive it through masonry walls.

1. To protect the valve during transportation, the cover cap should be screwed on hand tight and remain on until the cylinder is in place and ready for use.
2. Cylinders should never be rolled or dragged.
3. When moving large cylinders, they should be strapped to a properly designed wheeled cart to ensure stability.
4. Only one cylinder should be handled (moved) at a time.

4.6.5 Bibliography

Occupational Health and Safety Administration - Regulations (Standards - 29 CFR).

Compressed gases (general requirements), 1910.101.

4.7 Drum Handling Safety

4.7.1 Introduction

Accidents may occur during handling of drums and other hazardous waste containers. Hazards include detonations, fires, explosions, vapor generation, and physical injury resulting from moving heavy containers by hand and working around stacked drums, heavy equipment, and deteriorated drums. While these hazards are always present, proper work practices-such as minimizing handling and using equipment and procedures that isolate workers from hazardous substances-can minimize the risks to site personnel.

This section defines the safety of handling drums and other hazardous waste containers. It is intended to aid the team leader or safety manager in setting up a waste container handling program. In addition a general requirements and standards for storing, containing, and handling chemicals and containers, and for maintaining equipment used for handling materials. Containers are handled during characterization and removal of their contents and during other operations.

The appropriate procedures for handling drums depend on the drum contents. Thus, prior to any handling, drums should be visually inspected to gain as much information as possible about their contents. The inspection crew should look for:

- Symbols, words, or other marks on the drum indicating that its contents are hazardous, e.g., radioactive, explosive, corrosive, toxic, flammable.
- Symbols, words, or other marks on a drum indicating that it contains discarded laboratory chemicals, reagents, or other potentially dangerous materials in small-volume individual containers.
- Signs of deterioration such as corrosion, rust, and leaks.
- Signs that the drum is under pressure such as swelling and bulging.
- Drum type.
- Configuration of the drumhead.

Conditions in the immediate vicinity of the drums may provide information about drum contents and their associated hazards. Monitoring should be conducted around the drums using instruments such as a gamma radiation survey instrument, organic vapor monitors, and a combustible gas meter.

The results of this survey can be used to classify the drums into preliminary hazard categories, for example:

- Radioactive.
- Leaking/deteriorated.
- Bulging.
- Explosive/shock-sensitive.
- Contains small-volume individual containers of laboratory wastes or other dangerous materials.

As a precautionary measure, personnel should assume that unlabelled drums contain hazardous materials until their contents are characterized. Also, they should bear in mind that drums are frequently mislabeled-particularly drums that are reused.

4.7.2 Types of Drums

Closed-top drums are sealed and have small openings called bungs in the top through which liquids can be poured. Open-top drums have removable lids, and some do not have the small openings characteristic of the closed-top drum. Table 4.4. summarizes types of drums, their construction materials, and appropriate use.

4.7.2.1 Determine Drum Type

1. **Polyethylene or PVC-Lined Drums**: These often contain strong acids or bases. If the lining is punctured, the contents may quickly corrode the metal, resulting in a significant leak or spill.
2. **Metal Drum:** If a drum is made of a very expensive metal such as aluminum, nickel, or stainless steel, it may contain an extremely dangerous material.
3. **Laboratory Packs**: These are used to dispose of used chemicals and wastes from laboratories and

Table 4.4 Drum types.

Type of Drum	Construction	Contents
Closed-top (bung)	Metal	Non corrosive products in liquid form
Closed-top (bung)	Plastic or composite (plastic inside metal or cardboard)	Liquid, acid, or bases
Open-top	Metal	Corrosive solids or sludge
Open-top	Polyethylene	Corrosive solids or sludge
Special	Stainless steel, nickel, and Aluminum	Extremely hazardous chemicals
Overpack	Metal or plastic	Any drums listed above
Closed-top drums with fittings (bungs)	Fittings for pressurizing with inert gas	Reactive, flammable, or explosive liquids
Open-top	Plastic or metal	Lab packs of a variety of potentially dangerous and incompatible material.

hospitals. They may contain incompatible materials, radioisotopes, shock sensitive, volatile, corrosive, or toxic chemicals.

4.7.2.2 *Identifying Drum Contents*

Drum contents may be difficult to identify (characterize). Unknown contents are dangerous because they may be toxic, reactive, ignitable, shock sensitive, corrosive, radioactive, or a combination of these. Before handling a drum, always be sure to take the following step:

- Look for symbols, words, or colors on the drum indicating that its contents are hazardous.

4.7.3 Drum Inspection, Handling, and Staging Inspection

The appropriate procedures for handling drums depend on the drum contents. Thus, prior to any handling, drums should be visually inspected to gain as much information as possible about their contents. The inspection crew should look for:

- Symbols, words, or other marks on the drum indicating that its contents are hazardous.
- Symbols, words, or other marks on a drum indicating that it contains discarded laboratory chemicals, reagents, or other potentially dangerous materials in small-volume individual containers.
- Signs of deterioration such as corrosion, rust, and leaks.
- Signs that the drum is under pressure such as swelling and bulging.
- Drum type.
- Configuration of the drumhead.

Conditions in the immediate vicinity of the drums may provide information about drum contents and their associated hazards. Monitoring should be conducted around the drums using instruments such as a gamma radiation survey instrument, organic vapor monitors, and a combustible gas meter.

As a precautionary measure, personnel should assume that unlabelled drums contain hazardous materials until their contents are characterized.

The purpose of handling is:

1. To respond to any obvious problems that might impair worker safety, such as radioactivity, leakage, or the presence of explosive substances.
2. To unstack and orient drums for sampling.
3. To organize, if necessary, drums into different areas on site to facilitate characterization and remedial action.

Prior to handling, all personnel should be warned about the hazards of handling, and be instructed to minimize handling as much as possible. In all phases of handling, personnel should be alert for new information about potential hazards.

The major causes of leaks and spills at hazardous waste sites are:

- Swollen containers due to pressure of contents.
- Damage from rough handling during transport.
- Drum bungs that are not completely tightened.
- Corrosion from contact with the soil or from acids or chlorinated hydrocarbons in the drums.

Over pack drums (larger drums in which leaking or damaged drums are placed for storage or shipment) and an adequate volume of absorbent should be kept near areas where minor spills may occur.

Keep absorbent spill control materials (brooms, bulk solid absorbent such as vermiculite) available on site. There is some equipment which can be used to move drums:

- A drum grappler attached to a hydraulic excavator.
- A small front-end loader, which can be either loaded manually or equipped with a bucket sling.
- A rough terrain forklift.
- A roller conveyor equipped with solid rollers.
- Drum carts designed specifically for drum handling.

Drums are also sometimes moved manually. The drum grappler is the preferred piece of equipment for drum handling because it keeps the operator removed from the drums so that there is less likelihood of injury if the drums detonate or rupture.

If a drum is leaking, the operator can stop the leak by rotating the drum and immediately placing it into an over pack.

4.7.4 Leaking, Open, and Deteriorated Drums

If a drum containing a liquid cannot be moved without rupture, immediately transfer its contents to another drum using a pump designed for transferring that liquid. When transferring combustible liquids from one drum to another, be sure that the two containers are bonded with a heavy copper wire that is soldered and grounded. A copper wire from the container to the ground prevents sparks or shocks by dissipating the static charge to the ground.

Use protective equipment for your eyes, skin, and lungs when you open a drum, transfer the contents of a drum, or clean a spill.

Drum opening

There are three basic techniques available for opening drums at hazardous waste sites:

- Manual opening with non-sparking bung wrenches.
- Drum de-heading.
- Remote drum puncturing or bung removal.

The choice of drum opening techniques and accessories depends on the number of drums to be opened, their waste contents, and their physical condition.

Remote drum opening equipment should always be considered in order to protect worker safety.

Manual drum opening with bung wrenches or de-headers should be performed only with structurally sound drums and waste contents that are known to be non-shock sensitive, non-reactive, non-explosive, and non-flammable.

4.7.4.1 Bung wrench

A common method for opening drums manually is using a universal bung wrench.

These wrenches have fittings made to remove nearly all commonly encountered bungs. They are usually constructed of a non-sparking metal alloy (i.e. brass, bronze/manganese, aluminum) formulated to reduce the likelihood of sparks. The use of a "NON-SPARKING Wrench" does not completely eliminate the possibility of a spark being produced.

4.7.4.2 *Staging*

Although every attempt should be made to minimize drum handling, drums must sometimes be staged to facilitate characterization, remedial action, and to protect drums from potentially hazardous site conditions.

The number of staging areas necessary depends on site specific circumstances such as the scope of the operation, the accessibility of drums in their original positions, and perceived hazards. During staging, the drums should be physically separated into the following categories: those containing liquids, those containing solids, those containing lab packs, and those which are empty. This is done because the strategy for sampling and handling drums/containers in each of these categories will be different. This may be achieved by visual inspection of the drum and its labels, codes, etc. Solids and sludge are typically disposed of in open top drums. Closed head drums with a bung opening generally contain liquid. Where there is good reason to suspect that drums contain radioactive, explosive, or shock-sensitive materials, these drums should be staged in a separate isolated area. Placement of explosives and shock-sensitive materials in burned and fenced areas will minimize the hazard and the adverse effects of any premature detonation of explosives.

4.7.4.3 *Hazards Associated with Handling Drums*

Working with drums and other containers can present the following six hazards:

1. Unknown materials.
2. Spills, leaks and ruptures.
3. Physical exposure (through lungs, skin or mouth) to toxic chemicals.
4. Fires and explosions.
5. Unforeseen emergencies.
6. Back injuries.

Preventing physical exposure to toxic chemicals

When you work with drums it is important to protect your lungs, skin, and eyes from harmful contacts with chemicals. OSHA requires your employer to have a respiratory protection program. This program should state exactly what type of protection is necessary for each job where you work.

Always open bulging, corroded, dented, and otherwise damaged drums by remote handling tools. If the contents of a drum are known to be non-explosive and the drums are not damaged, use the following safety procedures for manual opening:

- Complete, appropriate protective equipment (respirators, splash aprons, eye protection, gloves).
- Drum should be positioned upright, bung up, or, if a side bung is used, drum on its side, bung up.
- Wrench bung plug open slowly and steadily; keep fire extinguishers available.
- If there is evidence of reactive, incompatible chemicals, pressurized contents or sudden release of toxic gases/vapors in high concentration, STOP, evacuate and finish the job using remote handling tools.

4.7.5 Preventing Back Injuries

Fires, explosions, and toxic gases are important and immediate dangers. However, the most common hazard for workers working with drums is back injury.

When handling drums, manual handling may mean lifting, lowering, pushing, pulling, carrying, moving, or holding drums of all sizes. If possible, use the following suggestions to avoid strain injuries:

- Introduce bulk handling to eliminate the use of drums.
- Provide and use mechanical handling equipment for drums.
- Palletize the drums and use mechanical lifting.
- Allocate sufficient space for handling the drums.
- Use a drum lifter.
- Ask your supplier to provide smaller sized drums.
- Clean up floor spills immediately.
- Provide unloading ramps.
- Provide and maintain even and non-slip floors.

4.8 Safe Welding Practices

4.8.1 Identifying Safe Welding Practices

Welding operations can pose significant hazards. The hazards of welding are summed up in three key areas: fires, fumes and your face.

Welding processes require heat and sometimes other substances to produce the weld. Byproducts resulting from the welding process include fumes and gases which can be serious health hazards to employees. Additionally, safety hazards can exist such as the potential for fire or explosion and injuries from arc radiation, electrical shock, or materials handling. This safety procedure affects to all employees who are exposed by this job duties to welding and torch or cutting operations.

Welding operations will be performed only by authorized and trained employees. When welding hazards exist that cannot be eliminated, then engineering practices, administrative practices, safe work practices, personal protective equipment, and proper training regarding welding will be implemented.

4.8.2 Definitions

- **Welder and Welding Operator** – is any operator of electric or gas welding and cutting work area.
- **Fire Hazards** – Move either the object you are welding or cutting or any movable fire hazards in the area to a safe place.
- **Guards** – If you can move neither of the above, then use guards to confine the heat, sparks and slag to protect the immovable fire hazards.
- **Combustible Materials** – Wherever there are floor openings or cracks in the flooring, close them or take precautions so that sparks will not drop through to combustible materials on the floor below. Use the same precautions with cracks or holes in walls, open doorways and open or broken windows.
- **Confined Space** – A space that is not designated for human occupancy, has limited openings for entry and exit, may lack adequate ventilation, and may contain or produce dangerous air contamination.

- **Hazardous** – Any act, condition, or substance which poses health and safety risks to employees.
- **Hot Work Permit** – A permit allowing employees to perform work involving welding, cutting, or any task that would deplete oxygen, create toxic fumes and vapor, or create the potential for fire or explosion.

4.8.3 General Provisions

4.8.3.1 Training

- Recognize the hazards associated with various welding operations.
- Know the safe work practices for welding operations.
- Understand the importance and requirements of hot work permits.
- Use the appropriate personal protective equipment (PPE) for the job.
- Recognize confined spaces and the requirements associated with them.
- Understand the importance of regular inspections of welding equipment, attachments, and accessories.
- Refresher training shall be provided upon the discretion of the supervisor or safety manager.

4.8.3.2 Types of Welding

Several types of welding operations are used. The most common welding operations are:

- Gas welding and cutting;
  - The gas welding process unites metals by heating. The gases commonly used as the fuel gas are oxygen and acetylene. The gas cutting process removes metals by a chemical reaction of the base metal with oxygen at an elevated temperature.
- Arc welding and cutting; and
  - The arc welding and cutting processes uses electric current and two welding leads. One welding lead is connected to the electric power supply while the other lead is attached to the work surface.

- Resistance welding:
  - Resistance welding is a metal – joining process where welding heat is generated at the joint by the resistance to the flow of electric current.

4.8.3.3 *Welding Hazards*

The hazards associated with welding include health and safety hazards. Health hazards are primarily respiratory hazards due to the generation of the fumes and gases. Safety hazards are generally physical hazards due to the work site and conditions and materials associated with the work site.

- Health Hazards
  - Toxic gases
  - Primary pulmonary gases
  - Non pulmonary gases
  - Particulate matter
  - Irritants and toxic inhalants
  - Air sampling may be required to identify the fumes and gases emitted from a specific operation.
- Safety Hazards
  - Fire
  - Proximity to combustible materials
  - Hazardous locations (rooms containing flammable or combustible vapors)
  - Closed containers that have held flammable liquids or other combustibles
  - Electric shock (arc welding)
  - Infrared and ultraviolet

4.8.3.4 *Safe Work Practice*

Safe work practices for all welding operations outlined below:

- Placing work at an optional height to avoid back strain or shoulder fatigue.
- Using fall protection equipment for work on elevated surfaces more than 6 feet above the floor or ground surface.

- Wearing personal protective equipment as applicable for the work conditions.
- Following special precautions when welding or cutting in a confined space.
- Posting warning signs to mark just- completed welding or cutting surfaces.
- Following safe house keeping principles.
- Using equipment as directed by the manufacturer instructions or practices.
- Following fire protection and prevention practices during the welding operation.
- Using proper ventilation techniques during welding operations.

4.8.3.5 *Electrical Hazards*

- The welding machine must be securely grounded.
- The electrode holder shall be specifically designed for its use, and have capacity capable of carrying the maximum-rated current required by the electrodes in use. The work leads must be of sufficient size also.
- The work leads will be checked for damaged insulation and secure attachment to the welding machine.
- The ground lead must be securely attached and close to the work to prevent unwanted arcing.
- Electrode holders left unattended shall not have a rod in them. Rod scraps shall be disposed of properly.
- No splices are allowed in the worklead with 25.4 centimeters of the electrode holder.
- Never dip an electrode holder in water to cool it.
- The power supply to a welding machine shall be turned off if it is not used for any appreciable length of time.

4.8.3.6 *Fire and Explosions*

Arc and welding produces intense heat. Temperatures up to 6649 degrees C are possible and special precautions need to be taken to prevent deadly fires and explosions.

- Never weld in an explosive atmosphere. If you suspect the presence of a gas, certain dust such as grain

and flour which are flammable contact management to checked it out before proceeding with any work.
- Never weld near stored ignitable materials or combustible debris. Never weld on a drum or a barrel unless it has been thoroughly cleaned of any previously contained material.
- Never weld on a compressed gas cylinder.
- When welding at a higher elevation, take precautions for falling sparks you produce.
- Always have adequate fire extinguishing equipment immediately available where you are welding.

4.8.3.7 Toxic gases and Fumes

The welding process produces various exhaust gases and fumes, depending on the materials you are working with. Precautions must be taken to avoid inhalation of toxic gases and fumes.

- Keep your head out of the fume path. Your welding helmet will also help protect your breathing.
- Provide ventilation, especially in welding booths, away from the welder.
- Some materials are known to be toxic or carcinogens. Respirators are required when working with them. They include; galvanized metals, cadmium plated, lead, mercury, chrome, nickel, and beryllium.
- Supplied air respirators may be required when welding in confined spaces.

4.8.3.8 Hot Work Permits

Hot work permits are a useful accountability tool to ensure that all the necessary precautions are taken prior to commencing welding. They also assure that employees are aware of and use the appropriate safeguards when performing welding operations. In confined spaces a hot work permit is required if any welding operations are performed in that space regardless of whether or not a confined space entry permit is required.

4.8.4 Inspections

All welding equipment including attachments and accessories will be inspected on a monthly basis by the supervisor or his designee.

A written record including the date, type of equipment, equipment number, and equipment serial number along with the signature of the employee performing the inspection will be maintained for a period of one year.

4.8.5 Ventilation Guidelines for Welding Operations

- Mechanical ventilation will be provided for welders and helpers when:
  - Welding is being performed in a space less than 283.17 cubic meters per welder.
  - A room ceiling height less than 1.83 meters.
  - A confined space or welding space contains partitions, balconies or other structural barriers to the extent that obstruct cross ventilation.
- The minimum rate for mechanical ventilation will be 56.63 cubic meters per minute per welder unless exhaust hoods or air supplied respirators are provided.
- When using local exhaust hoods, they will be as close to the operation as possible. The exhaust hood will provide a rate of 30.48 linear meters per minute of air flow in the welding zone.
- Air-supplied respirators will be used when mechanical ventilating is not possible or when materials such as beryllium and cadmium are used.
- Local exhaust ventilation or air-supplied respirators will be used when welding or torch cutting on coated metals indoors or in confined spaces. Outdoors operations shall be done using respiratory protective equipment.

5

Electrical Safety

5.1 Electric Shock and Lockout/Tagout

5.1.1 Introduction

Several occupational safety and health investigations have documented a number of fatalities whose circumstances suggest that the victims were unaware of the electrocution hazard from feedback electrical energy in power lines that were assumed to be de-energized. Occupational electrocutions from all causes continue to be a serious problem and take a very large toll on the workforce.

Safety laws are established to help provide safe working areas for electricians. Individuals can work safely on electrical equipment with today's safeguards and recommended work practices. In addition, an understanding of the principles of electricity is gained. It is good practice to always ask supervisors when in doubt about a procedure. Additionally, reporting any unsafe conditions, equipment, or work practices as soon as possible can help reduce the number of electrical shock accidents and fatalities.

5.1.2 Fuses[1]

Confirm that the switch for the circuit is open or disconnected before removing any fuse from a circuit. When removing fuses, use an approved fuse puller and break contact on the hot side of the circuit first. When replacing fuses, first install the fuse into the load side of the fuse clip, then into the line side.

5.1.3 GFCIs[2]

A groundfault circuit interrupter (GFCI) is an electrical device which protects personnel by detecting potentially hazardous ground faults and immediately disconnecting power from the circuit. Any current over 8 mA is considered potentially dangerous depending on the path the current takes, the amount of time exposed to the shock, as well as the physical condition of the person receiving the shock.

GFCls should be installed in such places as dwellings, hotels, motels, construction sites, marinas, receptacles near swimming pools and hot tubs, underwater lighting, fountains, and other areas in which a person may experience a ground fault.

A GFCI compares the amount of current in the ungrounded (hot) conductor with the amount of current in the neutral conductor (Figure 5.1). If the current in the neutral conductor becomes less than the current in the hot conductor, a ground fault condition exists. The missing current is returned to the source by some path other than the intended path (fault current). A fault current as low as 4 mA to 6 mA activates the GFCI and interrupts the circuit. Once activated, the fault condition is cleared and the GFCI manually resets before power may be restored to the circuit.

GFCI protection may be installed at different locations within a circuit. Direct-wired GFCI receptacles provide a ground fault protection at the point of installation. GFCI receptacles may also be connected to provide GFCI protection at all other receptacles installed downstream on the same circuit. GFCI CBs, when installed in a load center or panelboard, provide GFCI protection and conventional circuit overcurrent protection for all branch-circuit components connected to the CB.

---

1. Electricians Toolbox Etc., http://www.elec-toolbox.com/Safety/safety.htm
2. Ibid.

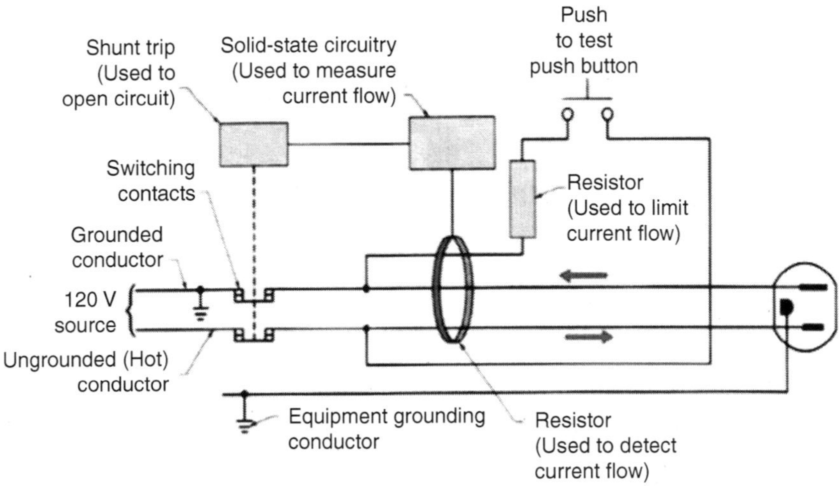

Figure 5.1 A GFCI compares the amount of current in the ungrounded (hot) conductor with the amount of current in the neutral conductor.

Plug-in GFCIs provide ground fault protection for devices plugged into them. Plug-in devices are generally utilized by personnel working with power tools in an area that does not include GFCI receptacles.

5.1.4 Electrical Shock[3]

Most fatal electrical shocks happen to people who are actually knowledgeable regarding electrical shock safety precautions. A majority of the electrical shock accidents occur when an employee in a rush disregards safety precautions that they know should be followed.

The following are some electromedical facts intended to make employees think twice before taking chances.

It's not the voltage but the current that kills. People have been killed by 100 volts AC in the home and with as little as 42 volts DC. The real measure of a shock's intensity lies in the amount of current (in milliamperes) forced through the body. Any electrical device used on a house wiring circuit can, under certain conditions, transmit a fatal amount of current.

3. Electricians Toolbox Etc., http://www.elec-toolbox.com/Safety/safety.htm

Currents between 100 and 200 milliamperes (0.1 ampere and 0.2 ampere) are fatal. Anything in the neighborhood of 10 milliamperes (0.01) is capable of producing painful to severe shock. Current values and their effects are summarized in Table 5.1 below.

The severity of the shock increases as the current increases. Below 20 milliamperes, breathing becomes labored, and it can cease completely even at values below 75 milliamperes. As the current approaches 100 milliamperes ventricular fibrillation occurs. This is an uncoordinated twitching of the walls of the heart's ventricles. Since you don't know how much current went through the body, it is necessary to perform artificial respiration to try to get the person breathing again. If the heart is not beating, cardio pulmonary resuscitation (CPR) is necessary.

Table 5.1 Current values and their effects.

	Readings	**Effects**
Safe Current Values	< than 1 mA	Causes No Sensation
	1 mA to 8 mA	Sensation of shock, not painful. Individual can let go at will since muscular control is not lost.
Unsafe Current Values	8 mA to 15 mA	Painful shock; individual can let go at will since muscular control is not lost.
	15 mA to 20 mA	Painful shock; control of adjacent muscles lost; victim can not let go.
	50 mA to 100mA	Ventricular fibrillation – a heart condition that can result in death is possible.
	100 mA to 200 mA	Ventricular fibrillation occurs.
	> than 200 mA	Severe burns, severe muscular contraction, so severe that chest muscles clamp the heart and stop it for the duration of the shock.

Source: Electricians Toolbox Etc., http://www.elec-toolbox.com/Safety/safety.htm

Electrical shock occurs when a person comes in contact with two conductors of a circuit or when the body becomes part of the electrical circuit. In either instance, a severe shock can cause the heart and lungs to stop functioning. Additionally, severe burns may occur where current enters and exits the body.

Prevention is the best medicine for electrical shock. Employees should respect all voltages and possess the knowledge of the principles of electricity. It is important to always follow safe work procedures. Do not take unnecessary chances. All electricians should be encouraged to take a basic course in CPR so they can aid a coworker in emergency situations.

Portable electric tools should always be well maintained and in safe operating condition. Make sure there is a third wire on the plug for grounding in case of shorts. The fault current should flow through the third wire to ground instead of through the operator's body to ground if electric power tools are grounded and if an insulation breakdown occurs.

5.1.5 Feedback Electrical Energy[4]

The potential for electrocution due to feedback electrical energy should never be underestimated. Workers may not be aware of the extreme hazard of electrocution and electrical injuries presented by feedback energy. The problem of feedback electrical energy in electrical transmission and distribution systems is always present and diligent efforts should be applied to safeguard against it. "Fuzzing" a line is a standard practice to test for the presence of high voltage in power lines. "Fuzzing" is accomplished by bringing a metallic object, such as a pair of lineman's pliers, close to a power line while watching for an arc and listening for a buzzing sound. This method does not reliably detect low voltages and should not be used as the only test for electrical energy in power lines. Once fuzzing has determined that high voltage is not present, low voltage testing equipment, such as a glowing neon light or a light-emitting diode, should be used to determine whether low voltage is present.

4. The National Institute for Occupational Safety and Health, Preventing Electrocutions by Undetected Feedback Electrical Energy Present in Power Lines, Publication No. 88-104, December 1987.

5.1.6 Universal Precautions[5]

- Exercise extreme caution when working on or in the vicinity of unverified de-energized power lines. All persons performing this work should treat all power lines as "hot" unless they positively know these lines are properly de-energized and grounded.
- "Fuzzing," although an accepted practice to check for high voltage in power lines, is not a reliable test method. "Fuzzing" should only be attempted after power lines have been de-energized and properly grounded. The person performing the work should personally ground all lines on both sides of the work area unless he or she is wearing the proper protective equipment.

5.1.7 Training Programs[6]

- Training programs for linemen should emphasize proper procedures for working with electrical transmission and distribution systems and their associated hazards.
- Training programs for linemen should include basic electrical theory that addresses electrical distribution systems and the identification, evaluation, and control of the hazards associated with these systems. Because the danger of feedback energy is always present, an improved method of de-energizing these systems should be stressed.

5.1.8 Protective Equipment and Work Practices[7]

- Power lines should not be repaired or otherwise accessed without adequate personal protective equipment unless the worker personally verifies that the line is de-energized and properly grounded.

---

5. Ibid.
6. Ibid.
7. Ibid.

- Workers must be specifically instructed to wear proper protective equipment, such as gloves and sleeves, required for the task to be performed.
- Linemen must be instructed to treat all power lines as energized unless they personally de-energize them by establishing a visible open point between the load and supply sides of the line to be repaired, by opening a fused disconnect, by opening a fused switch, or by removing a tap jumper if the load permits.
- Workers must verify that the power lines have been de-energized.
- Workers must provide proper grounding for the lines. Unless a power line is effectively grounded on both sides of a work area, it must be considered energized even though the line has been de-energized. Lines must be grounded to the system neutral.
- Grounds must be attached to the system neutral first and removed from the system neutral last. If work is being performed on a multiphase system, grounds must be placed on all lines.
- Lines should be grounded in sight of the working area and work should be performed between the grounds whenever possible. If work is to be performed out of sight of the point where the line has been de-energized, an additional ground should be placed on all lines on the source side of the work area.

5.1.9 Detection of Low Voltage[8]

- Appropriate safety and protective equipment should be provided to all persons working on or in the vicinity of power lines. All persons should also be trained in procedures that address all magnitudes of voltages. to which they may be exposed.
- Procedures should be established to perform a dual voltage check on the grounded load and supply sides of the open circuit. Once the "fuzzing" method has

8. The National Institute for Occupational Safety and Health, Preventing Electrocutions by Undetected Feedback Electrical Energy Present in Power Lines, Publication No. 88-104, December 1987.

determined that high voltage is not present, low voltage testing equipment, such as a glowing neon light or a light-emitting diode, should be used to determine if lower voltage is present.

5.1.10 Lockout/Tagout[9]

Electrical power must be removed when electrical equipment is inspected, serviced, or repaired. To ensure the safety of personnel working with the equipment, power is removed and the equipment must be locked out and tagged out.

Equipment should be locked out and tagged out before any preventive maintenance or servicing is performed (Figure 5.2). Lockout is the process of removing the source of electrical power and installing a lock which prevents the power from being turned on. Tagout is the process of placing a danger tag on the source of electrical power which indicates that the equipment may not be operated until the danger tag is removed.

Figure 5.2 Equipment must be locked out and tagged out before preventive maintenance or servicing is performed.

9. Electricians Toolbox Etc., http://www.elec-toolbox.com/Safety/safety.htm

A danger tag has the same importance and purpose as a lock and is used alone only when a lock does not fit the disconnect device. The danger tag shall be attached at the disconnect device with a tag tie or equivalent and shall have space for the worker's name, craft, and other required information. A danger tag must withstand the elements and expected atmosphere for as long as the tag remains in place. A lockout/tagout should be used when:

- Servicing electrical equipment that does not require power to be on to perform the service;
- Removing or bypassing a machine guard or other safety device. The possibility exists of being injured or caught in moving machinery; and
- Clearing jammed equipment. The danger exists of being injured if equipment power is turned on.

An approved procedure should be developed, implemented and followed when applying a lockout/tagout. Lockouts and tagouts are attached only after the equipment is turned off and tested to ensure that power is off. The lockout/tagout procedure is required for the safety of workers due to modern equipment hazards. An example of a standard procedure for equipment lockout/tagout is:

1. Prepare for machinery shutdown;
2. Machinery or equipment shutdown;
3. Machinery or equipment isolation;
4. Lockout or tagout application;
5. Release of stored energy; and
6. Verification of isolation.

A lockout/tagout should only be removed by the person that installed it, except in an emergency. In an emergency, the lockout/tagout may be removed only by authorized personnel. The authorized personnel shall follow approved procedures. The following lockout and tagout procedures and steps can effectively reduce the number of electrocutions in the workplace:

- Use a lockout and tagout whenever possible;
- Use a tagout when a lockout is impractical. A tagout is used alone only when a lock does not fit the disconnect device;

- Utilize a multiple lockout when individual employee lockout of equipment is impractical;
- Notify all employees affected before using a lockout/tagout;
- Remove all primary and secondary power sources; and
- Measure for voltage using a voltmeter to ensure that power is.

5.1.11 Lockout Devices[10]

Lockout devices are lightweight enclosures that allow the lockout of standard control devices. Lockout devices are available in various shapes and sizes that allow for the lockout of ball valves, gate valves, and electrical equipment such as plugs, disconnects, etc.

Lockout devices resist chemicals, cracking, abrasion, and temperature changes. They are available in colors to match ANSI pipe colors. Lockout devices are sized to fit standard industry control device sizes. Locks used to lock out a device may be color coded and individually keyed. The locks are rust-resistant and are available with various size shackles.

Danger tags provide additional lockout and warning information. Various danger tags are available. Danger tags may include warnings such as "Do Not Start," "Do Not Operate," or may provide space to enter worker, date, and lockout reason information. Tag ties must be strong enough to prevent accidental removal and must be self-locking and nonreusable.

5.1.12 Specific Procedures for Logout/Tagout

1. **Preparation for Lockout/Tagout[11]**
    - Prepare and conduct a survey to locate and identify all isolating devices to be certain which switches, valves, or other energy isolating devices apply to the equipment to be locked or tagged out. It is possible

10. Electricians Toolbox Etc., http://www.elec-toolbox.com/Safety/safety.htm
11. North Carolina State University, Environmental Health and Public Safety, http://www.ncsu.edu/ehs/www99/right/handsMan/workplace/lockout.html

that more than one energy source (electrical, mechanical, stored energy, or others) may be involved.

2. **Sequence of Lockout or Tagout System Procedure**[12]
   - Notify affected employees that a lockout or tagout system is going to be utilized the reason for using the system.
   - The authorized employee shall know the type and magnitude of energy that the machine or equipment utilizes and shall understand the potential dangers and hazards.
   - If the machine or equipment is operating, shut it down by the normal stopping procedure (depress stop button, open toggle switch, etc.).
   - Operate the switch, valve, or other energy isolating devices in order to isolate the equipment from its energy sources.
   - Stored energy must be dissipated or restrained by methods such as repositioning, blocking, bleeding down, etc.
   - Lockout/tagout the energy isolating devices with assigned individual locks and/or tags.
   - After ensuring that no personnel are exposed, and to serve as a check on having disconnected the energy sources, operate the push button or other normal operating controls to make sure the equipment will not operate. *CAUTION:* Always return operating controls to neutral or off position after the test.
   - The equipment is now locked out or tagged out.
3. **Restoring Machines or Equipment to Normal Operations**[13]
   - Once the servicing and/or maintenance is complete and the equipment is ready for normal production operations, safeguard the area around the machines or equipment to ensure that no one is exposed.
   - Lockout or tagout devices should only be removed after all tools have been removed from the machine

12. Ibid.
13. North Carolina State University, Environmental Health and Public Safety, http://www.ncsu.edu/ehs/www99/right/handsMan/workplace/lockout.html

or equipment, guards have been reinstalled, and employees are in the clear. Operate the energy isolating devices to restore energy to the machine or equipment.

4. **Procedure Involving More Than One Person**[14]
   If more than one individual is required to lockout or tagout equipment, each shall place his/her own personal lockout/tagout device on the energy isolating devices. A multiple lockout or tagout device may be used when an energy isolating device cannot accept multiple locks or tags. If lockout is used, a single lock may be used to lockout the machine or equipment with the key being placed in a lockout box or cabinet which allows the use of multiple locks to secure it. Each employee will then use his/her own lock to secure the box or cabinet. As each person no longer needs to maintain his/her lockout protection, that person will remove his/her lock from the box or cabinet.
5. **Temporary Removal of Lockout/Tagout Devices**[15]
   In situations where lockout/tagout devices must be temporarily removed from the energy isolating device and the machine or equipment energized to test or position the machine, equipment or component thereof, the following sequence of actions should be followed:
   1. Remove non-essential items and ensure that machine or equipment components are operationally intact.
   2. Notify affected employees that lockout/tagout devices have been removed and
   3. Ensure that all employees have been safely positioned or removed from the area.
   4. Have employees who applied the lockout/tagout devices remove the lockout/tagout devices.
   5. Energize and proceed with testing or positioning.
   6. De-energize all systems and reapply energy control measures.

14. Ibid.
15. Sonoma State University, Environmental Health & Safety, http://www.sonoma.edu/ehs/esp/LOTOp.shtml

6. **Maintenance Requiring Undisrupted Energy Supply**[16]
   Where maintenance, repairing, cleaning, servicing, adjusting, or setting up operations cannot be accomplished with the prime mover or energy source disconnected, such operations may only be performed under the following conditions:
   - The operating station (e.g. external control panel) where the machine may be activated must at all times be under the control of a qualified operator.
   - All participants must be in clear view of the operator or in positive communication with each other.
   - All participants must be beyond the reach of machine elements which may move rapidly and present a hazard.
   - Where machine configuration or size requires that the operator leave the control station to install tools, and where there are machine elements which may move rapidly, if activated, such elements must be separately locked out.
   - During repair procedures where mechanical components are being adjusted or replaced, the machine shall be de-energized or disconnected from its power source.

5.1.13 Bibliography

Electrical Safety-Related Work Practices Program, 29 Code of Federal Regulations 1910.331-335, Washington, DC: U.S. Government Printing Office, Office of the Federal Register, 1985.

Electricians Toolbox Etc., http://www.elec-toolbox.com/Safety/safety.htm.

Occupational Safety and Health Act of 1984, Fatalgram: Reports of OSHA Fatality/Catastrophe Investigations, Washington, DC: U.S. Department of Labor, Occupational Safety and Health, 1984.

North Carolina State University, Environmental Health and Public Safety, http://www.ncsu.edu/ehs/www99/right/handsMan/workplace/lockout.html.

16. Sonoma State University, Environmental Health & Safety, http://www.sonoma.edu/ehs/esp/LOTOp.shtml

Occupational Safety and Health Administration, An Illustrated Guide to Electrical Safety, U.S. Department of Labor, 1983.
Sonoma State University, Environmental Health & Safety, http://www.sonoma.edu/ehs/esp/LOTOp.shtml.

5.2 Linemen General Safety Practices

5.2.1 Introduction[17]

The subject of electrical hazards analysis has been recognized by a small segment of the electrical industry for many years. However, the electrical industry, most commonly at the user level, has largely ignored the subject, essentially reacting to catastrophic accidents, rather than proactively trying to predict and prevent them. Recent changes in consensus standards, along with a better general understanding of the seriousness of electrical hazards have resulted in a renewal of interest in the subject.

5.2.2 Scope[18]

Key definitions that are crucial to understanding shock and arc-flash hazard protection are:

- **Limited Approach Boundary** – "An approach limit at a distance from an exposed live part within which a shock hazard exists."
- **Restricted Approach Boundary** – "An approach limit at a distance from an exposed live part within which there is an increased risk of shock, due to electric arc over combined with inadvertent movement, for personnel working in close proximity to the live part."
- **Prohibited Approach Boundary** – "An approach limit at a distance from an exposed live part within which work is considered the same as making contact with the live parts."

17. National Fire Protection Agency, 70E-2004, "Standard for Electrical Safety in the Workplace."
18. Ibid.

- **Flash Protection Boundary** – "An approach limit at a distance from exposed live parts within which a person could receive a second degree burn if an electrical arc flash were to occur."

Additional electrical safety-related work practices must be used if circuits, operating at 50 volts or more, are not de-energized (placed in an electrically safe work condition). Work practices must protect the employee from an arc flash, as well as inadvertent contact with live parts operating at 50 volts or more. Analyses must be performed before an employee approaches exposed live parts, within the limited approach boundary.

5.2.3 Shock Hazard Analysis[19]

Inadvertent contact with energized conductors can injure or kill employees. There are several contributing factors to the causes of numerous shock-related injuries and fatalities including:

- Poor or faulty insulation;
- Improper or inadequate grounding;
- Loose connections;
- Defective or outdated parts;
- Ground faults in equipment;
- Unguarded live parts;
- Failure to de-energize electrical equipment when it is being repaired or inspected;
- Intentional use of obviously defective and unsafe tools; or
- Use of tools or equipment too close to energized parts.

Evaluate the procedures or work practices that will be involved to appropriately assess the electrical shock hazard associated with any type of maintenance or repair work. Evaluate the practices against both regulatory and consensus standards requirements as well as recognized good practice within the industry.

19. Ibid.

5.2.4 Regulatory Issues[20]

- Place all equipment in a de-energized state prior to any maintenance or repair work. (Limited exceptions exist).
- A qualified person must verify the de-energized state prior to beginning any work.
- Maintain the de-energized state through the consistent use of locks and tags, and in some cases, grounding.
- Perform energized work in accordance with written procedures.[21]

5.2.5 Standard Requirements[22]

- The shock hazard analysis must establish the:
  - Limited approach boundary,
  - Restricted approach boundary, and the
  - Prohibited approach boundary.
- This applies to all exposed live parts operating at 50 volts or more
- Only qualified persons are permitted within these boundaries.
- An unqualified person may not enter these boundaries unless the conductors and equipment have been placed in an electrically safe work condition.

5.2.6 Test Equipment Industry Recognized Good Practices[23]

- Plan every job.
- Anticipate unexpected results and the required action for these results.
- Use procedures as tools.
- Identify all hazards.

20. National Fire Protection Agency, 70E-2004, "Standard for Electrical Safety in the Workplace."
21. IEEE Conference Record of 2006 Annual Pulp and Paper Industry Technical Conference, June 18–23, 2006.
22. National Fire Protection Agency, 70E-2004, "Standard for Electrical Safety in the Workplace."
23. Ibid.

- Keep unqualified workers away from hazards.
- Assess employee's abilities and qualifications.

Include an assessment of the physical condition of the electrical system. As part of the assessment, identify the proper personal protective equipment (PPE) for shock protection, which would include, but not be limited to, rubber insulating gloves with leather protectors, rubber blankets and mats, and insulated hand tools.

Insure that equipment covers and guards are in place; that access to exposed conductors is limited to electrically qualified personnel; and overcurrent protective devices are operable and of appropriate interrupting rating.

5.2.7 Flash Hazard Analysis[24]

A large number of all serious electrical injures are related to electrical arcs created during short circuits and switching procedures. An electrical arc is basically an electrical current passing through ionized air. This current flow releases a tremendous amount of energy as both radiated light and convected heat. Human errors and equipment malfunctions contribute to the initiation of an electrical arc. Engineering design and construction of arc resistant equipment as well as requirements for safe work practices are continuing to target the risk of electrical arc-flash hazards.[25]

The principle factors used in the determination of the flash hazards to personnel are:

- Available short circuit current at the arc location;
- Duration of the electrical arc;
- Distance from the arc to personnel;
- The arc gap; and
- Environmental conditions and surroundings at the arc location.

24. National Fire Protection Agency, 70E-2004, "Standard for Electrical Safety in the Workplace."

25. IEEE Conference Record of 2006 Annual Pulp and Paper Industry Technical Conference, June 18-23, 2006.

Minimizing or eliminating the hazard is the first choice; however, when this is not possible flame-resistant rated clothing and personal protective equipment must be utilized.

In addition to flame-resistant clothing and personal protective equipment, there are other safe work practices that can be adopted to minimize or eliminate the hazards including:

- lockout/tagout along with temporary grounding;
- body positioning;
- clothing;
- insulated tools; and
- carefully scrutinized to insure that the risk to employees is minimized.

5.2.8 Blast Hazard Analysis[26]

During an electrical arc, both the conducting material and the surrounding air are heated to extremely high temperatures. The resulting expansion of the air and vaporized conductive material creates a concussive wave surrounding the arc. Pressures may becomce excessive, destroying equipment enclosures and throwing debris great distances. The pressure created during an electrical explosion is directly proportional to the available short circuit at the arc location. With a current short circuit study available, the anticipated blast pressure can be estimated from tables or charts.

Utilize blast pressure calculations to determine whether enclosures will withstand an internal fault if sufficient manufacturer's data is available. Recognize the magnitude of the hazard so that appropriate safety practices, such as correct body positioning, can be incorporated into work procedures. Blasts can severely injure or kill a person if the blast hazard is high, or if it is in a limited space. Serious consideration should be given to not allowing personnel in the area during specific equipment operations if these conditions are present.

26. Ibid.

5.2.9 Selection of Electrical Protective Equipment[27]

The requirements for the hazard analysis and selection of protective clothing must first be defined.

Assess the workplace to determine if hazards are present, or are likely to be present, which require the use of personal protective equipment. If such hazards are determined, the employer should select and have each employee use, the type of personal productive equipment that will protect the affected employee from the hazards identified in the hazard assessment.

Train the employee to be knowledgeable with the following issues and scenarios:

- When personal productive equipment is necessary;
- What personal productive equipment is necessary;
- How to properly don, doff, adjust, and wear personal productive equipment;
- The limitations of the personal productive equipment; and
- The proper care, maintenance, useful life, and disposal of personal productive equipment.

Include shock, arc, and blast assessments in the hazard analyses. Identify the selection, inspection, and use requirements for electrical personal productive equipment. Specify the type of clothing that is prohibited.

Utilize protective shields, protective barriers, or insulating materials to protect each employee from shock, burns, or other electrically related injuries while that employee is working near exposed energized parts which might be accidentally contacted or where dangerous electric heating or arcing might occur.

Protective clothing, including a complete multilayered flash suit with hood and face shield, may be required for the operation, insertion, or removal of a circuit breaker.

Calculate the incident energy (in cal/cm^2) available at the work site in order to determine and the protective clothing required

27. IEEE Conference Record of 2006 Annual Pulp and Paper Industry Technical Conference, June 18–23, 2006.

for the specific task. Additionally, determine a "Flash protection boundary" for all energized work. At this boundary, exposed flesh must not receive a second-degree burn or worse.

After determining the type, purchase the necessary protective clothing and train employees on how to properly wear the gear.

5.2.10 Exterior Safety Rules[28]

5.2.10.1 Rope[29]

Follow these precautions when using rope:

- Ensure that any rope used near live conductors or equipment is free of metal strands.
- Turn in the rope for replacement when the strands become frayed.
- Keep the rope clean so that sharp particles of solder or other substances cannot cut your hands or rubber gloves.
- Keep your hands at a safe distance when taking hold of a rope passing through a block.
- Do not place tape, string, or marlin cord on a rope to repair defects.
- Never loop a rope over your hand or arm because it could be a tripping hazard while climbing.
- Never leave uncoiled wire, rope, or other material unguarded on streets, roadways, or alleys because of the potential tripping hazard.
- Never hang safety lines, ropes, or wires from a pole because they could interfere with passing traffic.

5.2.10.2 Body Belts and Safety Straps[30]

Inspect all body belts and safety straps during regular tool inspections to determine the condition of all parts. Repair or replace suspect of faulty straps or belts.

---

28. http://www.globalsecurity.org/military/library/policy/army/fm/5-424/chap13.htm
29. Ibid.
30. Ibid.

Ensure that the leather loops holding the D-rings are not worn or crushed enough at the edges to weaken the leather or cause it to tear. If the straps are otherwise sound, utilize them until they are not less than 3.2 milimeters thick in any portion other than the doubled part of the strap. In this portion, the leather may wear slightly less than 3.2 milimeters because a double portion is approximately twice as strong as a single portion.

5.2.10.3 Rubber Protective Equipment[31]

Follow the safety rules below for inspection, use, and care of rubber protective equipment:

Inspection

- Ensure that each worker personally tests his rubber gloves before starting work on or near live conductors or equipment carrying between 300 and 5,000 volts. Conduct a thorough visual inspection for holes and thin spots. Immediately tag and turn in rubber gloves or other protective equipment that is believed to be defective.
- Perform semiannual electrical test on rubber gloves. Give other rubber protective equipment an electrical test annually. Make arrangements with the servicing utility to conduct these tests.

Use

- Wear rubber gloves when working on or within reach of energized conductors or equipment operating at voltages between 300 and 5,000 volts.
- Wear leather protective gloves over rubber gloves to protect them from mechanical damage. Pin holes or thin spots caused by using gloves without protectors may be hazardous to workers.
- Use approved live-line tools without using rubber gloves for work on lines that are energized at more than 5,000 volts.

31. Ibid.

- Wear rubber gloves while cutting supposedly dead cables and while testing supposedly burned-out transformers on both primary and secondary sides.
- Do not remove rubber gloves until you are entirely out of reach of energized equipment.
- Do not work with the gauntlet of rubber gloves rolled down.
- Wear rubber gloves when tending reels when wire is being strung or removed near other wires energized between 300 and 5,000 volts, and if working near someone who is near conductors or equipment energized at voltages between 300 and 5,000 volts.
- Use a hand line to raise or lower all rubber protective equipment in a canvas tool bag or a bucket.

Care

Issue rubber gloves to each lineman and inform the lineman of the following safety precautions and procedures:

- Store rubber gloves in a canvas bag when they are not in use.
- Give rubber gloves an air test each day before starting work and when encountering an object that may have damaged the gloves.
- Wash gloves with warm water and mild detergent.
- Be careful when applying or removing rubber protective equipment. Do not work over energized conductors or equipment or get in a position that may result in unwanted contact with the equipment.
- Dry rubber protective equipment before putting it away if possible.
- Do not carry rubber protective equipment in compartments on trucks or in tool bags with tools or other equipment.
- Roll, never fold, rubber blankets when putting them away.
- Place rubber hoods and hoses flat in truck compartments.

5.2.10.4 Gaffs[32]

Take the following safety measures to ensure that gaffs are the proper length and sharpness:

- Do not sharpen gaffs on the underside, except to make the shoulder, because it changes the angle to which they are set and renders the climber unsafe.
- Make the underside straight when removing metal to make the shoulder.
- Do not sharpen the gaff with a grindstone or an emery wheel.
- Place the climber in a vise with the gaff uppermost.
- Sharpen the gaff on the two outer surfaces with a file. Take long strokes from the heel to the point of the gaff. Remove only enough material to make a good point.
- Never sharpen the gaff to a needlepoint. Leave a shoulder about 3.2 milimeters back from the point. The distance across the gaff at the shoulder should be about 4.0 milimeters. Sharpening the gaff in this manner prevents the point from sinking too far into the pole.

5.2.10.5 Climbers[33]

Take the following precautions and safety measures when using climbers:

- Do not use climbers after gaffs are worn shorter than 3.2 centimeters on the underside.
- Remove climbers when moving from job to job or when working on the ground.
- Remove climbers before working on a ladder, standing on a rubber blanket, or riding in a vehicle.

32. http://www.globalsecurity.org/military/library/policy/army/fm/5-424/chap13.htm
33. http://www.globalsecurity.org/military/library/policy/army/fm/5-424/chap13.htm

5.2.10.6 Hot-Line Tools[34]

Follow these safety rules when working with hot-line tools:

- Do not perform hot-line work when rain or snow is threatening or when heavy dew, fog, or other excessive moisture is present. Exceptions to this rule are when conducting switching operations, fusing, or clearing damaged equipment that presents a hazard to the public or to troops.
- Remain alert. If rain or snow starts to fall or an electrical storm appears while a job is in progress, complete the work as quickly as possible to allow safe, temporary operation of the line until precipitation or lightning ceases. Judgment of safe weather conditions for hot-line work is the foreman's responsibility.
- Perform hot-line work during daylight if possible. In emergency situations, work under artificial light if all conductors and equipment being worked on are made clearly visible.
- Do not wear rubber gloves with hot-line tools because they make detection of brush discharges impossible.
- Avoid holding outer braces or other metal attachments.
- Avoid unnecessary conversation.
- Maintain close cooperation among everyone on the job.
- Treat wooden pole structures the same as steel towers.
- Be careful with distribution primaries. When they are located on the same pole with high-tension lines, cover them with rubber protective equipment before climbing through or working above them.
- Do not change your position on the pole without first looking around and informing others.
- Never use your hands to hold a live line clear of a lineman on a pole. Secure the line with live-line tools and lock it in a clamp.
- Stay below the live wire when moving it until it is thoroughly secured in a safe working position.

34. Ibid.

- Take special precautions on poles having guy lines.
- Do not use a rope on conductors carrying more than 5,000 volts unless the rope is insulated from the conductor with an insulated tension link stick.

5.2.10.7 *Vehicles*[35]

Follow the safety rules below when operating or working with vehicles:

- Never hang coils of wire or rope outside a truck because they could easily catch on passing vehicles.
- Take care when using materials or tools that extend 1.2 meters or more behind a truck body. Attach a danger flag near the end of the projection during the day, and use a red lantern at night.
- Provide a red lantern or taillight with any equipment that is being pulled by a truck at night.

5.2.11 Exterior Working Practices[36]

5.2.11.1 *Excavation*[37]

- Place necessary barriers, red flags, and warning signs to protect traffic and guard workers when making excavations in or near highways.
- Place red lights or flares at night to give ample warning to approaching traffic.
- Place signs between 30.5 meters and 91.4 meters from the outside edge of the work area.
- Display a "MEN AT WORK" sign or another suitable warning when opening manholes briefly in daytime for inspections, tests, troubleshooting, and so forth.
- Notify the person in charge when you find obstructions in digging.

35. http://www.globalsecurity.org/military/library/policy/army/fm/5-424/chap13.htm
36. Ibid.
37. Ibid.

5.2.11.2 *Manholes*[38]

Rules when working at manholes:

- Make a gas check before opening a manhole cover.
- Remove or replace manhole covers with cover hooks or other approved tools.
- Do not enter a manhole until after a gas check has been made.
- Use a mine safety lamp or a chemical detector to make checks several times throughout the day.
- Carefully check manholes near gasoline storage tanks. If gas exists, ventilate the manhole with a power blower.
- Use only gas-proof lights in a manhole if you suspect explosive gases.
- Assign someone with a red flag or lantern to warn pedestrians and vehicular traffic when a manhole is opened during daylight.
- At night, mark open manholes.
- Ensure the safety of workers in manholes if they cannot be seen by fellow workers on the surface.

5.2.11.3 *Potheads*[39]

- Open the neutral pothead last and close it first when opening and closing circuits with potheads.
- Do not stand on a concrete floor or on the ground while operating potheads in manholes or vaults. Use an insulated stool, a crossarm, or other dry lumber. Cover all grounded framework safely.
- Keep your body clear of all grounded equipment when working on or near potheads.

5.2.11.4 *Poles*[40]

- Warn persons nearby when moving a pole by hand, with a pole cart, or with the truck derrick.

38. Ibid.
39. http://www.globalsecurity.org/military/library/policy/army/fm/5-424/chap13.htm
40. Ibid.

- Station someone with a red flag to warn or stop traffic as necessary.
- Report poles that appear unsafe for climbing to the foreman so that they can be properly braced or guyed.
- Do not remove wires or supporting guy lines from a pole until the condition of the butt has been checked.
- Support unsound poles with pike poles or guy lines before starting excavation or removing wires.
- Remove all the equipment around the pole being removed before starting excavation.
- Climb to a working position by the safest path and take the safest position possible, keeping all body parts clear of grounded or live portions of equipment.
- Do not use pins, braces, or guy lines as handholds.
- Never carry any tools or material in your hands when climbing.
- Watch for hazardous and dangerous conditions such as rusted hardware and defective guy lines, braces, and arms. Repair such conditions immediately.
- When more than one person is ascending or descending, ensure that the first person reaches his working position or the ground before the second person starts to ascend or descend.

5.2.11.5 Climbing[41]

- Exercise caution when climbing past another worker to avoid damaging the safety straps or gaffs.
- Ensure that the safety strap does not catch on the pole steps when climbing. When wearing a body belt and a safety strap, snap both ends of the strap into the same D-ring when not in the working position.
- Never put both of the snaps for the safety strap in one D-ring while in the working position on a pole. Ensure that one snap is fastened in each D-ring.
- Avoid pressure on the tongue snap when leaning against arms, pins, braces, and wires to prevent the safety snap from becoming disengaged from the D-ring.

41. Ibid.

5.2.12 Electrical Safety Rules[42]

5.2.12.1 Voltage Low[43]

Electricians may work on energized conductors and equipment operating at 300 volts or less if all adjacent energized or grounded conductors and equipment are covered with insulating material or approved rubber protective equipment.

Safety precautions and procedures for working on energized conductors and equipment operating at 300 volts or less are as follows:

- Tape or cover all bare or exposed places on one conductor before exposing another one.
- Open switches and remove fuses when working on building wiring, motors, belting, shafting, blowers, or other machinery (including shop machine tools). Attach hold cards to the switches.
- Do not touch steam or water pipes when handling portable electric tools or light cords.
- Never leave joints or loose ends of wire untaped unless otherwise protected.
- Never turn on current in a building where a fire has occurred and electric wiring or apparatus has been affected until a careful safety inspection is made by the fire department or a safety inspector.

5.2.12.1.1 Intermediate[44]

De-energize ground lines and equipment operating between 300 and 5,000 volts before starting any work. The only exception to this is when working on overhead lines. Wear rubber gloves when working on overhead lines.

5.2.12.1.2 Checking Energized Conductors[45]

Consider conductors and equipment normally operating between 300 and 5,000 volts to be energized until positively proved otherwise. If confirmation cannot be determined by a visual check, use

42. http://www.globalsecurity.org/military/library/policy/army/fm/5-424/chap13.htm
43. Ibid.
44. Ibid.
45. Ibid.

an approved voltage detector. Check the detector on a conductor that is known to be energized and note a positive indication before using it. Repeat this check on a known live conductor after making the test on a dead conductor. If rechecking the voltage detector on a live conductor is impossible, use two voltage detectors, one as a check against the other.

5.2.12.1.3 Insulation[46]

When working within reach of conductors or equipment operating between 300 and 5,000 volts, all energized and grounded conductors or equipment must be covered with rubber line hose, insulator hoods, line protectors (pigs), or blankets or must be isolated with suitable barriers. Cover or barricade any energized or grounded conductors and equipment coming within reach before starting new work.

Utilize the following safety precautions when working on equipment and conductors operating at 300 to 5,000 volts:

- De-energize by opening sectionalizing switches and effectively ground the section of line being worked on.
- Treat line wires being installed on poles as if they were energized when those poles are already carrying energized conductors.
- Use rope to pull line wires into position.
- Remove wire from the underside of reels.
- Ground the truck when using a pay-out reel to string wires near energized wires of this classification.
- Exercise caution when stringing wire near other wires energized with 300 to 5,000 volts. Ensure that the worker tending the pay-out reel wears rubber gloves and does not allow any part of his body to come in contact with the wire being strung.
- Apply the brake to the rim of the pay-out reel on the side opposite the direction the wire is being pulled. A crossarm or sidearm may be used as a brake.
- Hold the loose ends of jumpers in the clear, or have someone at each end hold them if necessary.

46. Ibid.

- Keep the tie wire coiled during installation or removal to prevent accidental contact with an energized conductor.
- Test for voltage before removing a primary or secondary wire. The wire should be rechecked by the foreman to ensure that it is dead.
- Treat the spans of wire to be removed as energized. Attach a rope to each end and lower the spans one at a time.
- Make all possible connections to dead wires and equipment with only the final connection made to the energized conductor.
- Never open the ground connection at a ground pipe or a bus bar unless it has first been disconnected at the point of contact with the equipment it is intended to ground.
- Make the ground pipe connection first whenever a ground wire on a pole is to be connected.
- Treat coil wires in the take-up reel that was used to coil the wires as energized when removing the reel from the line arms. Attach a rope to the wires at the end farthest from the take-up reel, and wear rubber gloves when tending the reel.
- Ensure proper direction of phase rotation when restoring three-phase power service.
- Identify cable while in the manhole before anyone cuts a cable or makes an opening in the lead sheath or sleeve. Check its duct location with that shown on the working print. Check the working print against the installation engineer's map records. Additionally, check the cable identity by listening with an exploring coil for the pulsating beat of the interrupter signal.
- Remove a 7.6 centimeter strip of lead around the cable and test it with two voltage detectors after the cable has been identified and grounded. Do this one at a time, at two or more points around the cable and at the center of the exposed insulation.
- Place a hacksaw on the exposed cable insulation adjacent to and touching the grounded lead sheet when cutting cable.
- Open the switch before working on an oil circuit breaker that has an operating bus feed switch.

- Treat series circuits, series lamps, and devices in series circuits as energized unless they are opened by disconnects.
- Never open a series circuit at the point where work is in progress, such as at lamps or other series-circuit devices. Always bridge the point of the circuit being working on with a jumper.

5.2.12.4 High Voltage[47]

Lines and equipment operating at 5,000 volts or more must be de-energized and grounded before any work is started. An exception is overhead lines, which can be worked on with live-line tools.

Follow these rules and procedures when working on high-voltage equipment operating at 5,000 volts or more:

- Do not work on more than one conductor at a time with live-line tools.
- Keep live-line tools dry and free from dirt.
- Dry and test tools subjected to damp weather before using them again.
- Protect live-line tools carried on line trucks to prevent scars and abrasions. Waterproof canvas bags, compartments with padded hooks, or bins built into the truck may be used for this purpose.
- Perform all work on de-energized circuits between two sets of grounds. Place one set on the first pole or structure toward the source of energy and the other on the pole nearest the load.
- Do not ground a supposedly dead circuit without first feeling out the circuit for a static discharge with a switch pole.
- Net off energized equipment properly at the substation before working near it, and install suitable barriers and warning signs.
- Remove substation apparatus from service when wiping, cleaning, repairing, or performing other maintenance. Place hold cards on the main and control

47. http://www.globalsecurity.org/military/library/policy/army/fm/5-424/chap13.htm

switches before starting work. Use a red light to illuminate hold cards if they are not distinctly visible.
- Do not perform work on any apparatus until the electrical foreman has proved it to be dead and safe.
- Do not approach or touch reactors and connected equipment unless they have been disconnected from all live lines and have been grounded.
- Do not raise, move, or lower cables that are energized at voltages up to 12 kilovolts more than 45.7 centimeters. Do not move cables that are energized at voltages over 12 kilovolts.
- Discharge electrolytic and oxide-film lightning arresters by grounding and shorting the horn gaps before touching the arresters.

Table 5.2 gives the minimum body clearance from energized lines and equipment.[48]

5.2.12.3 Circuits[49]

5.2.12.3.1 De-energized[50]

Perform all electrical installation work on a de-energized (dead) circuit if possible. Rip the main service-entrance switch, remove

Table 5.2 Minimum body clearance distance.

Operating Voltage (Kilovolts)	Minimum Distance (Meters)
5 to 7.5	.30
7.5 to 12	.61
12 to 33	.91
33 to 66	1.2
66 to 132	1.5
132 to 220	2.4

48. http://www.globalsecurity.org/military/library/policy/army/fm/5-424/chap13.htm
49. Ibid.
50. Ibid.

the fuses, and tag the switch. The tag should list the electrician's name, the date, and the time that he turned the switch off. If existing conditions do not permit this procedure, de-energize the circuit being repaired or modified by removing the branch circuit fuses or tripping the circuit breaker connected to the branch circuits to off.

5.2.12.3.2 HOT[51]

Working on hot circuits is not recommended except in emergency situations. Only experienced personnel who have been cautioned on the dangers involved should work on these circuits. AC voltages as low as 67 volts and DC voltages as low as 110 volts have caused death by electrocution.

When working on a hot line, closely follow the procedures listed below:

- Insulate yourself to prevent your body from becoming a conductor path for the current flow.
- Work on only one side of the circuit at a time.
- Use insulating gloves at all times and stand on non-conducting materials.
- Ensure that all equipment, tools, and wearing apparel is dry, because dry materials offer more resistance to the flow of electricity than do moistened or wet items.

5.2.12.3.3 Protection[52]

Fuses or circuit breakers are installed as overload and short-circuit protection for circuit components and connected loads. Limit their selection with regard to ampere rating to the maximum value allowable for the smallest conductor or equipment used in the circuit.

5.2.12.3.4 Overload[53]

Determine the cause of trouble and rewire overloaded circuits by increasing the wire size of the run or dividing the connected load into several circuits.

---

51. Ibid.
52. http://www.globalsecurity.org/military/library/policy/army/fm/5-424/chap13.htm
53. Ibid.

5.2.12.3.5 Bypassed[54]

Do not use jumpers to bypass the fuse, and do not remove the circuit-breaker protection from the circuit except when testing for short circuits. These primary safety devices are of vital importance in a circuit installation.

5.2.13 Transformers and Circuit Breakers[55]

Take the following safety precautions when working on transformers and circuit breakers:

- Prevent moisture from entering when removing covers from oil-filled transformers.
- Do not allow tools, bolts, nuts, or similar objects to drop into the transformers.
- Tie tools or parts with suitable twine.
- Have workers empty their pockets of lose articles such as knives, keys, and watches.
- Remove all oil from transformer covers, the floor, and the scaffold to eliminate slipping hazards.
- Exhaust gaseous vapor with an air blower before allowing work in large transformer cases because they usually contain some gaseous fumes and are not well ventilated.
- Stay away from the base of the pole or structure while transformers are being raised or lowered.
- Ensure that anyone working on a pole or structure takes a position above or well clear of transformers while the transformers are being raised or supported with blocks.
- Ground the secondary side of a transformer before energizing it except when the transformer is part of an ungrounded delta bank.
- Make an individual secondary-voltage test on all transformers before connecting them to secondary mains. On banks of three transformers connected Y-delta, bring in the primary neutral and leave it connected until the secondary connections have been completed to get a true indication on the lamp tests.

---

54. Ibid.
55. Ibid.

- Disconnect secondary-phase leads before opening primary cutouts when taking a paralleled transformer out of service. Do not disconnect secondary neutral or ground connections until you have opened the primary cutouts.
- Do not stand on top of energized transformers unless absolutely necessary and then only with the permission of the foreman and after all possible precautions have been taken. These precautions include placing a rubber blanket protected with a rubbish bag over the transformer cover. Do not wear climbers.
- Treat the grounded case of a connected transformer the same as any grounded conductor. Treat the ungrounded case of a connected transformer the same as any energized conductor because the case may become energized if transformer windings break down.
- Ensure that the breaker cannot be opened or closed automatically before working on an oil circuit breaker and that it is in the open position or the operating mechanism is blocked.
- Ensure that metal-clad switching equipment is de-energized before working on it.
- Ensure that regulators are off the automatic position and set in the neutral position before doing any switching on a regulated feeder.
- Do not break the charging current of a regulator or large substation transformer by opening disconnect switches because a dangerous arc may result. Use oil or air brake switches unless special instructions to do otherwise have been issued by the proper authority.
- Do not operate outdoor disconnecting switches without using the disconnect pole provided for this purpose.
- Ensure that all contacts are actually open and that safe clearance is obtained on all three phases each time an air brake switch is opened. Do not depend on the position of the operating handle as evidence that the switch is open.
- Do not operate switches or disconnect switches without proper authority and then only if thoroughly familiar with the equipment.

- Remove potential transformer fuses with wooden tongs. Wear rubber gloves and leather over gloves.
- Do not open or remove disconnect switches when carrying load. First open the oil circuit breaker in series with the switches. Open disconnect switches slowly and reclose immediately if an arc is drawn.

5.2.14 Wire Markers[56]

Wire markers are a distinct aid to installation. Use them as a guide, and thoroughly check the condition of each conductor with a test lamp or voltmeter before working on the circuit.

5.2.15 Adequacy and Effectiveness of the Training Program[57]

Ensure adequacy and effectiveness of the electrical safe work practices program and training of qualified electrical personnel by performing the following steps:

- Conduct a comprehensive job task analysis.
- Complete a task hazard assessment including:
  - Shock hazard,
  - Arc flash hazard (using current short circuit and coordination studies),
  - Blast hazard, and
  - Other hazards (Slip, fall, struck-by, environmental, etc.).
- Analyze task for the personal protective equipment needed.
- Conduct training needs assessment for qualified and non-qualified electrical workers.

5.2.16 Bibliography

GlobalSecurity.org, website, http://www.globalsecurity.org/military/library/policy/army/fm/5-424/chap13.htm.

---

56. http://www.globalsecurity.org/military/library/policy/army/fm/5-424/chap13.htm
57. Ibid.

IEEE Conference Record of 2006 Annual Pulp and Paper Industry Technical Conference, June 18-23, 2006, http://www.electricaltrainingservices.com/common/documents/avotraining/Electrical%20Hazards%20Analysis.pdf.

National Fire Protection Association, 70E-2004, Standard for Electrical Safety in the Workplace, 2004.

Occupational Safety and Health Administration, 29 CFR 1910, Electrical Standards, Federal Register Vol. 46, No. 11, Friday, January 16, 1981.

Occupational Safety and Health Administration, 29 CFR 1910.147, Control of Hazardous Energy Source), September 1, 1989.

Occupational Safety and Health Administration, 29 CFR 1910.331-.335, Electrical Safety-Related Work Practices, August 6, 1990.

Occupational Safety and Health Administration, 29 CFR 1910.269, Electric Power Generation, Transmission, and Distribution, January 31, 1994.

5.3 Electrical Safe Work Practices Plan

5.3.1 Introduction

This safety procedure provides guidelines for safely working around electrical hazards. It includes provisions for training, lockout requirements, and specific types of work practices and the required precautionary practices when using portable electric equipment. The reader may refer to other sections of the standards for specific guidelines on the use of safety equipment and various safe work practices.

5.3.2 Training

It is the responsibility of each exposed employee's immediate supervisor to ensure that the employee has received the training necessary to safely perform his or her duties. This training will be given via classroom and on-the-job instruction and is to be documented. Exposed employees shall be trained in and be familiar with the safety related work as they pertain to their respective job assignments. Additional training requirements for qualified persons are also mandated.

Employees will be trained in specific hazards associated with their potential exposure. This training will include isolation of energy, hazard identification, premises wiring, connection to supply, generation, transmission, distribution installations, clearance distances, use of personal protective equipment and insulated tools, and emergency procedures.

5.3.3 Qualified Person

Those persons who are permitted to work on or near exposed energized parts and are trained in the applicable electrical safe work practices. Qualified persons shall, at a minimum, be trained in and familiar with:

- The skills and techniques necessary to distinguish exposed live parts from other parts of electric equipment.
- The skills and techniques necessary to determine the nominal voltage of exposed live parts.
- The clearance distances specified in Table 5.3 and the corresponding voltage to which the qualified person will be exposed.

5.3.4 Safe Work Practices

5.3.4.1 Lock-out/Tag-out and Safety Related Work Practices

All electrical energy sources must be locked out when any employee is exposed to direct or indirect contact with parts of fixed electrical equipment or circuits.

Safety related work practices will be used to prevent electric shock or other injuries resulting from either direct or indirect

Table 5.3 Approach distance for qualified employees.

Alternating Current	Distance
300V and less	Avoid contact
Over 300V, but less than 750V	30.5 cm
Over 750V, but less than 2kV	46 cm
Over 2kV, but less than 15kV	61 cm
Over 15kV, but less than 37kV	91 cm
Over 37kV, but less than 87.5kV	107 cm
Over 87.5kV, but less than 121kV	122 cm
Over 121kV, but less than 140kV	137 cm

electrical contacts. Safety related work practices will be consistent with the nature and extent of the associated electrical hazards. Specific types of work practices covered by this safety procedure include:

- Working with de-energized parts;
- Working with energized parts;
- Vehicular and mechanical equipment near overhead lines and underground lines;
- Illumination;
- Conductive materials and equipment;
- Portable ladders; and
- Housekeeping.

5.3.4.2 *Portable Equipment*

All portable electric equipment will be handled in such a manner that will not damage or reduce service life. Flexible cords connected to equipment may not be used for raising or lowering equipment and will not be used if damage to the outer insulation is present. Additionally, visual inspections are required and unauthorized alterations of the grounding protection are not allowed to ensure the safety of employees.

Prior to each shift, a visual inspection will be performed for external defects and for possible internal damage. Attachment plugs and receptacles may not be connected or altered in a manner that would prevent proper continuity of the equipment grounding conductor.

In addition, these devices may not be altered to allow the grounding pole of a plug to be inserted into slots intended for connection to the current-carrying conductors.

Portable electric equipment and flexible cords used in highly conductive work locations or in job locations where employees are likely to contact water or conductive liquids shall be approved by the manufacturer for those locations. The hazardous locations that employees should be aware of include, wet locations and locations where combustible or flammable atmospheres are present.

For wet locations, employees' hands will not be wet when plugging and unplugging energized equipment. Energized plug and receptacle connections will be handled only with protective equipment if the condition could provide a conductive path to the employee's hand (if, for example, a cord connector is wet from being immersed in water). In addition, ground-fault circuit interrupter (GFCI) protection is required for some equipment/locations and is also recommended for use in all wet or highly conductive locations.

For combustible/flammable atmospheres, all electric equipment and wiring systems in classified locations must meet electric code requirements for that particular classification.

5.3.4.3 *Personal Protective Equipment*

Employees working in confined areas such as electrical vaults or any other area where there are potential electrical hazards will be provided with and use protective equipment that is appropriate for the work to be performed. Examples of personal protective equipment (PPE) which might be needed for protection against electric shock include but are not limited to:

- Nonconductive hard-hats, gloves, and foot protection or insulating mats;
- Eye and face protection whenever there is danger from electric arcs or flashes;
- Insulated tools or handling equipment; and
- Protective shields and barriers to protect against electrical shock and burns.

Additionally, other ways of protecting employees from the hazards of electrical shock will be implemented, including insulation and guarding of live parts. The insulation must be appropriate for the voltage and the insulating material must be undamaged, clean, and dry.

Guarding prevents the employee from coming too close to energized parts. It can be in the form of a physical barricade or it can be provided by installing the live parts out of reach from the working surface.

5.3.4.4 *Conductive Materials and Equipment*

Conductive materials and equipment (e.g., hand tools) will be handled to prevent contact with exposed energized conductors or circuit parts. Conductive articles of jewelry and clothing (such as watch bands, bracelets, rings, key chains, necklaces, metalized aprons, cloth with conductive thread, or metal headgear) will not be worn.

5.3.4.5 *De-energized Parts*

All electrical parts exceeding 50 volts will be de-energized before an employee works on or near equipment unless:

- The de-energizing creates a more hazardous situation;
- The equipment, by design, cannot be shut down; and/or
- The decision to work without de-energizing shall be made by management and documented before work begins.

When any employee is exposed to direct or indirect contact with parts of fixed electrical equipment or circuits which have been de-energized, the electrical energy source will be locked out.

5.3.4.6 *Energized Parts*

If work must be performed while equipment is energized or if de-energizing is not feasible, additional safety measures will be taken to ensure the safety of the qualified employee and any other persons who may be exposed. Protection from energized parts will be suitable for the type of hazard involved. Exposed energized parts in areas accessible to the public shall be continuously protected by an authorized attendant. In areas not accessible to the public, employees shall be protected from exposed energized parts by the use of signs or tags. In addition to signs or tags, barricades shall be used where necessary to limit access to areas with exposed energized parts.

Only qualified persons will be allowed to perform work directly on energized parts or equipment. Qualified persons will be capable of working safely on energized circuits and will be familiar with special precautionary techniques, personal protective equipment, insulating and shielding materials and insulated tools. Qualified persons must also have received the training required in this safety procedure.

5.3.4.7 *Illumination*

Employees will be provided with adequate light to work on energized equipment or equipment will be relocated to ensure adequate light is available.

5.3.4.8 *Portable Ladders*

Portable ladders will have nonconductive surfaces if they are used where the employee or the ladder could be exposed to electrical shock hazards.

5.3.4.9 *Reclosing Circuits*

If circuits are tripped using a protective device such as ground fault circuit interrupter (GFCI), power will not be restored until the reason for the interruption is determined and corrected. Fuses or breakers will not be replaced or reset until it is determined that the circuit is safe to operate. Fuses will not be replaced with higher rated fuses or with makeshift devices to bypass circuit protection as designed. Problems will be identified and promptly repaired by a qualified person.

5.3.4.10 *Vehicular and Mechanical Equipment Near Overhead Power Lines*

Overhead power lines will be de-energized and grounded before any work is performed by any vehicle or mechanical equipment near the energized overhead power lines. If the overhead lines can not be de-energized, then the vehicle or mechanical equipment will be operated so that a clearance of 3.1 meters is maintained. If the voltage of the overhead line exceeds 50 kV, the distance will be increased 10.1 centimeters for every 10 kV increase in power. If lines are protected with properly rated insulating devices, the distance may be decreased.

If the equipment is an aerial lift insulated for the voltage involved and if the work is performed by a qualified person, the clearance may be reduced to a distance given in Table 5.3. If protective measures such as guarding or isolation are provided, these measures must protect the employee from contacting such lines directly with any part of the body or indirectly through conductive materials, tools, or equipment.

5.3.4.11 *Electrical Equipment/Machinery*

All electrical equipment and machinery must be grounded effectively so that there is no potential difference between the metal enclosures. Use the voltage detector to find discrepancies and other test equipment to determine the corrective action required. Disconnects should be easily identified with the specific machinery they shut off. Disconnects should also be accessible near the machinery for use in an emergency. The disconnects should be activated periodically to be sure they are operable. All electrical connections to the equipment must be secure so that no cord or cable tension will be transmitted to the electrical terminals within the equipment. The wiring installation should be such that it is protected from damage at all times.

5.3.4.12 GFCI Protection

Generally, GFCI protection is not required by the NEC on a retroactive basis. Where there is an employee exposure to potential line-to-ground shock hazards, GFCI protection should be provided. This is especially important in work areas where portable electrical equipment is being used in wet or damp areas in contact with earth or grounded conductive surfaces.

5.3.4.13 Wiring

Temporary wiring that is being used on a permanent basis should be replaced with fixed wiring. Conduit and/or cable systems must be protected from damage by vehicles or other mobile equipment. All fittings and connections to junction boxes and other equipment must be secure. No exposed wiring can be allowed. Check for missing knockouts and cover plates. Jerry-rigged splices on flexible cords and cables should be correctly repaired. Electrical equipment should be installed in a neat and professional manner. Check for damaged insulation on flexible cords and pendant drop cords.

5.4 Electrical Equipment

5.4.1 Introduction

Ensure that all individuals are protected from accidental or unexpected activation of mechanical and/or electrical equipment during the maintenance, repairing, cleaning, servicing, and/or adjusting of machinery or equipment.

5.4.2 Electrical Safety Facts[58]

Risks associated with the use of electricity include electrical shock and electrical fires are caused by shorts and overloaded circuits or wiring. Sparks from electrical equipment can act as a source of ignition for flammable or explosive vapors or combustible materials.

Rules and procedures to reduce the electrical hazards in the workplace:

1. Know the location of electrical panels and disconnect switches in or near your work area so that power can be quickly shut down in the event of a fire or electrical accident. To enhance safety, post the location of the electrical panel on the equipment it services.
2. Label electrical panels to indicate what equipment or power source they control. A minimum 91 meter clearance must be maintained around electrical panels at all times to permit ready and safe operation and maintenance of such equipment.
3. Do not overload circuits or wiring. Overloading can lead to overheated wires and arcing, which can cause fires and electrical shock injuries.
4. Inspect all electrical equipment before use to ensure that cords and plugs are in good condition – not worn, twisted, frayed, abraded, corroded, or with exposed wires or missing ground pins. Live parts must be effectively insulated or physically guarded. Equipment with damaged or defective cords or plugs should be taken out of service immediately and repaired by qualified personnel.
5. Ensure that all electrical outlets have a grounding connection requiring a three-pronged plug. All electrical equipment should have three-pronged, grounded plugs or be double-insulated.
6. Ensure that electrical outlets, wiring, and other electrical equipment integral to the building are only

58. U.S. Occupational Safety and Health Administration, Standard 29 CFR 1910, Subpart S - Electrical http://www.osha.gov/pls/oshaweb/owadisp.show_document?p_id= 9911&p_table=STANDARDS

serviced and repaired by qualified trades personnel or other qualified electricians.

7. Work on electrical equipment must be done only after the power has been disconnected. The power cord must be unplugged and under the exclusive control of the person performing the work so that the equipment cannot be accidentally turned on by someone else. The main disconnect device or circuit breaker must be shut off and locked and tagged with a special padlock and tag. Service and/or repair work on hard-wired equipment may only be carried out by authorized individuals who have received lockout/tag out training.
8. Limit the use of extension cords. In all other cases, request the installation of a new electrical outlet. Do not use extension cords as substitution for fixed receptacle outlets. Long-term use of extension cords is a violation. The long-term use of multi-outlet power strips is also illegal, except for use with computer equipment.
9. Ensure that all extension cords used are carefully placed, visible, and not subject to damage. Cords must not run across aisles or corridors where they might be damaged or create a tripping hazard. Cords must not run through doors, walls or partitions, under rugs, or above dropped ceilings. They must not be tied in knots, draped overhead, or attached to walls.
10. Ensure that the wire size of an extension cord is adequate for the current to be carried. Failure to do so can lead to electrical fires. Cords used for 110-120 volt service should be UL listed with a polarized three prong plug. Extension cords must never be linked together—use the proper length extension cord needed for the application.
11. Keep corrosive chemicals and organic solvents away from electrical cord—these can easily erode the insulation on wires.
12. Keep flammable materials away from electrical equipment.
13. Keep electrical equipment away from wet or damp locations or potential water spillage, unless specifically rated for use under such conditions.

14. Do not handle electrical equipment when hands, feet, or body are wet or perspiring or when standing on a wet floor.
15. Do not sure adapters which interrupt continuity of equipment grounding connections.
16. Ensure that only authorized workers perform testing work on electric circuits or equipment that may expose them to shock or arc-flash hazard.
17. Do not reset a circuit breaker that has tripped unless it has been determined that the equipment and circuit can be safely energized.
18. Do not remove covers from electrical equipment that could expose energized un-insulated parts. Keep space clear around electrical; ensure that nothing is stored in this dedicated space. Marking the floor to indicate the following dimensions is encouraged:
    a. 91.4 centimeters for 150 volts to ground or less;
    b. 106.7 centimeters for 151 volts to 600 volts;
    c. Width of 76.2 centimeters or width of equipment, whichever is greater.

5.4.3 Vehicular and Mechanical Equipment[59]

Maintain a clearance of 305cm for all vehicles or mechanical equipment capable of having parts or its structure elevated near energized overhead lines such as cranes, mobile scaffolds, elevating platforms, dump trucks, lift trucks, flatbed trailer cranes, etc.

Clearance is increased 10 cm for every 10 kV over 50kV if the voltage of the overhead line is more than 50 kV.

Under any of the following conditions, the above clearance may be reduced:

- For vehicles in transit with its structure lowered, the clearance may be reduced to 122 cm for 50 kV or 122 cm plus 10.2 cm for every 10 kV over 50 kV.

59. U.S. Occupational Safety and Health Administration, Standard 29 CFR 1910, Subpart S - Electrical http://www.osha.gov/pls/oshaweb/owadisp.show_document?p_id= 9911&p_table=STANDARDS

- The clearance may be reduced to distance within the designed working dimensions of the insulating barrier if the insulating barriers are installed to avoid contact with the lines and if the barriers are rated for the voltage of the line being guarded and are not a part of an attachment to the vehicle or its elevated structure.
- The clearance between the un-insulated portion of the lift and the power line may be reduced to distance given in the above if the equipment is an aerial lift insulated for the voltage involved, and if the work is performed by a qualified person.
- When working on the ground, avoid contact with the vehicle or mechanical equipment or any of its attachments, unless.
- Using protective equipment rated for the voltage; or if the equipment is positioned so that no un-insulated part may provide a conductive part to individuals on the ground can come closer to the line than 3.05 meters or 3.15 meters for every ten (10) kV over 50 kV.
- Individuals may not work on the ground when any vehicle or mechanical equipment has elevated part near energized overhead lines and is grounded due to possible overhead line contact. Additional precautionary measures should be taken as necessary to protect individuals from hazardous ground conditions which can develop within a few feet from the grounding point such as barricades or insulation.

5.4.4 Use of Equipment[60]

Handling

Portable equipment can not be handled in a way which will cause damage. Flexible cords connected to the equipment can not be used for raising or lowering equipment. The flexible cords can not

60. U.S. Occupational Safety and Health Administration, Standard 29 CFR 1910, Subpart S - Electrical http://www.osha.gov/pls/oshaweb/owadisp.show_document?p_id= 9911&p_table=STANDARDS

be fastened with staples or hung in a manner which would cause damage to the outer jacket or the insulation.

Visual Inspection

1. Inspect portable cord-and plugged-connected equipment and flexible cord sets (extension cords) before use on any shift for external defects such as loose parts, deformed and missing pins, or damage to the outer jacket or insulation and for possible internal damage such as pinched or crushed outer jacket.
2. Remove the defected or damaged item from service which might expose employees to injury are detected, and prevent employees from using it until it is repaired and tested to ensure it is safe to use.
3. Ensure the plug and receptacle contacts match prior to connecting an attachment to a receptacle.

Grounding-type Equipment

1. Flexible cords used with grounding-type equipment must contain an equipment grounding conductor.
2. Attachment plugs and receptacles are not connected or altered in a manner which prevents proper continuity of the ground the point where plugs are attached to receptacles.
3. Plugs and receptacle are not altered to allow the grounding pole to be inserted into current connector slots.
4. Adapters which interrupt the continuity of the equipment grounding connection are not used.

Conductive Work Locations

All portable electric equipment and flexible cords used in highly conductive work locations, such as those with water or other conductive liquids, or in places where employees are likely to contact water or conductive liquids, require approval for those locations.

Connecting Attachment Plugs

1. Ensure that hands are not wet when plugging and unplugging flexible cords and cord-and-plug-connected equipment, if energized equipment is involved.
2. Handle the he energized plug and receptacle connectors with insulated protective equipment if the cord connectors are wet from immersion in water or if the condition of the connection could provide a conducting path.

5.4.5 Test Equipment[61]

1. Ensure only qualified individuals are permitted to perform testing work on electric circuits or equipment.
2. Visually inspect all test equipment and all associated test leads, cables, power cords, probes for external defects and damage prior to use.
3. Remove defected or damaged equipment from service that might pose cause injury to an employee and prohibit use until the equipment is repaired and tested.
4. Test equipment and accessories, and make sure they are rated for the circuits for the environment in which they are utilized.

5.4.6 Bibliography

Environmental Health & Safety, Lockout/Tagout Procedure, Sonoma State University, http://www.sonoma.edu/ehs/esp/LOTOp.shtml.

National Fire Protection Association, 70E-2004, Standard for Electrical Safety in the Workplace, 2004.

U.S. Occupational Safety and Health Administration, Standard 29 CFR 1910, Subpart S - Electrical http://www.osha.gov/pls/oshaweb/owadisp.show_document?p_table=STANDARDS&p_id=10135.

61. U.S. Occupational Safety and Health Administration, Standard 29 CFR 1910, Subpart S - Electrical http://www.osha.gov/pls/oshaweb/owadisp.show_document?p_id= 9911&p_table=STANDARDS

5.5 Safe Work Practices Near Power Lines

5.5.1 Introduction

Trees and electricity are both beneficial, but when the two come together, problems arise. Add a third element, a human, and a life may be in jeopardy. Workers pruning trees near energized power lines face serious hazards from contact with those lines or related equipment. A worker's body may come into direct contact with an energized line. Tools or other conductive material, such as severed tree limbs, may touch the lines, resulting in an indirect, but equally dangerous, contact. Either way, people are exposed to electrical shock by becoming a path to the ground - between an energized conductor and grounded equipment, tools or trees - or by becoming a path between two conductors.

The electrical current passing through the body can cause third degree burns, irregular heart action or stoppage, or the stoppage of breathing. In addition, you may be exposed to thermal burns or flash burns from arcing.

If you're a tree worker, your first concern should be to protect yourself from such hazards.

5.5.2 Plan Ahead

Before any work begins, the job site should be inspected to locate overhead power lines and electrical equipment. If there appears to be any conflict, contact the electric utility company so measures can be taken to ensure a safe work environment.

Remember, the distance the tree is away from the energized lines can be deceiving. In some cases, it may be just as dangerous to trim a tree 3.05 to 6.1 meters from energized lines, as it is to trim one much closer, because a long, large branch can fall into the lines and conduct electricity back to the trimmer who believes he has maintained more than adequate clearance.

Unqualified individuals should never trim trees around energized lines. Where electrical hazards exist, only qualified line clearance tree trimmers should do the work. Climb the side of the tree away from any powerlines: the crotch of the tree selected for tying in should be over the work area, but located in such a way that a slip or fall would swing away from energized lines.

Ladders and aerial equipment, including insulated aerial equipment, should not be placed in contact with electrical conductors.

If an aerial device does contact an electrical line or conductor, the truck supporting the device and any other attached equipment should be considered energized. Anyone touching the equipment could be injured or even killed.

5.5.3 Safe Work Practices

5.5.3.1 General

Determine whether you and your crews can maintain the 3.05 meter clearance from power lines. If the line exceeds 50,000 volts, clearance requirements increase.

Mechanized equipment must maintain a minimum 3.05 meter distance from powerlines and conductors rated at 50,000 volts or less. The equipment includes load, cables, tag lines and any other attachments. If the powerline is rated more than 50,000 volts, the clearance requirement exceeds 3.05 meters.

Consider these clearance requirements for work near overhead powerlines as your "circle of safety." These working clearances are minimum safety clearances. Whenever possible, even greater clearances should be maintained.

If your work is going to bring you closer than the above clearances allow, notify the electric utility company promptly. No work should be performed near energized conductors until danger of contact with those conductors has been effectively guarded against.

Inspect the work site carefully prior to climbing or working on a tree to determine whether electrical lines pass through the tree or within your reach or that of other workers. All lines on utility poles should be viewed as possibly being energized, whether you think they are cable or telephone lines.

Branches hanging on energized powerlines should be removed only by qualified line clearance tree trimmers in the employ of the electric utility company and using approved insulated equipment.

5.5.3.2 Safe Operations of Cranes

Electrocutions of crane operators and crew members working near overhead power lines is a serious problem. Workers are killed each year when cranes contact overhead power lines. Employers, supervisors, and workers may not be fully aware of the hazards of operating cranes near overhead power lines or may not implement the proper safety procedures for controlling these hazards.

Each year an average of 15 electrocutions were caused by contact between cranes or similar boomed vehicles and energized, overhead power lines. The actual number of workers who died from crane contact with energized power lines is higher than reported by NTOF because methods for collecting and reporting these data tend to underestimate the total number of deaths [NIOSH 1993a].

A study conducted by the U.S. Occupational Safety and Health Administration (OSHA) showed that 377 (65%) of 580 work-related electrocutions occurred in the construction industry during the period 1985-89 [OSHA 1990]. Nearly 30% (113) of these electrocutions involved cranes.

Employers should take precautions when cranes and boomed vehicles are operated near overhead power lines (Figure 5.3). Any overhead power line should be considered energized unless the owner of the line or the electric utility company indicates that it has been de-energized and it is visibly grounded. The following are recommended regulations which are based on OSHA standards:

Figure 5.3 Improper operation of boomed vehihcle near powerlines.

- Employers shall ensure that overhead power lines are de-energized or separated from the crane and its load by implementing one or more of the following procedures:
  - De-energize and visibly ground electrical distribution and transmission lines.
  - Use independent insulated barriers to prevent physical contact with the power lines.
  - Maintain minimum clearance between energized power lines and the crane and its load.
- Where it is difficult for the crane operator to maintain clearance by visual means, a person shall be designated to observe the clearance between the energized power lines and the crane and its load.
- The use of cage-type boom guards, insulating links, or proximity warning devices shall not alter the need to follow required precautions. These devices are not a substitute for de-energizing and grounding lines or maintaining safe line clearances.

The American National Standards Institute (ANSI) has published a standard for mobile and locomotive cranes that includes operation near overhead power lines [ANSI 1994]. This consensus standard (B30.5-1994) contains guidelines for preventing contact between cranes and electrical energy. The standard addresses the following issues:

- Considering any overhead wire to be energized unless and until the person owning the line or the utility authorities verify that the line is not energized.
- De-energizing power lines before work begins, erecting insulated barriers to prevent physical contact with the energized lines, or maintaining safe clearance between the energized lines and boomed equipment.
- Limitations of cage-type boom guards, insulating links, and proximity warning devices.
- Notifying line owners before work is performed near power lines.
- Posting warnings on cranes cautioning the operators to maintain safe clearance between energized power lines and their equipment.

The Construction Safety Association of Ontario, Canada (CSA) recommends safe work practices in addition to those addressed in the OSHA and ANSI standards [CSA 1982]. These recommendations include the following:

- Operate the crane at a slower-than-normal rate in the vicinity of power lines.
- Exercise caution near long spans of overhead power lines, since wind can cause the power lines to sway laterally and reduce the clearance between the crane and the power line.
- Mark safe routes where cranes must repeatedly travel beneath power lines.
- Exercise caution when traveling over uneven ground that could cause the crane to weave or bob into power lines.
- Keep all personnel well away from the crane whenever it is close to power lines.
- Prohibit persons from touching the crane or its load until a signal person indicates that it is safe to do so.

5.5.3.3 *Procedures to Follow If Contact Occurs*

To protect against electrical shock injury in the event of contact between a crane and an energized line, the CSA recommends the following:

- The crane operator should remain inside the cab.
- All other personnel should keep away from the crane, ropes, and load, since the ground around the machine might be energized.
- The crane operator should try to remove the crane from contact by moving it in the reverse direction from that which caused the contact.
- If the crane cannot be moved away from contact, the operator should remain inside cab until the lines have been de-energized.

5.5.3.4 *Recommended Practices*

Employers should comply with recommended standards and take the following measures to protect workers and operators of cranes and other boomed vehicles from contacting energized overhead power lines.

- Train workers to comply with the recommended standards. These regulations require workers and employers to consider all overhead power lines to be energized until (1) the owner of the lines or the electric utility indicates that they are not energized, and (2) they have been visibly grounded.
- Employers shall ensure that overhead power lines are de-energized or separated from the crane and its load by implementing one or more of the following:
  - De-energize and visibly ground electrical distribution and transmission lines at the point of work.
  - Use insulated barriers that are not a part of the crane to prevent contact with the lines.
  - If the power lines are not de-energized, operate cranes in the area only if a safe minimum clearance is maintained as follows:
    - At least 3.05 meters for lines rated 50 kilovolts or below,
    - At least 3.05 meters plus 10.2 milimeters for each kilovolt above 50 kilovolts; or maintain twice the length of the line insulator (but never less than 3.05m).
- Where it is difficult for the crane operator to maintain safe clearance by visual means, designate a person to observe the clearance and to give immediate warning when the crane approaches the limits of safe clearance.
- Do not use cage-type boom guards, insulating links, or proximity warning devices as a substitute for de-energizing and grounding lines or maintaining safe clearance.
- Follow ANSI Guidelines – Train workers to follow ANSI guidelines for operating cranes near overhead power lines (ANSI Standard B30.5-1994, 5-3.4.5)[ANSI 1994]. These guidelines recommend posting signs at the operator's station and on the outside of the crane warning that electrocution may occur if workers do not maintain safe minimum clearance that equals or exceeds OSHA requirements shown in Table 5.4.

Table 5.4. OSHA requirements for safe minimum clearance.

Power Line Voltage Phase To Phase (Kv)	Minimum Safe Clearance (Meters)
50 or below	3.1
Above 50 to 200	4.6
Above 200 to 350	6.1
Above 350 to 500	7.6
Above 500 to 750	10.7
Above 750 to 1,000	13.7

- Notify power line owners – Before beginning operations near electrical lines, notify the owners of the lines or their authorized representatives and provide them with all pertinent information: type of equipment (including length of boom) and date, time, and type of work involved. Request the cooperation of the owner to de-energize and ground the lines or to help provide insulated barriers. NIOSH encourages employers to consider de-energization (where possible) as the primary means of preventing injury from contact between cranes and power lines.
- Develop safety programs – Develop and implement written safety programs to help workers recognize and control the hazards of crane contact with overhead power lines.
- Evaluate jobsites – Evaluate jobsites before beginning work to determine the safest areas for material storage, the best placement for machinery during operations, and the size and type of machinery to be used.
- Know the location and voltage of all overhead power lines at the jobsite before operating or working with any crane. Research has shown that it is difficult to judge accurately the distance to an overhead object such as a power line [Middendorf 1978]. Therefore, NIOSH recommends that no other duties or responsibilities

be assigned when workers are designated to observe clearance during crane movement or operation.

- Evaluate alternative work methods – Evaluate alternative work methods that do not require the use of cranes. For example, it may be possible to use concrete pumping trucks instead of crane-suspended buckets for placing concrete near overhead power lines. Alternative methods should be carefully evaluated to ensure that they do not introduce new hazards into the workplace.

5.5.3.5 Tree Spraying

Spraying operations near energized electrical wires or equipment may create safety problems. Spraying directed into overheated lines may damage equipment or cause arcing between wires. Electricity also may be conducted down the hose stream, creating a hazard for workers.

5.5.3.6 Electrical Emergencies

Strong winds, ice or unintentional contact with equipment may cause trees or tree limbs to fall into powerlines. This may cause wires to break and fall to the ground. Should this happen, notify the electric utility company immediately.

A fallen wire can create hazards for workers and the general public. Objects touched by a fallen wire – fences, vehicles, buildings or even the surrounding ground – must be considered energized and should not be touched. It is impossible to tell simply by looking whether a downed wire is energized. Consider all downed wires energized and dangerous until the electric utility personnel notify you otherwise.

Where a power line has fallen across a vehicle, occupants should remain within the vehicle. If they must leave the vehicle because of a life-threatening situation, such as fire or potential explosion, they should jump clear of the vehicle without touching either the vehicle or wire and the ground at the same time. Once clear of the vehicle, they should shuffle away, taking small steps and keeping both feet in contact with the ground.

Remember, electricity can be transmitted from the victim to you. If a switch is accessible, shut off the power immediately. Otherwise, stand on a dry surface and pull the victim away with a dry board or rope. Do not use your hands or anything metal.

Use a CO_2 or dry chemical extinguisher to put out an electrical fire. Water should be used only by trained firefighting personnel.

In an emergency involving power lines or electrical equipment, call the electric utility company immediately.

5.5.3.7 Training Workers

Ensure that workers assigned to operate cranes and other boomed vehicles are specifically trained in safe operating procedures. Also ensure that workers are trained (1) to understand the limitations of such devices as boom guards, insulated lines, ground rods, nonconductive links, and proximity warning devices, and (2) to recognize that these devices are not substitutes for de-energizing and grounding lines or maintaining safe clearance. Workers should also be trained to recognize the hazards and use proper techniques when rescuing coworkers or recovering equipment in contact with electrical energy. CSA guidelines list techniques that can be used when equipment contacts energized power lines [CSA 1982].

All employers and workers should be trained in cardiopulmonary resuscitation (CPR).

Refer to Sec.2(q) (Crane Operation Safety) for additional safe work practices.

5.5.4 Bibliography

ANSI [1994]. American National standard for mobile and locomotive cranes. New York, NY: American National Standards Institute, ANSI B30.5-1994.

CFR. Code of Federal regulations. Washington, DC: U.S. Government Printing Office, Office of the Federal Register.

CSA (Construction Safety Association) [1982]. Mobile crane manual. Toronto, Ontario, Canada: Construction Safety Association of Ontario.

Middendorf L [1978]. Judging clearance distances near overhead power lines. In: Proceedings of the Human Factors Society, 22nd annual meeting. Santa Monica, CA: Human Factors Society, Inc.

NIOSH [1985]. NIOSH Alert: request for assistance in preventing electrocutions from contact between cranes and power lines. Cincinnati, OH: U.S. Department of Health and Human Services, Public Health Service, Centers for Disease Control, National Institute for Occupational Safety and Health, DHHS (NIOSH) Publication No. 85-111.

NIOSH [1990a]. Foreman electrocuted and lineman injured after truck-mounted crane boom contacts 7,200-volt overhead power line in Virginia. Morgantown, WV: U.S. Department of Health and Human Services,

Public Health Service, Centers for Disease Control, National Institute for Occupational Safety and Health, Fatal Accident Circumstances and Epidemiology (FACE) Report No. 90-39.

NIOSH [1990b]. Laborer touching suspended cement bucket electrocuted when crane cable contacts 7,200-volt power line in North Carolina. Morgantown, WV: U.S. Department of Health and Human Services, Public Health Service, Centers for Disease Control, National Institute for Occupational Safety and Health, Fatal Accident Circumstances and Epidemiology (FACE) Report No. 90-29.

NIOSH [1990c]. Well driller electrocuted when pipe on crane cable contacts 12,000-volt overhead power line in Virginia. Morgantown, WV: U.S. Department of Health and Human Services, Public Health Service, Centers for Disease Control, National Institute for Occupational Safety and Health, Fatal Accident Circumstances and Epidemiology (FACE) Report No. 90-38.

NIOSH [1991]. Construction laborer is electrocuted when crane boom contacts overhead 7,200-volt power line in Kentucky. Morgantown, WV: U.S. Department of Health and Human Services, Public Health Service, Centers for Disease Control, National Institute for Occupational Safety and Health, Fatal Accident Circumstances and Epidemiology (FACE) Report No. 91-21.

NIOSH [1993a]. Fatal Injuries to workers in the United States, 1980-1989: a decade of surveillance. Cincinnati, OH: U.S. Department of Health and Human Services, Public Health Service, Centers for Disease Control and Prevention, National Institute for Occupational Safety and Health, DHHS (NIOSH) Publication No. 93-108.

NIOSH [1993b]. Truck driver and company president electrocuted after crane boom contacts power line – West Virginia. Morgantown, WV: U.S. Department of Health and Human Services, Public Health Service, Centers for Disease Control and Prevention, National Institute for Occupational Safety and Health, Fatality Assessment and Control Evaluation (FACE) Report No. 93-14.

OSHA [1990]. Analysis of construction fatalities – the OSHA database 1985-1989. Washington, DC: U.S. Department of Labor, Occupational Safety and Health Administration.

5.6 Functional Safety for Electric Power Transmission

5.6.1 Introduction

This section presents the main types of hazards for personnel, equipment and electric power systems which should be taken into consideration in the design of computer-based systems. An emphasis is put on description of the hazard for safety of an electric power system, whose nature is fundamentally different from the nature of those occurring in the process industry and in all other sectors of industry. To illustrate the problems with hazards and risks identification, a power substation software interlocking case study is described. The case study description is preceded by a short presentation of the results of an extensive review of available publications with the aim of identifying the current state of practice in assuring functional safety of computer systems applied in the sector.

Safety as a crucial issue of computer applications in such fields as nuclear power stations, space industry, aircraft industry, chemical industry or railways is well recognized by wide circles of specialists of these fields as well as by specialists in software engineering and computer science.

Contemporary electric power systems (EPSs) are very complex and highly technologically advanced systems. The vast, highly interconnected North American electric power system has been called the "greatest machine ever created". Similarly one could say the same about contemporary electric power systems of other countries.

Safety issues are important in designing computer-based control systems for EPSs. At the same time application of computer-based systems in this sector offers researchers and practitioners a wide range of issues, which are both basic for the future of the sector.

5.6.2 Structure of Electric Power Systems

The electric power industry consists of many different companies involved in generating electric power, bulk transmission of energy from power stations to loading centers and its distribution to customers. In order to perform their functions and to attain suitable effectiveness, all power stations, substations, power lines forming power grids, the related control centers and other components are interconnected, although they belong to various companies, forming an electric power system (EPS). However an increasing tendency can be observed, for example in Europe, aiming to build up more and more stronger connections between the separate EPS-s existing in the individual countries. Energy related policy of the European Union aims at building Trans-European networks in order to provide a sound basis for free and competitive electricity market in Europe and strengthening security of energy supply.

In the most general terms, an EPS can be partitioned into generating stations and high-voltage power networks known as grids. Power networks consist of transmission networks, called extra-high voltage (EHV) networks, which are used to transmit power from generating stations to main load centres and distribution networks of lower voltages, also known as high-voltage (HV) networks, which are used to transmit power to customers. Both, transmission and distribution networks consist of power lines, substations and control centers.

Substations form vital nodes in the HV and EHV networks because they make possible modifications in the configuration of networks during the operation of the system by means of switching devices that can be controlled by computer-based control systems applied in the substations. Initiation of the control procedure may be performed locally by substation operators or remotely from the EPS control canters. They also are the points in which power is leaded out of the generation stations and supplied to the consumers.

The power network of generation, transmission and distribution subsystems forming an EPS is coupled with telecommunication and telecontrol systems that are used for communication and transmission of data between power generating stations, substations and control centers for remote operation and remote real-time signaling, metering, control and fault protection. Development of these EPS communication systems goes toward an integrated EPS telecommunication network. In the last decade due to the process of computerization of the EPS-s the data network is used on increasingly larger scale for transmitting the data that are critical to safe and reliable operation of an EPS and for the power grid - related financial transactions.

5.6.3 Hazards in Electric Power Systems

In an EPS, most hazards connected with current application of computer-based systems concern mainly some applications in power stations, substations and control centers.

Most hazards, which appear in power stations and substations, are hazardous for people and equipment and are connected with the technological process of energy production or switching operations performed at substations. Those kinds of hazards are approximate to the kind of hazards appearing in other sectors of industry. Their nature is rather obvious and due to space limitation in this paper they will not be analyzed except of hazards connected with switching operations performed in substations.

Further on in this section an emphasis is put on description of concept of safety of an electric power system, whose nature is fundamentally different from the nature of those occurring in the process industry sector and all other sectors of industry.

5.6.3.1 Hazards In Substations

Electric power substations consist of two essential parts:

- main (high or extra-high voltage) circuits, also called primary circuits; and
- auxiliary circuits also called secondary circuits.

Main circuit of a substation is composed of a busbar system (or busbar systems) and connections of power lines, power transformers, etc to the busbar system through switching devices. A busbar system it is set of three electric conductors of very low impedance that serves as a common connection for individual power lines, power transformers, etc. Substations are divided into bays. The Bay of a substation it is a part of substation containing high (or extra-high) voltage switching devices and connections of individual power lines, individual power transformers, etc to the substation busbar system (or busbar systems) as well as protection, control, and measurement devices for the individual power line, the individual power transformer, etc. If it is a bay used to connect a power line to busbars is called a line bay, if it is used for connection a power transformer it is called a transformer bay, etc. Normally a substation contains a few or more line and transformer bays and sometimes also other bays. All bays are similar to the line bay.

Auxiliary circuits are electrical circuits contain measurement, signaling, control and protection devices.

The Main hazard for substation staff and equipment is connected with the fact, that for design reasons, disconnectors used in substations are not able to switch on or off the current (e.g. to switch on or off the loaded line) and are used only to ensure the required isolation clearance between disconnected elements which due to design restrictions cannot be achieved by a circuit breaker. In order to take these limitations into consideration switching off a line for example must be performed according to the following sequence:

- breaking of the current using a circuit-breaker;

- opening of the line disconnector in order to achieve isolation clearance between the disconnected line and the circuit-breaker; and
- opening of the busbar disconnector, depending on which busbars, "1A" or "2", the line is connected to in order to achieve the isolation clearance between the circuit-breaker and busbars.

If this sequence is carried out in an incorrect way, and for example the signal for opening would be sent at first to the disconnector (e.g. busbar disconnector), an electric arc will arise between the contacts of the disconnector accompanied by high rate optic and acoustic phenomena, spraying melted metal etc. The arc would travel to neighboring phases, resulting in an inter-phase short circuit. It would look like an explosion. This failure would cause considerable material losses because of complete destruction of the disconnector and partially or complete destruction also other components in the substation, disturbance in substation operation and interruption of energy supply to consumers. Sprayed melted metal could seriously injure personnel if accidentally someone of the personnel is near an exploding disconnector.

Depending on the situation in a given EPS, the incorrect sequence of switching operation could also cause even large power system failure and collapse a part of the EPS, i.e. it could create a hazard for the described below safety of the EPS.

Note that in high voltage networks transmission lines have three wires, one for each phase. It concerns also busbar systems in substations. Each busbar system in a substation consist of three bars.

5.6.3.2 *Security*

In electric power systems engineering safety of an EPS is described by means of the word "security," although in less formal descriptions the word "safety" is also used. Because in computer science and software engineering the word "security" is used in a different meaning to avoid misunderstanding, the codes refer to the safety of EPS by means of the word "safety" which according to widely accepted definition means the ability of the bulk power electric system to withstand sudden disturbances such as electric short circuits or unanticipated loss of system components.

Generally speaking, if all generators connected to an EPS work synchronously and the voltage and the frequency in the EPS is within the required limits, then this is a normal, stable state of an EPS. In case of sudden disturbances, e.g. sudden increase of the load, or switching off an important transmission line due to a failure in a substation computer control or protection system, the stable state of operation is disturbed. Each EPS is designed with a certain stability margin. When the disturbance is greater than the stability margin it results in loss of stability and collapse of the EPS, which may be total, and this is called a black-out, or partial.

The loss of stability is understood in the sense applied in dynamical systems, it means generators connected to the EPS fall-out of synchronism. It causes an emergency shutdown of the generators by automatic protection devices in order to protect them against destruction and emergency stop of the power stations. Economic and social results of the loss of stability by an EPS are always very serious and can be catastrophic. This is the reason why maintaining stability of an EPS is the main requirement for the EPS planning, operation and control. Hazard for stability is equivalent to hazard for safety of an EPS. Major blackouts are rare events, but as it has been stated in one publication, their impact can be catastrophic.

A large system failure, which happened in 1977 in New York, blacked out the city and left it without an electricity supply. Losses were estimated at 310 million USDs. The restart of an industrial plant after occurring a power failure with duration time longer than critical one lasts on average 17.4 hours.

In the history of the electric utility industry, safety as it is understood today, is a relatively recent concept, which emerged after the famous 1965 blackout that left almost 30 million of people of north-eastern United States and Ontario, Canada, without electricity.

Through the first two-thirds of the 20th century, the term "safety" was not used and the safety of an EPS was subsumed with its reliability and implemented into the system in the system planning process by providing a robust system that could withstand any "credible" sudden disturbances without a serious disruption. Perhaps the epitome of this approach was mid-century American Electric Power system, which in 1974 withstood five simultaneous major tornadoes, losing eleven 345 kV lines, one 500 kV line, two 765 kV lines, and three major switching substations, without interruption of service to customers. Such practices even if technically feasible are no longer considered economically or environmentally feasible.

The focus in the safety concept was shifted from system robustness, which was designed into the system at the planning stage, onto risk aversion, which is based on automatic protection and control devices, and still to a considerably high degree on intervention of a system operator in real time in an emergency state.

5.6.4 Assuring Functional Safety

Since electric power and telecommunications infrastructures are considered the most vulnerable to sabotage and cyber-attack, in specifying of functional safety requirements for substation automation systems and for computer-based systems used in power stations and control centers interaction of safety, security and availability requirements should be taken into account.

The automation of substations, similarly like automation of equipment and plants in other sectors of industry, is more and more based not only on hardware but also on software solutions. Software engineers are not able to design protection and automation systems used in electric power sector without co-operation with electric power systems engineers. It especially concerns requirements specification and safety analysis phases. It seems that these facts should have influence on curricula of electrical engineering faculties. Curricula should at least include requirements engineering (lectures, classes and laboratory) to the extent which enables graduates to carry out specification of requirements on their own, co-operate with software engineers during validation of models used in a safety analysis, carry out the safety analysis, evaluate of results, and co-operate with software engineers.

5.6.5 Recommended References

IEEE Recommended Practice For The Design Of Reliable Industrial And Commercial Power Systems, ANSI/IEEE Std 493-1990, May 1995.

Balu, N., Bertram, T., et al.: On-Line Power System Security Analysis. Proceedings of the IEEE, Vol. 80, No. 2, pp. 262-280, 1992.

Electric Power Research Institute (EPRI): Electricity Technology Roadmap: Powering Progress - 1999 Summary and Synthesis. EPRI, Palo Alto, 1999.

Electric Power Research Institute (EPRI): Issues and Solutions: North American Grid Operations (2000-2005). EPRI, 1999.

Executive Summary: EPRI/DOD Initiative On Complex Interactive Networks/Systems, www.epri.com, 1999.

Martin O., Messie M., and de Labrouhe G.: The digital monitoring and control of substations, in Proc. CIGRE Conf., paper no. 23/13-02, 1994.

Design and maintenance practice for substation secondary systems, CIGRE Working Group 23.05 Tech. Rep., April 1994.

Zurakowski Z.: Task B1b: Identification and Preparation of Case Studies – Extra-High Voltage Substation Software Interlocking Case Study, EC COPERNICUS JRP 1594, Technical Report TR ISAT 97/8, Institute of Power Systems Automation, December 1996, Updated: February and April 1997.

Zurakowski, Z.: Safety and Security Issues in Electric Power Industry. In Proceedings of the 19th Conference SAFECOMP 2000, Rotterdam, The Netherlands, October 2000.

6

Worker Safety Rules

6.1 Critical Incident Stress

6.1.1 Responding to Emergency Events

Workers responding to emergency events and or disasters will see and experience events that will strain their ability to function. These events, which include having to witness or experience tragedy, death, serious injuries and threatening situations are called "Critical Incidents." The physical and psychological well-being of those experiencing this stress, as well as their future ability to function through a prolonged response, will depend upon how they manage this stress. Post traumatic stress disorder differs from critical incident stress by lasting longer than four weeks after the event triggering the emotional, mental or physical response. Most instances of critical incident stress last between two days and four weeks.

The signs and symptoms of critical incident stress can be physical, emotional, cognitive, or behavioral. Individuals express stress in different ways and therefore manifest different reactions. Table 6.1 below is not exhaustive but will help supervisors to identify workers who are exhibiting stress reactions.

Table 6.1 Signs and symptoms of stress.

Physical	Cognitive	Emotional	Behavioral
Fatigue Chills Unusual thirst Chest pain Headaches Dizziness	Uncertainty Confusion Nightmares Poor attention/ decision making ability Poor concentration, memory Poor problem solving ability	Grief Fear Guilt Intense anger Apprehension and depression Irritability Chronic anxiety	Inability to rest Withdrawal Antisocial behavior Increased alcohol consumption Change in communications Loss/increase in appetite

During the emergency phase of the response, monitoring of employees by simple conversation and observation may help to identify early signs for some responders. The following steps can help to reduce significant stress detected early in the response:

- Limit exposure to noise and odors.
- Dictate an immediate 15 minute rest break.
- Provide non-caffeinated fluids to drink.
- Provide low sugar and low fat food.
- Get the person to talk about his or her feelings.
- Do not rush the person back to work.

6.1.2 Critical Incident Stress Management

Critical incident stress management is system of education, prevention and mitigation of the effects from exposure to highly stressful critical incidents. It is handled most effectively by specially trained individuals, such as crisis intervention specialists.

6.1.3 Critical Incident Stress Debriefing

Critical incident stress debriefing (CISD) is a facilitator-led group process conducted soon after a traumatic event with individuals considered to be under stress from trauma exposure. When structured, the process usually (but not always) consists of seven steps:

- Introduction;
- Fact Phase;
- Thought Phase;
- Reaction Phase;
- Symptom Phase;
- Teaching Phase; and
- Re-entry Phase.

During the group process, participants are encouraged to describe their experience of the incident and its aftermath, followed by a presentation on common stress reactions and stress management. This early intervention process supports recovery by providing group support and linking employees to further counseling and treatment services if they become necessary.

6.2 Toxic Industrial Chemicals

Toxic industrial chemicals are industrial chemicals that are manufactured, stored, transported, and used throughout the world. Toxic industrial chemicals can be in the gas, liquid, or solid state. They can be chemical hazards (e.g., carcinogens, reproductive hazards, corrosives, or agents that affect the lungs or blood) or physical hazards (e.g., flammable, combustible, explosive, or reactive). Table 6.2 lists the most common TICs listed by their hazard index.

Table 6.2 TICs listed by Hazard index.

High	Medium	Low
Ammonia (CAS# 7664-41-7)	Acetone cyanohydrin (CAS# 75-86-5)	Allyl isothiocyanate (CAS# 57-06-7)
Arsine (CAS# 7784-42-1)	Acrolein (CAS# 107-02-8)	Arsenictrichloride (CAS# 7784-34-1)
Boron trichloride (CAS#10294-34-5)	Acrylonitrile (CAS# 107-13-1)	Bromine (CAS# 7726-95-6)

Table 6.2 (cont.) TICs listed by Hazard index.

High	Medium	Low
Boron trifluoride (CAS#7637-07-2)	Allyl alcohol (CAS# 107-18-6)	Bromine chloride (CAS# 13863-41-7)
Carbon disulfide (CAS# 75-15-0)	Allylamine (CAS# 107-11-9)	Bromine pentafluoride (CAS# 7789-30-2)
Chlorine (CAS# 7782-50-5)	Allyl chloro carbonate (CAS# 2937-50-0)	Bromine trifluoride (CAS# 7787-71-5)
Diborane (CAS# 19287-45-7)	Boron tribromide (CAS# 10294-3-4)	Carbonyl fluoride (CAS# 353-50-4)
Ethylene oxide (CAS# 75-21-8)	Carbon monoxide (CAS# 630-08-0)	Chlorine pentafluoride (CAS# 13637-63-3)
Fluorine (CAS# 7782-41-4)	Carbonyl sulfide (CAS# 463-58-1)	Chlorine trifluoride (CAS# 7790-91-2)
Formaldehyde (CAS# 50-00-0)	Chloroacetone (CAS# 78-95-5)	Chloroacetaldehyde (CAS# 107-20-0)
Hydrogen bromide (CAS# 10035-10-6)	Chloroacetonitrile (CAS# 7790-94-5)	Chloroacetyl chloride (CAS# 79-04-9)
Hydrogen chloride (CAS# 7647-01-0)	Chlorosulfonic acid (CAS# 779094-5)	Crotonaldehyde (CAS# 123-73-9)
Hydrogen cyanide (CAS#74-90-8)	Diketene (CAS# 674-82-8)	Cyanogen chloride (CAS# 506-77-4)
Hydrogen fluoride (CAS# 7664-39-3)	1,2-Dimethylhy drazine (CAS# 54073-8)	Dimethyl sulfate (CAS# 77-78-1)
Hydrogen sulfide (CAS# 7783-0604)	Ethylene dibromide (CAS# 106-93-4)	Diphenylmethane-4.4′-diisocyanate (CAS# 101-68-8)
Nitric acid, fuming (CAS# 7697-37-2)	Hydrogen selenide (CAS# 7783-07-5)	Ethyl chlroroformate (CAS# 541-41-3)
Phosphorus trichloride (CAS#7719-12-2)	Methyl bromide (CAS# 74-83-9)	Ethyl phosphonothioic dichloride (CAS# 993-43-1)

Table 6.2 (cont.) TICs listed by Hazard index.

High	Medium	Low
Sulfur dioxide (CAS# 7446-09-5)	Methyl chloroformate (CAS# 79-22-1)	Ethyl phosphonic dichloride (CAS# 1066-50-8)
Sulfuric acid (CAS# 7664-93-9)	Methyl chlorosilane (CAS# 993-00-0)	Ethyleneimine (CAS# 151-56-4)
Tungsten hexafluoride (CAS# 7783-82-6)	Methyl hydrazine (CAS# 60-34-4)	Hexachloro cyclopentadiene (CAS# 77-47-4)
	Methyl isocyanate (CAS# 624-83-9)	Hydrogen iodide (CAS# 10034-85-2)
	Methyl mercaptan (CAS# 74-93-1)	Iron pentacarbonyl (CAS# 13463-40-6)
	Nitrogen dioxide (CAS# 10102-44-0)	Isobutyl chloroformate (CAS# 543-27-1)
	Phosphine (CAS# 7803-51-2)	Isopropyl chloroformate (CAS# 108-23-6)
	Phosphorus oxychloride (CAS# 10025-87-3)	Isopropyl isocyanate (CAS# 1795-48-8)
	Phosphorus pentafluoride (CAS# 7647-19-0)	n-Butyl chloroformate (CAS# 592-34-7)
	Selenium hexafluoride (CAS# 7783-79-1)	n-Butyl isocyanate (CAS# 111-36-4)
	Silicon tetrafluoride (CAS# 7783-61-1)	Nitric oxide (CAS# 10102-43-9)
	Stibine (CAS# 7803-52-3)	n-Propyl chloroformate (CAS# 109-61-5)
	Sulfur trioxide (CAS# 7446-11-9)	Parathion (CAS#: 56-38-2)

Table 6.2 (cont.) TICs listed by Hazard index.

High	Medium	Low
	Sulfuryl fluoride (CAS# 2699-79-8)	Perchloromethyl mercaptan (CAS# 594-42-3)
	Tellurium hexafluoride (CAS# 7783-80-4)	sec-Butyl chloroformate (CAS# 17462-58-7)
	n-Octyl mercaptan (CAS# 111-88-6)	tert-Butyl isocyanate (CAS# 1609-86-5)
	Titanium tetrachloride (CAS# 7550-45-0)	Tetraethyl lead (CAS# 78-00-2)
	Tricholoroacetyl chloride (CAS# 76-02-8)	Tetraethyl pyrophosphate (CAS# 107-49-3)
	Trifluoroacetyl chloride (CAS# 354-32-5)	Tetramethyl lead (CAS# 75-74-1)
		Toluene 2.4-diisocyanate (CAS# 584-84-9)
		Toluene 2.6-diisocyanate (CAS# 91-08-7)

Source: Guide for the Selection of Chemical and Biological Decontamination Equipment for Emergency First Responders. National Institute of Justice Guide 103-00 (Volume I), (2001, October). This guide for emergency first responders provides information about the selection and use of chemical and/or biological decontamination equipment for various applications.

The time that these agents persist in the environment is dependent on many variables such as physical state (solid, liquid, or gas), weather conditions (wind speed, rain or snow, air or surface temperature), indoor or outdoor release, chemical stability, method of release (vapor or aerosol), or quantity released.

Many toxic industrial chemicals are highly toxic and may rapidly affect exposed individuals. Toxic industrial chemicals (whether as a gas, aerosol, or liquid) enter the body through inhalation, through the skin, or through digestion. The time that it takes for a toxic industrial chemical to begin working is dependent mainly on the route that the agent enters the body. Generally poisoning occurs more quickly if a chemical enters through the lungs (because of the ability of the agent to rapidly diffuse throughout the body). Information related to how the chemicals affect humans and symptoms of exposure to specific chemicals can be found in material safety data sheets (MSDS) or chemical information cards.

If you are exposed to a toxic industrial chemicals attack, get away from the impacted area quickly without passing through the contaminated area, if possible. If available, a good way to protect yourself from a toxic industrial chemicals is to wear suitable chemical protective clothing and respiratory protection. However, it must be stressed that this protective equipment does not always work against toxic industrial chemicals. The effectiveness is determined by the materials of construction, the type and level of exposure, and duration of exposure. If you have been exposed to a toxic industrial chemical, consult with your physician as soon as possible. Personal decontamination can be performed by removing contaminated clothing and washing exposed skin with soap and water.

"Shelter-in-place" means to go indoors, close up the building, and wait for the danger to pass. If you are advised to shelter in place, close all doors and windows; turn off fans, air conditioners, and forced-air heating units that bring in fresh air from the outside; only re-circulate air that is already in the building; move to an inner room or basement; and keep your radio turned to the emergency response network or local news to find out what else you need to do.

First responders should consider the possible impact of a release of toxic industrial chemicals and address these in their health and safety plans (HASPs). The health and safety plan should include guidelines such as monitoring, detection, awareness training, personal protective equipment, decontamination, and medical surveillance of acutely exposed workers.

There are a wide variety of direct reading instruments and analytical sampling and analysis procedures available for detecting toxic industrial chemicals. Direct reading instruments include detector tubes, flame ionization detector (FID), photoionization detector (PID), and specialized gas chromatographs.

During a chemical release, or after a release has occurred with no information available about the duration of the release or the airborne concentration of toxic industrial chemicals, at a minimum, Level B protection should be used. The requirement of OSHA's hazardous waste operations and emergency response (HAZWOPER) standard (29 CFR 1910.120(q)) provides additional information for responding to hazardous substance releases including toxic industrial chemicals.

6.3 Electrical Protective Devices

6.3.1 Design Requirements

Insulating blankets, matting, covers, line hose, gloves, and sleeves made of rubber shall meet the following requirements:

- Blankets, gloves, and sleeves shall be produced by a seamless process.

6.3.2 Manufacture and Markings

- Each item shall be clearly marked as follows:
  - Class 0 equipment shall be marked Class 0.
  - Class 1 equipment shall be marked Class 1.
  - Class 2 equipment shall be marked Class 2.
  - Class 3 equipment shall be marked Class 3.
  - Class 4 equipment shall be marked Class 4.
- Non-ozone-resistant equipment other than matting shall be marked Type I.
- Ozone-resistant equipment other than matting shall be marked Type II.
- Other relevant markings, such as the manufacturer's identification and the size of the equipment, may also be provided.
- Markings shall be non-conducting and shall be applied in such a manner as not to impair the insulating qualities of the equipment.
- Markings on gloves shall be confined to the cuff portion of the glove.

6.3.3 Electrical Requirements

Equipment shall be capable of withstanding the a-c proof-test voltage specified in Table 6.3 or the d-c proof-test voltage specified in Table 6.3.

The proof test shall reliably indicate that the equipment can withstand the voltage involved.

The test voltage shall be applied continuously for 3 minutes for equipment other than matting and shall be applied continuously for 1 minute for matting.

Gloves shall also be capable of withstanding the a-c proof-test voltage specified in Table 6.3 after a 16-hour water soak.

When the a-c proof test is used on gloves, the 60-hertz proof-test current may not exceed the values specified in Table 6.3 at any time during the test period.

If the a-c proof test is made at a frequency other than 60 hertz, the permissible proof-test current shall be computed from the direct ratio of the frequencies.

For the test, gloves (right side out) shall be filled with tap water and immersed in water to a depth that is in accordance with Table 6.5. Water shall be added to or removed from the glove, as necessary, so that the water level is the same inside and outside the glove.

After the 16-hour water soak specified in paragraph 1,2,9,3 of this section, the 60-hertz proof-test current may exceed the values given in Table 6.3 by not more than 2 milliamperes.

Table 6.3 A-C proof-test requirements.

Class of Equipment	Proof-test Voltage rms V	Maximum Proof-test Current, mA (Gloves Only)			
		267-mm Glove	356-mm Glove	406-mm Glove	457-mm Glove
0	5,000	8	12	14	16
1	10,000	—	14	16	18
2	20,000	—	16	18	20
3	30,000	—	18	20	22
4	40,000	—	—	22	24
		—			

Equipment that has been subjected to a minimum breakdown voltage test may not be used for electrical protection.

Material used for Type II insulating equipment shall be capable of withstanding an ozone test, with no visible effects. The ozone test shall reliably indicate that the material will resist ozone exposure in actual use. Any visible signs of ozone deterioration of the material, such as checking, cracking, breaks, or pitting, is evidence of failure to meet the requirements for ozone-resistant material.

6.3.4 Workmanship and Finish

Equipment shall be free of harmful physical irregularities that can be detected by the tests or inspections required under this section.

Surface irregularities that may be present on all rubber goods because of imperfections on forms or molds or because of inherent difficulties in the manufacturing process and that may appear as indentations, protuberances, or imbedded foreign material are acceptable under the following conditions:

- The indentation or protuberance blends into a smooth slope when the material is stretched.
- Foreign material remains in place when the insulating material is folded and stretches with the insulating material surrounding it.

The following are important reference materials:

- American Society for Testing and Materials (ASTM) D 120-87, Specification for Rubber Insulating Gloves.
- ASTM D 178-93 (or D 178-88), specification for rubber insulating Matting.
- ASTM D 1048-93 (or D 1048-88a), specification for rubber insulating Blankets.
- ASTM D 1049-93 (or D 1049-88), specification for rubber insulating Covers.
- ASTM D 1050-90, specification for rubber insulating Line Hose.
- ASTM D 1051-87, specification for rubber insulating Sleeves.

These standards contain specifications for conducting the various tests required in paragraph 1 of this section. For example, the a-c and d-c proof tests, the breakdown test, the water soak procedure, and the ozone test mentioned in this paragraph are described in detail in the ASTM standards.

6.3.5 In-Service Care and Use

Electrical protective equipment shall be maintained in a safe, reliable condition.

The following specific requirements apply to insulating blankets, covers, line hose, gloves, and sleeves made of rubber:

- Maximum use voltages shall conform to those listed in Tables 6.3, 6.4, and 6.5.
- Insulating equipment shall be inspected for damage before each day's use and immediately following any incident that can reasonably be suspected of having caused damage. Insulating gloves shall be given an air test, along with the inspection.
- Insulating equipment with any of the following defects may not be used:
  - A hole, tear, puncture, or cut;
  - Ozone cutting or ozone checking (the cutting action produced by ozone on rubber under mechanical stress into a series of interlacing cracks);

Table 6.4 D-C Proof-test requirements.

Class of Equipment	Proof-test Voltage
0	20,000
1	40,000
2	50,000
3	60,000
4	70,000

Note: The d-c voltages listed in this table are not appropriate for proof testing rubber insulating line hose or covers. For this equipment, d-c proof tests shall use a voltage high enough to indicate that the equipment can be safely used at the voltages listed in ASTM D Standards 1050-90 and ASTM D 1049-88.

Table 6.5 Glove tests – water level.[1,2]

Class of Glove	AC Proof-test mm.	DC Proof-test mm.
0	38	38
1	38	51
2	64	76
3	89	102
4	127	153

1. The water level is given as the clearance from the cuff of the glove to the water line, with a tolerance of + or -13 mm. (+ or -0.5 in.).

2. If atmospheric conditions make the specified clearances impractical, the clearances may be increased by a maximum of 25 mm. (1 in.).

- An embedded foreign object;
- Any of the following texture changes: swelling, softening, hardening, or becoming sticky or inelastic;
- Any other defect that damages the insulating properties.

Insulating equipment found to have other defects that might affect its insulating properties shall be removed from service and returned for testing.

Insulating equipment shall be cleaned as needed to remove foreign substances.

Insulating equipment shall be stored in such a location and in such a manner as to protect it from light, temperature extremes, excessive humidity, ozone, and other injurious substances and conditions.

Protector gloves shall be worn over insulating gloves, except as follows:

- Protector gloves need not be used with Class 0 gloves, under limited-use conditions, where small equipment and parts manipulation necessitate unusually high finger dexterity.

- *Note:* Extra care is needed in the visual examination of the glove and in the avoidance of handling sharp objects.

Any other class of glove may be used for similar work without protector gloves if the employer can demonstrate that the possibility of physical damage to the gloves is small and if the class of glove is one class higher than that required for the voltage involved. Insulating gloves that have been used without protector gloves may not be used at a higher voltage until they have been tested.

Electrical protective equipment shall be subjected to periodic electrical tests. Test voltages and the maximum intervals between tests shall be in accordance with Tables 6.6 and 6.7.

The test method used shall reliably indicate whether the insulating equipment can withstand the voltages involved.

Table 6.6 Rubber insulating equipment voltage requirements.

Class of Equipment	Maximum Use Voltage[1] a – c – rms	Retest Voltage[2] a – c – rms	Retest Voltage[2] d – c – avg
0	1,000	5,000	20,000
1	7,000	10,000	40,000
2	17,000	20,000	50,000
3	26,000	30,000	60,000
4	36,000	40,000	70,000

1. The maximum use voltage is the a-c voltage (rms) classification of the protective equipment that designates the maximum nominal design voltage of the energized system that may be safely worked. The nominal design voltage is equal to the phase-to-phase voltage on multiphase circuits. However, the phase-to-ground potential is considered to be the nominal design voltage:

   1. If there is no multiphase exposure in a system area and if the voltage exposure is limited to the phase-to-ground potential, or
   2. If the electrical equipment and devices are insulated or isolated or both so that the multiphase exposure on a grounded wire circuit is removed.

2. The proof-test voltage shall be applied continuously for at least 1 minute, but no more than 3 minutes.

Table 6.7 Rubber insulating equipment test intervals.

Type of Equipment	When to Test
Rubber insulating line hose	Upon indication that insulating value is suspect.
Rubber insulating covers	Upon indication that insulating value is suspect.
Rubber insulating blankets	Before first issue and every 12 months thereafter.[1]
Rubber insulating gloves	Before first issue and every 6 months thereafter.[1]
Rubber insulating sleeves	Before first issue and every 12 months thereafter.[1]

1. If the insulating equipment has been electrically tested but not issued for service, it may not be placed into service unless it has been electrically tested within the previous 12 months.

Note: Standard electrical test methods considered as meeting this requirement are given in the following national consensus standards:

- American Society for Testing and Materials (ASTM) D 120-87, specification for rubber insulating gloves.
- ASTM D 1048-93, specification for rubber insulating blankets.
- ASTM D 1049-93, specification for rubber insulating covers.
- ASTM D 1050-90, specification for rubber insulating line hose.
- ASTM D 1051-87, specification for rubber insulating sleeves.
- ASTM F 478-92, specification for in-service care of insulating line hose and covers.
- ASTM F 479-93, specification for in-service care of insulating blankets.
- ASTM F 496-93b, specification for in-service care of insulating gloves and sleeves.

Insulating equipment failing to pass inspections or electrical tests not be used by employees, except as follows:

- Rubber insulating line hose may be used in shorter lengths with the defective portion cut off.
- Rubber insulating blankets may be repaired using a compatible patch that results in physical and electrical properties equal to those of the blanket.
- Rubber insulating blankets may be salvaged by severing the defective area from the undamaged portion of the blanket. The resulting undamaged area may not be smaller than 560 mm by 560 mm for Class 1, 2, 3, and 4 blankets.

Rubber insulating gloves and sleeves with minor physical defects, such as small cuts, tears, or punctures, may be repaired by the application of a compatible patch. Also, rubber insulating gloves and sleeves with minor surface blemishes may be repaired with a compatible liquid compound. The patched area shall have electrical and physical properties equal to those of the surrounding material. Repairs to gloves are permitted only in the area between the wrist and the reinforced edge of the opening.

Repaired insulating equipment shall be retested before it may be used by employees.

The employer shall certify that equipment has been tested. The certification shall identify the equipment that passed the test and the date it was tested.

Note: Marking of equipment and entering the results of the tests and the dates of testing onto logs are two acceptable means of meeting this requirement.

6.4 Hand Protection

Employers shall select and require employees to use appropriate hand protection when employees' hands are exposed to hazards such as those from skin absorption of harmful substances; severe cuts or lacerations; severe abrasions; punctures; chemical burns; thermal burns; and harmful temperature extremes.

Employers shall base the selection of the appropriate hand protection on an evaluation of the performance characteristics of the hand protection relative to the task(s) to be performed, conditions present, duration of use, and the hazards and potential hazards identified.

6.5 Hazard Assessment and Personal Protective Equipment Selection

6.5.1 General Guidance

The following guidelines will assist employers and employees in implementing requirements for a hazard assessment and the selection of personal protective equipment.

1. Controlling hazards. PPE devices alone should not be relied on to provide protection against hazards, but should be used in conjunction with guards, engineering controls, and sound manufacturing practices.
2. Assessment and selection. It is necessary to consider certain general guidelines for assessing the foot, head, eye and face, and hand hazard situations that exist in an occupational or educational operation or process, and to match the protective devices to the particular hazard. It should be the responsibility of the safety officer to exercise common sense and appropriate expertise to accomplish these tasks.
3. Assessment guidelines. In order to assess the need for PPE the following steps should be taken:
    a. Survey. Conduct a walk-through survey of the areas in question. The purpose of the survey is to identify sources of hazards to workers and co-workers. Consideration should be given to the basic hazard categories:
        i. Impact;
        ii. Penetration;
        iii. Compression (roll-over);
        iv. Chemical;
        v. Heat;
        vi. Harmful dust; and
        vii. Light (optical) radiation.
    b. Sources. During the walk-through survey the safety officer should observe:
        i. sources of motion; i.e., machinery or processes where any movement of tools, machine elements or particles could exist, or movement of personnel that could result in collision with stationary objects;

ii. sources of high temperatures that could result in burns, eye injury or ignition of protective equipment, etc.;
iii. types of chemical exposures;
iv. sources of harmful dust;
v. sources of light radiation, i.e., welding, brazing, cutting, furnaces, heat treating, high intensity lights, etc.;
vi. sources of falling objects or potential for dropping objects;
vii. sources of sharp objects which might pierce the feet or cut the hands;
viii. sources of rolling or pinching objects which could crush the feet; and
ix. layout of workplace and location of co-workers. In addition, injury/accident data should be reviewed to help identify problem areas.

c. Organize data. Following the walk-through survey, it is necessary to organize the data and information for use in the assessment of hazards. The objective is to prepare for an analysis of the hazards in the environment to enable proper selection of protective equipment.
d. Analyze data. Having gathered and organized data on a workplace, an estimate of the potential for injuries should be made. Each of the basic hazards (paragraph 3.a.) should be reviewed and a determination made as to the type, level of risk, and seriousness of potential injury from each of the hazards found in the area. The possibility of exposure to several hazards simultaneously should be considered.

4. Selection guidelines. After completion of the procedures in paragraph 3, the general procedure for selection of protective equipment is to:
   a. Become familiar with the potential hazards and the type of protective equipment that is available, and what it can do; i.e., splash protection, impact protection, etc.;
   b. compare the hazards associated with the environment; i.e., impact velocities, masses, projectile shape, radiation intensities, with the capabilities of the available protective equipment;

   c. select the protective equipment which ensures a level of protection greater than the minimum required to protect employees from the hazards; and
   d. fit the user with the protective device and give instructions on care and use of the PPE. It is very important that end users be made aware of all warning labels for and limitations of their PPE.
5. Fitting the device. Careful consideration must be given to comfort and fit. PPE that fits poorly will not afford the necessary protection. Continued wearing of the device is more likely if it fits the wearer comfortably. Protective devices are generally available in a variety of sizes. Care should be taken to ensure that the right size is selected.
6. Devices with adjustable features. Adjustments should be made on an individual basis for a comfortable fit that will maintain the protective device in the proper position. Particular care should be taken in fitting devices for eye protection against dust and chemical splash to ensure that the devices are sealed to the face. In addition, proper fitting of helmets is important to ensure that it will not fall off during work operations. In some cases a chin strap may be necessary to keep the helmet on an employee's head. (Chin straps should break at a reasonably low force, however, so as to prevent a strangulation hazard). Where manufacturer's instructions are available, they should be followed carefully.
7. Reassessment of hazards. It is the responsibility of the safety officer to reassess the workplace hazard situation as necessary, by identifying and evaluating new equipment and processes, reviewing accident records, and reevaluating the suitability of previously selected PPE.
8. Selection chart guidelines for eye and face protection. Some occupations (not a complete list) for which eye protection should be routinely considered are: carpenters, electricians, machinists, mechanics and repairers, millwrights, plumbers and pipe fitters, sheet metal workers and tinsmiths, assemblers, sanders, grinding machine operators, lathe and milling machine operators, sawyers, welders, laborers, chemical process operators

and handlers, and timber cutting and logging workers. Table 6.8 chart provides general guidance for the proper selection of eye and face protection to protect against hazards associated with the listed hazard "source" operations.

Table 6.8 Eye and face protection chart.

Source	**Assessment of Hazard**	**Protection**
IMPACT - Chipping, grinding machining, masonry work, woodworking, sawing, drilling, chiseling, powered fastening, riveting, and sanding.	Flying fragments, objects, large chips, particles sand, dirt, etc.	Spectacles with side protection, goggles, face shields. See notes 1, 3, 5, 6, 10. For severe exposure, use faceshield.
HEAT - Furnace operations, pouring, casting, hot dipping, and welding.	Hot sparks Splash from molten metals High temperature exposure	Faceshields, goggles, spectacles with side protection. For severe exposure use faceshield. See notes 1, 2, 3. Faceshields worn over goggles. See notes 1, 2, 3. Screen face shields, reflective face shields. See notes 1, 2, 3.
CHEMICALS - Acid and chemicals handling, degreasing plating.	Splash Irritating mists	Goggles, eyecup and cover types. For severe exposure, use face shield. See notes 3, 11. Special-purpose goggles.
DUST - Woodworking, buffing, general dusty conditions.	Nuisance dust	Goggles, eyecup and cover types. See note 8.

Table 6.8 Eye and face protection chart.

Source	Assessment of Hazard	Protection
LIGHT and/or RADIATION Welding: Electric arc Welding: Gas Cutting, Torch brazing, Torch soldering Glare	Optical radiation Optical radiation Optical radiation Poor vision	Welding helmets or welding shields. Typical shades: 10-14. See notes 9, 12. Welding goggles or welding face shield. Typical shades: gas welding 4-8, cutting 3-6, brazing 3-4. See note 9. Spectacles or welding face shield. Typical shades, 1, 5-3. See notes 3, 9. Spectacles with shaded or special-purpose lenses, as suitable. See notes 9, 10.

Notes to Eye and Face Protection Selection Chart:

1. Care should be taken to recognize the possibility of multiple and simultaneous exposure to a variety of hazards. Adequate protection against the highest level of each of the hazards should be provided. Protective devices do not provide unlimited protection.

2. Operations involving heat may also involve light radiation. As required by the standard, protection from both hazards must be provided.

3. Face shields should only be worn over primary eye protection (spectacles or goggles).

4. As required by the standard, filter lenses must meet the requirements for shade designations in 1910.133.1.5. Tinted and shaded lenses are not filter lenses unless they are marked or identified as such.

5. As required by the standard, persons whose vision requires the use of prescription (Rx) lenses must wear either protective devices fitted with prescription (Rx) lenses or protective devices designed to be worn over regular prescription (Rx) eyewear.

6. Wearers of contact lenses must also wear appropriate eye and face protection devices in a hazardous environment. It should be recognized that dusty and/or chemical environments may represent an additional hazard to contact lens wearers.

Table 6.8 (cont.) Eye and face protection chart.

7. Caution should be exercised in the use of metal frame protective devices in electrical hazard areas.

8. Atmospheric conditions and the restricted ventilation of the protector can cause lenses to fog. Frequent cleansing may be necessary.

9. Welding helmets or face shields should be used only over primary eye protection (spectacles or goggles).

10. Non-side shield spectacles are available for frontal protection only, but are not acceptable eye protection for the sources and operations listed for "impact."

11. Ventilation should be adequate, but well protected from splash entry. Eye and face protection should be designed and used so that it provides both adequate ventilation and protects the wearer from splash entry.

12. Protection from light radiation is directly related to filter lens density. See note 4. Select the darkest shade that allows task performance.

6.5.2 Head Protection

Head protection (helmets) is designed to provide protection from impact and penetration hazards caused by falling objects. Head protection is also available which provides protection from electric shock and burn. When selecting head protection, knowledge of potential electrical hazards is important. Class A helmets, in addition to impact and penetration resistance, provide electrical protection from low-voltage conductors (they are proof tested to 2,200 volts). Class B helmets, in addition to impact and penetration resistance, provide electrical protection from high-voltage conductors (they are proof tested to 20,000 volts). Class C helmets provide impact and penetration resistance (they are usually made of aluminum which conducts electricity), and should not be used around electrical hazards.

Where falling object hazards are present, helmets must be worn. Some examples include: working below other workers who are using tools and materials which could fall; working around or under conveyor belts which are carrying parts or materials; working below machinery or processes which might cause material or objects to fall; and working on exposed energized conductors. Some examples of occupations for which head protection should be routinely considered are: carpenters, electricians, linemen, mechanics and repairers, plumbers and pipe fitters, assemblers, packers, wrappers, sawyers, welders, laborers, freight handlers, timber cutting and logging, stock handlers, and warehouse laborers.

6.5.3 Foot Protection

Safety shoes and boots which meet the ANSI Z41-1991 Standard provide both impact and compression protection. Where necessary, safety shoes can be obtained which provide puncture protection. In some work situations, metatarsal protection should be provided, and in other special situations electrical conductive or insulating safety shoes would be appropriate.

Safety shoes or boots with impact protection would be required for carrying or handling materials such as packages, objects, parts or heavy tools, which could be dropped; and, for other activities where objects might fall onto the feet. Safety shoes or boots with compression protection would be required for work activities involving skid trucks (manual material handling carts) around bulk rolls (such as paper rolls) and around heavy pipes, all of which could potentially roll over an employee's feet. Safety shoes or boots with puncture protection would be required where sharp objects such as nails, wire, tacks, screws, large staples, scrap metal etc., could be stepped on by employees causing a foot injury.

Some occupations (not a complete list) for which foot protection should be routinely considered are: shipping and receiving clerks, stock clerks, carpenters, electricians, machinists, mechanics and repairers, plumbers and pipe fitters, structural metal workers, assemblers, drywall installers and lathers, packers, wrappers, craters, punch and stamping press operators, sawyers, welders, laborers, freight handlers, gardeners and grounds-keepers, timber cutting and logging workers, stock handlers and warehouse laborers.

6.5.4 Hand Protection

Gloves are often relied upon to prevent cuts, abrasions, burns, and skin contact with chemicals that are capable of causing local or systemic effects following dermal exposure. OSHA is unaware of any gloves that provide protection against all potential hand hazards, and commonly available glove materials provide only limited protection against many chemicals. Therefore, it is important to select the most appropriate glove for a particular application and to determine how long it can be worn, and whether it can be reused.

It is also important to know the performance characteristics of gloves relative to the specific hazard anticipated; e.g., chemical hazards, cut hazards, flame hazards, etc. These performance characteristics should be assessed by using standard test procedures. Before purchasing gloves, the employer should request documentation from the manufacturer that the gloves meet the appropriate test standard(s) for the hazard(s) anticipated. Other factors to be considered for glove selection in general include:

1. As long as the performance characteristics are acceptable, in certain circumstances, it may be more cost effective to regularly change cheaper gloves than to reuse more expensive types; and
2. The work activities of the employee should be studied to determine the degree of dexterity required, the duration, frequency, and degree of exposure of the hazard, and the physical stresses that will be applied.

With respect to selection of gloves for protection against chemical hazards:

1. The toxic properties of the chemical(s) must be determined; in particular, the ability of the chemical to cause local effects on the skin and/or to pass through the skin and cause systemic effects;
2. Generally, any "chemical resistant" glove can be used for dry powders;
3. For mixtures and formulated products (unless specific test data are available), a glove should be selected on the basis of the chemical component with the shortest breakthrough time, since it is possible for solvents to carry active ingredients through polymeric materials; and
4. Employees must be able to remove the gloves in such a manner as to prevent skin contamination.

6.5.5 Cleaning and Maintenance

It is important that all PPE be kept clean and properly maintained. Cleaning is particularly important for eye and face protection where dirty or fogged lenses could impair vision. PPE should be inspected, cleaned, and maintained at regular intervals so that the PPE provides the requisite protection. It is also important to ensure that contaminated PPE which cannot be decontaminated is disposed of in a manner that protects employees from exposure to hazards.

6.6 Sanitation

6.6.1 Definitions

Non-water carriage toilet facility means a toilet facility not connected to a sewer.

Number of employees, unless otherwise specified, means the maximum number of employees present at any one time on a regular shift.

Personal service room means a room used for activities not directly connected with the production or service function performed by the establishment. Such activities include, but are not limited to, first-aid, medical services, dressing, showering, toilet use, washing, and eating.

Potable water means water which meets the quality standards prescribed in the U.S. Public health service drinking water standards, published in 42 CFR Part 72, or water which is approved for drinking purposes by the state or local authority having jurisdiction.

Toilet facility means a fixture maintained within a toilet room for the purpose of defecation or urination, or both.

Toilet room means a room maintained within or on the premises of any place of employment, containing toilet facilities for use by employees.

Toxic material means a material in concentration or amount which exceeds the applicable limit established by a standard, such as 1910.1000 and 1910.1001 or, in the absence of an applicable standard, which is of such toxicity so as to constitute a recognized hazard that is causing or is likely to cause death or serious physical harm.

Urinal means a toilet facility maintained within a toilet room for the sole purpose of urination.

Water closet means a toilet facility maintained within a toilet room for the purpose of both defecation and urination and which is flushed with water.

Wet process means any process or operation in a workroom which normally results in surfaces upon which employees may walk or stand becoming wet.

6.6.2 Housekeeping

All work places shall be kept clean to the extent that the nature of the work allows.

The floor of every workroom shall be maintained, so far as practicable, in a dry condition. Where wet processes are used, drainage shall be maintained and false floors, platforms, mats, or other dry standing places shall be provided, where practicable, or appropriate waterproof footgear shall be provided.

To facilitate cleaning, every floor, working place, and passageway shall be kept free from protruding nails, splinters, loose boards, and unnecessary holes and openings.

6.6.3 Waste Disposal

Any receptacle used for putrescible solid or liquid waste or refuse shall be so constructed that it does not leak and may be thoroughly cleaned and maintained in a sanitary condition. Such a receptacle shall be equipped with a solid tight-fitting cover, unless it can be maintained in a sanitary condition without a cover. This requirement does not prohibit the use of receptacles which are designed to permit the maintenance of a sanitary condition without regard to the aforementioned requirements.

All sweepings, solid or liquid wastes, refuse, and garbage shall be removed in such a manner as to avoid creating a menace to health and as often as necessary or appropriate to maintain the place of employment in a sanitary condition.

Vermin control. Every enclosed workplace shall be so constructed, equipped, and maintained, so far as reasonably practicable, as to prevent the entrance or harborage of rodents, insects, and other vermin. A continuing and effective extermination program shall be instituted where their presence is detected.

6.6.4 Water Supply

3.6.4.1 Potable Water

Potable water shall be provided in all places of employment, for drinking, washing of the person, cooking, washing of foods, washing of cooking or eating utensils, washing of food preparation or processing premises, and personal service rooms.

Portable drinking water dispensers shall be designed, constructed, and serviced so that sanitary conditions are maintained, shall be capable of being closed, and shall be equipped with a tap.

Open containers such as barrels, pails, or tanks for drinking water from which the water must be dipped or poured, whether or not they are fitted with a cover, are prohibited.

A common drinking cup and other common utensils are prohibited.

6.6.4.2 *Non-potable Water*

Outlets for non-potable water, such as water for industrial or fire-fighting purposes, shall be posted or otherwise marked in a manner that will indicate clearly that the water is unsafe and is not to be used for drinking, washing of the person, cooking, washing of food, washing of cooking or eating utensils, washing of food preparation or processing premises, or personal service rooms, or for washing clothes.

Construction of non-potable water systems or systems carrying any other non-potable substance shall be such as to prevent back-flow or back-siphonage into a potable water system.

Non-potable water shall not be used for washing any portion of the person, cooking or eating utensils, or clothing. Non-potable water may be used for cleaning work premises, other than food processing and preparation premises and personal service rooms: Provided that this non-potable water does not contain concentrations of chemicals, fecal coliform, or other substances which could create unsanitary conditions or be harmful to employees.

6.6.5 Toilet Facilities

Except as otherwise indicated in toilet rooms separate for each sex, shall be provided in all places of employment in accordance with Table 6.9 of this section. The number of facilities to be provided for each sex shall be based on the number of employees of that sex for whom the facilities are furnished. Toilet rooms are to be occupied by no more than one person at a time, can be locked from the inside, and contain at least one water closet, separate toilet rooms for each sex need not be provided. Where such single-occupancy rooms have more than one toilet facility, only one such facility in each toilet room shall be counted for the purpose of Table 6.9.

The requirements of this section do not apply to mobile crews or to normally unattended work locations so long as employees

Table 6.9 Minimum number of water closets.

Number of Employees	Minimum Number of Water Closets.[1]
1 to 15	1
16 to 35	2
36 to 55	3
56 to 80	4
81 to 110	5
111 to 150	6
Over 150	2

1. Where toilet facilities will not be used by women, urinals may be provided instead of water closets, except that the number of water closets in such cases shall not be reduced to less than 2/3 of the minimum specified.

2. One additional fixture for each additional 40 employees.

working at these locations have transportation immediately available to nearby toilet facilities which meet the other requirements of this subparagraph.

The sewage disposal method shall not endanger the health of employees.

6.6.6 Construction of Toilet Rooms

Each water closet shall occupy a separate compartment with a door and walls or partitions between fixtures sufficiently high to assure privacy.

6.6.7 Washing Facilities

Washing facilities shall be maintained in a sanitary condition.

Lavatories shall be made available in all places of employment. The requirements of this subdivision do not apply to mobile crews or to normally unattended work locations if employees working at these locations have transportation readily available to nearby washing facilities which meet the other requirements of this paragraph.

Each lavatory shall be provided with hot and cold running water, or tepid running water.

Hand soap or similar cleansing agents shall be provided.

Individual hand towels or sections thereof, of cloth or paper, warm air blowers or clean individual sections of continuous cloth toweling, convenient to the lavatories, shall be provided.

6.6.8 Showers

Whenever showers are required the following guidelines are recommended:

- One shower shall be provided for each 10 employees of each sex, or numerical fraction thereof, who are required to shower during the same shift.
- Body soap or other appropriate cleansing agents convenient to the showers shall be provided.
- Showers shall be provided with hot and cold water feeding a common discharge line.
- Employees who use showers shall be provided with individual clean towels.

6.6.9 Change Rooms

Whenever employees are required by a particular standard to wear protective clothing because of the possibility of contamination with toxic materials, change rooms equipped with storage facilities for street clothes and separate storage facilities for the protective clothing shall be provided.

6.6.10 Clothes Drying Facilities

Where working clothes are provided by the employer and become wet or are washed between shifts, provision shall be made to ensure that such clothing is dry before reuse.

6.6.11 Consumption of Food and Beverages

No employee shall be allowed to consume food or beverages in a toilet room, nor in any area exposed to a toxic material.

6.6.12 Waste Disposal Containers

Receptacles constructed of smooth, corrosion resistant, easily cleanable, or disposable materials, shall be provided and used for the disposal of waste food. The number, size, and location of such receptacles shall encourage their use and not result in overfilling. They shall be emptied not less frequently than once each working day, unless unused, and shall be maintained in a clean and sanitary condition. Receptacles shall be provided with a solid tight-fitting cover unless sanitary conditions can be maintained without use of a cover.

6.6.13 Sanitary Storage

No food or beverages shall be stored in toilet rooms or in an area exposed to a toxic material.

6.6.14 Food Handling

All employee food service facilities and operations shall be carried out in accordance with sound hygienic principles. In all places of employment where all or part of the food service is provided, the food dispensed shall be wholesome, free from spoilage, and shall be processed, prepared, handled, and stored in such a manner as to be protected against contamination.

6.7 Safety Color Code for Marking Physical Hazards

The color Red shall be the basic color for the identification of fire protection equipment and apparatus.

Safety cans or other portable containers of flammable liquids having a flash point at or below 80 degrees Fahrenheit, table containers of flammable liquids (open cup tester), excluding shipping containers, shall be painted red with some additional clearly visible identification either in the form of a yellow band around the can or the name of the contents conspicuously stenciled or painted on the can in yellow. Red lights shall be provided at barricades and at temporary obstructions, as specified in ANSI Safety Code for Building Construction, A10.2-1944, which is incorporated by reference as specified in Sec. 1910.6. Danger signs shall be painted red.

Emergency stop bars on hazardous machines such as rubber mills, wire blocks, flat work ironers, etc., shall be red. Stop buttons or electrical switches which letters or other markings appear, used for emergency stopping of machinery shall be red.

Yellow shall be the basic color for designating caution and for marking physical hazards such as: Striking against, stumbling, falling, tripping, and "caught in between."

6.8 Specifications for Accident Prevention Signs and Tags

6.8.1 General

These specifications apply to the design, application, and use of signs or symbols intended to indicate and, in so far as possible, to define specific hazards of a nature such that failure to designate them may lead to accidental injury to workers or the public, or both, or to property damage. These specifications are intended to cover all safety signs except those designed for streets, highways, railroads, and marine regulations. These specifications do not apply to plant bulletin boards or to safety posters.

All new signs and replacements of old signs shall be in accordance with these specifications.

As used in this section, the word "sign" refers to a surface prepared for the warning of, or safety instructions of, industrial workers or members of the public who may be exposed to hazards.

Excluded from this definition, however, are news releases, displays commonly known as safety posters, and bulletins used for employee education.

6.8.2 Classification of Signs

Classification of signs according to use:

Danger signs - There shall be no variation in the type of design of signs posted to warn of specific dangers and radiation hazards.

All employees shall be instructed that danger signs indicate immediate danger and that special precautions are necessary.

Caution signs - Caution signs shall be used only to warn against potential hazards or to caution against unsafe practices.

All employees shall be instructed that caution signs indicate a possible hazard against which proper precaution should be taken.

Safety instruction signs - Safety instruction signs shall be used where there is a need for general instructions and suggestions relative to safety measures.

6.8.3 Sign Design

All signs shall be furnished with rounded or blunt corners and shall be free from sharp edges, burrs, splinters, or other sharp projections. The ends or heads of bolts or other fastening devices shall be located in such a way that they do not constitute a hazard.

6.8.3.1 Danger Signs

The colors red, black, and white shall be those of opaque glossy samples as specified in Table 1 of Fundamental Specification of Safety Colors for CIE Standard Source "C", American National Standard Z53.1-1967, which is incorporated by reference as specified in Sec. 1910.6.

6.8.3.2 Caution Signs

Standard color of the background shall be yellow; and the panel, black with yellow letters. Any letters used against the yellow background shall be black. The colors shall be those of opaque glossy samples as specified in Table 1 of American National Standard Z53.1-1967.

Safety instruction signs - Standard color of the background shall be white; and the panel, green with white letters. Any letters used against the white background shall be black. The colors shall be those of opaque glossy samples as specified in Table 1 of American National Standard, Z53.1-1967.

6.8.3.3 Sign Wordings

The wording of any sign should be easily read and concise. The sign should contain sufficient information to be easily understood. The wording should make a positive, rather than negative suggestion and should be accurate in fact.

6.8.4 Biological Hazard Signs

The biological hazard warning shall be used to signify the actual or potential presence of a biohazard and to identify equipment, containers, rooms, materials, experimental animals, or combinations thereof, which contain, or are contaminated with, viable hazardous agents. For the purpose of this subparagraph the term "biological hazard," or "biohazard," shall include only those infectious agents presenting a risk or potential risk to the well-being of man.

6.8.5 Accident Prevention Tags

This paragraph applies to all accident prevention tags used to identify hazardous conditions and provide a message to employees with respect to hazardous conditions.

The following definitions are used:

- **"Biological hazard"** or **"BIOHAZARD"** mean those infectious agents presenting a risk of death, injury or illness to employees.
- **"Major message"** means that a portion of a tag's inscription that is more specific than the signal word and that indicates the specific hazardous condition or the instruction to be communicated to the employee. Examples include: "High Voltage," "Close Clearance," "Do Not Start," or "Do Not Use" or a corresponding pictograph used with a written text or alone.
- **"Pictograph"** means a pictorial representation used to identify a hazardous condition or to convey a safety instruction.
- **"Signal word"** means that portion of a tag's inscription that contains the word or words that are intended to capture the employee's immediate attention.
- **"Tag"** means a device usually made of card, paper, pasteboard, plastic or other material used to identify a hazardous condition.

Tags shall be used as a means to prevent accidental injury or illness to employees who are exposed to hazardous or potentially hazardous conditions, equipment or operations which are out of the ordinary, unexpected or not readily apparent. Tags shall be used until such time as the identified hazard is eliminated or the hazardous operation is completed. Tags need not be used where signs, guarding or other positive means of protection are being used.

All required tags shall meet the following criteria:

- Tags shall contain a signal word and a major message.
- The signal word shall be either "Danger," "Caution," or "Biological Hazard," "BIOHAZARD," or the biological hazard symbol.

The major message shall indicate the specific hazardous condition or the instruction to be communicated to the employee.

The signal word shall be readable at a minimum distance of 1.52 m or such greater distance as warranted by the hazard.

The tag's major message shall be presented in either pictographs, written text or both.

The signal word and the major message shall be understandable to all employees who may be exposed to the identified hazard.

All employees shall be informed as to the meaning of the various tags used throughout the workplace and what special precautions are necessary.

Tags shall be affixed as close as safely possible to their respective hazards by a positive means such as string, wire, or adhesive that prevents their loss or unintentional removal.

Danger tags shall be used in major hazard situations where an immediate hazard presents a threat of death or serious injury to employees. Danger tags shall be used only in these situations.

Caution tags shall be used in minor hazard situations where a non-immediate or potential hazard or unsafe practice presents a lesser threat of employee injury. Caution tags shall be used only in these situations.

Warning tags may be used to represent a hazard level between "Caution" and "Danger," instead of the required "Caution" tag, provided that they have a signal word of "Warning," an appropriate major message, and otherwise meet the general tag criteria of paragraph 6.4 of this section.

Biological hazard tags shall be used to identify the actual or potential presence of a biological hazard and to identify equipment, containers, rooms, experimental animals, or combinations thereof, that contain or are contaminated with hazardous biological agents.

6.9 Permits for Confined Spaces

6.9.1 Definitions

This section contains additional information and requirements for practices and procedures to protect employees from the hazards of entry into permit-required confined spaces. The following are definitions derived from the U.S. OSHA standards to help orient workers and employees.

Acceptable entry conditions means the conditions that must exist in a permit space to allow entry and to ensure that employees involved with a permit-required confined space entry can safely enter into and work within the space.

Attendant means an individual stationed outside one or more permit spaces who monitors the authorized entrants and who performs all attendant's duties assigned in the employer's permit space program.

Authorized entrant means an employee who is authorized by the employer to enter a permit space.

Blanking or blinding means the absolute closure of a pipe, line, or duct by the fastening of a solid plate (such as a spectacle blind or a skillet blind) that completely covers the bore and that is capable of withstanding the maximum pressure of the pipe, line, or duct with no leakage beyond the plate.

Confined space means a space that:

1. Is large enough and so configured that an employee can bodily enter and perform assigned work; and
2. Has limited or restricted means for entry or exit (for example, tanks, vessels, silos, storage bins, hoppers, vaults, and pits are spaces that may have limited means of entry.); and
3. Is not designed for continuous employee occupancy.

Double block and bleed means the closure of a line, duct, or pipe by closing and locking or tagging two in-line valves and by opening and locking or tagging a drain or vent valve in the line between the two closed valves.

Emergency means any occurrence (including any failure of hazard control or monitoring equipment) or event internal or external to the permit space that could endanger entrants.

Engulfment means the surrounding and effective capture of a person by a liquid or finely divided (flowable) solid substance that can be aspirated to cause death by filling or plugging the respiratory system or that can exert enough force on the body to cause death by strangulation, constriction, or crushing.

Entry means the action by which a person passes through an opening into a permit-required confined space. Entry includes ensuing work activities in that space and is considered to have occurred as soon as any part of the entrant's body breaks the plane of an opening into the space.

Entry permit (permit) means the written or printed document that is provided by the employer to allow and control entry into a permit space and that contains information specified in this section.

Entry supervisor means the person (such as the employer, foreman, or crew chief) responsible for determining if acceptable entry conditions are present at a permit space where entry is planned, for authorizing entry and overseeing entry operations, and for terminating entry as required by this section. An entry supervisor also may serve as an attendant or as an authorized entrant, as long as that person is trained and equipped as required by this section for each role he or she fills. Also, the duties of entry supervisor may be passed from one individual to another during the course of an entry operation.

Hazardous atmosphere means an atmosphere that may expose employees to the risk of death, incapacitation, and impairment of ability to self-rescue (that is, escape unaided from a permit space), injury, or acute illness from one or more of the following causes:

1. Flammable gas, vapor, or mist in excess of 10 percent of its lower flammable limit (LFL).
2. Airborne combustible dust at a concentration that meets or exceeds its LFL; this concentration may be approximated as a condition in which the dust obscures vision at a distance of 1.52 m or less.
3. Atmospheric oxygen concentration below 19.5 percent or above 23.5 percent.
4. Atmospheric concentration of any substance for which a dose or a permissible exposure limit which could result in employee exposure in excess of safe levels of exposure. For exposure limits see Part II .5. An atmospheric concentration of any substance that

is not capable of causing death, incapacitation, and impairment of ability to self-rescue, injury, or acute illness due to its health effects is not covered by this provision.

5. Any other atmospheric condition that is immediately dangerous to life or health. For air contaminants which do not have established doses or permissible exposure limits, other sources of information, such as Material Safety Data Sheets, published information, and internal documents can provide guidance in establishing acceptable atmospheric conditions.

Hot work permit means the employer's written authorization to perform operations (for example, riveting, welding, cutting, burning, and heating) capable of providing a source of ignition.

Immediately dangerous to life or health (IDLH) means any condition that poses an immediate or delayed threat to life or that would cause irreversible adverse health effects or that would interfere with an individual's ability to escape unaided from a permit space. Some materials -- hydrogen fluoride gas and cadmium vapor, for example -- may produce immediate transient effects that, even if severe, may pass without medical attention, but are followed by sudden, possibly fatal collapse 12-72 hours after exposure. The victim "feels normal" from recovery from transient effects until collapse. Such materials in hazardous quantities are considered to be "immediately" dangerous to life or health.

Inerting means the displacement of the atmosphere in a permit space by a noncombustible gas (such as nitrogen) to such an extent that the resulting atmosphere is noncombustible. It is important to note that inerting produces an IDLH oxygen-deficient atmosphere.

Isolation means the process by which a permit space is removed from service and completely protected against the release of energy and material into the space by such means as: blanking or blinding; misaligning or removing sections of lines, pipes, or ducts; a double block and bleed system; lockout or tagout of all sources of energy; or blocking or disconnecting all mechanical linkages.

Line breaking means the intentional opening of a pipe, line, or duct that is or has been carrying flammable, corrosive, or toxic material, an inert gas, or any fluid at a volume, pressure, or temperature capable of causing injury.

Non-permit confined space means a confined space that does not contain or, with respect to atmospheric hazards, have the potential to contain any hazard capable of causing death or serious physical harm.

Oxygen deficient atmosphere means an atmosphere containing less than 19.5 percent oxygen by volume.

Oxygen enriched atmosphere means an atmosphere containing more than 23.5 percent oxygen by volume.

Permit-required confined space (permit space) means a confined space that has one or more of the following characteristics:

1. Contains or has a potential to contain a hazardous atmosphere;
2. Contains a material that has the potential for engulfing an entrant;
3. Has an internal configuration such that an entrant could be trapped or asphyxiated by inwardly converging walls or by a floor which slopes downward and tapers to a smaller cross-section; or
4. Contains any other recognized serious safety or health hazard.

Permit-required confined space program (permit space program) means the employer's overall program for controlling, and, where appropriate, for protecting employees from, permit space hazards and for regulating employee entry into permit spaces.

Permit system means the employer's written procedure for preparing and issuing permits for entry and for returning the permit space to service following termination of entry.

Prohibited condition means any condition in a permit space that is not allowed by the permit during the period when entry is authorized.

Rescue service means the personnel designated to rescue employees from permit spaces.

Retrieval system means the equipment (including a retrieval line, chest or full-body harness, wristlets, if appropriate, and a lifting device or anchor) used for non-entry rescue of persons from permit spaces.

Testing means the process by which the hazards that may confront entrants of a permit space are identified and evaluated. Testing includes specifying the tests that are to be performed in the permit space. Testing enables employers both to devise and implement adequate control measures for the protection of authorized entrants and to determine if acceptable entry conditions are present immediately prior to, and during, entry.

6.9.2 General Comments and Requirements

The employer shall evaluate the workplace to determine if any spaces are permit-required confined spaces.

If the workplace contains permit spaces, the employer shall inform exposed employees, by posting danger signs or by any other equally effective means, of the existence and location of and the danger posed by the permit spaces.

A sign reading **DANGER – PERMIT-REQUIRED CONFINED SPACE, DO NOT ENTER** or using other similar language would satisfy the requirement for a sign.

If the employer decides that its employees will not enter permit spaces, the employer shall take effective measures to prevent its employees from entering the permit spaces.

If the employer decides that its employees will enter permit spaces, the employer shall develop and implement a written permit space program that complies with this section. The written program shall be available for inspection by employees and their authorized representatives.

An employer whose employees enter a permit space need not comply with the recommendations of this section provided that:

- The employer can demonstrate that the only hazard posed by the permit space is an actual or potential hazardous atmosphere;
- The employer can demonstrate that continuous forced air ventilation alone is sufficient to maintain that permit space safe for entry;
- The employer develops monitoring and inspection data that supports the demonstrations required in this section;
- The determinations and supporting data required are documented by the employer and are made available to each employee who enters the permit space; and
- Entry into the permit space is performed in accordance with the requirements of this section.

The following requirements apply to entry into permit spaces that meet the conditions set forth in this section.

- Any conditions making it unsafe to remove an entrance cover shall be eliminated before the cover is removed.
- When entrance covers are removed, the opening shall be promptly guarded by a railing, temporary cover, or other temporary barrier that will prevent an accidental fall through the opening and that will protect each employee working in the space from foreign objects entering the space.
- Before an employee enters the space, the internal atmosphere shall be tested, with a calibrated direct-reading instrument, for oxygen content, for flammable gases and vapors, and for potential toxic air contaminants, in that order. Any employee who enters the space, or that employee's authorized representative, shall be provided an opportunity to observe the pre-entry testing required by this paragraph.

6.9.3 Oxygen Content, Flammable Gases and Vapors, and Potential Toxic Air Contaminants

There may be no hazardous atmosphere within the space whenever any employee is inside the space.

Continuous forced air ventilation shall be used, as follows:

- An employee may not enter the space until the forced air ventilation has eliminated any hazardous atmosphere.
- The forced air ventilation shall be so directed as to ventilate the immediate areas where an employee is or will be present within the space and shall continue until all employees have left the space.
- The air supply for the forced air ventilation shall be from a clean source and may not increase the hazards in the space.
- The atmosphere within the space shall be periodically tested as necessary to ensure that the continuous forced air ventilation is preventing the accumulation

of a hazardous atmosphere. Any employee who enters the space, or that employee's authorized representative, shall be provided with an opportunity to observe the periodic testing required by this paragraph.

- If a hazardous atmosphere is detected during entry.
- Each employee shall leave the space immediately.
- The space shall be evaluated to determine how the hazardous atmosphere developed; and
- Measures shall be implemented to protect employees from the hazardous atmosphere before any subsequent entry takes place.

The employer shall verify that the space is safe for entry and that the pre-entry measures required of this section have been taken, through a written certification that contains the date, the location of the space, and the signature of the person providing the certification. The certification shall be made before entry and shall be made available to each employee entering the space or to that employee's authorized representative.

When there are changes in the use or configuration of a non-permit confined space that might increase the hazards to entrants, the employer shall reevaluate that space and, if necessary, reclassify it as a permit-required confined space.

A space classified by the employer as a permit-required confined space may be reclassified as a non-permit confined space under the following procedures:

- If the permit space poses no actual or potential atmospheric hazards and if all hazards within the space are eliminated without entry into the space, the permit space may be reclassified as a non-permit confined space for as long as the non-atmospheric hazards remain eliminated.
- If it is necessary to enter the permit space to eliminate hazards, such entry shall be performed under the recommendations of this section. If testing and inspection during that entry demonstrate that the hazards within the permit space have been eliminated, the permit space may be reclassified as a non-permit confined space for as long as the hazards remain eliminated. Control of atmospheric hazards through forced air ventilation does not constitute elimination of the hazards.

The employer shall document the basis for determining that all hazards in a permit space have been eliminated, through a certification that contains the date, the location of the space, and the signature of the person making the determination. The certification shall be made available to each employee entering the space or to that employee's authorized representative.

If hazards arise within a permit space that has been declassified to a non-permit space, each employee in the space shall exit the space. The employer shall then reevaluate the space and determine whether it must be reclassified as a permit space, in accordance with other applicable provisions of this section.

When an employer (host employer) arranges to have employees of another employer (contractor) perform work that involves permit space entry, the host employer shall:

- Inform the contractor that the workplace contains permit spaces and that permit space entry is allowed only through compliance with a permit space program meeting the requirements of this section;
- Apprise the contractor of the elements, including the hazards identified and the host employer's experience with the space, that make the space in question a permit space;
- Apprise the contractor of any precautions or procedures that the host employer has implemented for the protection of employees in or near permit spaces where contractor personnel will be working;
- Coordinate entry operations with the contractor, when both host employer personnel and contractor personnel will be working in or near permit spaces; and
- Debrief the contractor at the conclusion of the entry operations regarding the permit space program followed and regarding any hazards confronted or created in permit spaces during entry operations.

In addition to complying with the permit space requirements that apply to all employers, each contractor who is retained to perform permit space entry operations shall:

- Obtain any available information regarding permit space hazards and entry operations from the host employer;

- Coordinate entry operations with the host employer, when both host employer personnel and contractor personnel will be working in or near permit spaces; and
- Inform the host employer of the permit space program that the contractor will follow and of any hazards confronted or created in permit spaces, either through a debriefing or during the entry operation.

6.9.4 Permit Space Program

6.9.4.1 General Details

The employer shall:

- Implement the measures necessary to prevent unauthorized entry;
- Identify and evaluate the hazards of permit spaces before employees enter them;
- Develop and implement the means, procedures, and practices necessary for safe permit space entry operations, including, but not limited to, the following:
  - Specifying acceptable entry conditions providing each authorized entrant or that employee's authorized representative with the opportunity to observe any monitoring or testing of permit spaces;
  - Isolate the permit space;
  - Perform the following: Purging, inerting, flushing, or ventilating the permit space as necessary to eliminate or control atmospheric hazards;
  - Provide pedestrian, vehicle, or other barriers as necessary to protect entrants from external hazards; and
  - Verify that conditions in the permit space are acceptable for entry throughout the duration of an authorized entry.

The employer shall provide the following equipment at no cost to employees, maintain that equipment properly, and ensure that employees use that equipment properly:

- Testing and monitoring equipment needed to comply with this section;

- Ventilating equipment needed to obtain acceptable entry conditions;
- Communications equipment necessary for compliance with this section;
- Personal protective equipment insofar as feasible engineering and work practice controls do not adequately protect employees;
- Lighting equipment needed to enable employees to see well enough to work safely and to exit the space quickly in an emergency;
- Barriers and shields as required by this section;
- Equipment, such as ladders, needed for safe ingress and egress by authorized entrants;
- Rescue and emergency equipment needed to comply with this section, except to the extent that the equipment is provided by rescue services; and
- Any other equipment necessary for safe entry into and rescue from permit spaces.

The employer shall evaluate permit space conditions as follows when entry operations are conducted:

- Test conditions in the permit space to determine if acceptable entry conditions exist before entry is authorized to begin, except that, if isolation of the space is infeasible because the space is large or is part of a continuous system (such as a sewer), pre-entry testing shall be performed to the extent feasible before entry is authorized and, if entry is authorized, entry conditions shall be continuously monitored in the areas where authorized entrants are working;
- Test or monitor the permit space as necessary to determine if acceptable entry conditions are being maintained during the course of entry operations; and
- When testing for atmospheric hazards, test first for oxygen, then for combustible gases and vapors, and then for toxic gases and vapors.

The employer shall provide each authorized entrant or that employee's authorized representative an opportunity to observe the pre-entry and any subsequent testing or monitoring of permit spaces.

The employer shall reevaluate the permit space in the presence of any authorized entrant or that employee's authorized representative who requests that the employer conduct such reevaluation because the entrant or representative has reason to believe that the evaluation of that space may not have been adequate.

The employer shall immediately provide each authorized entrant or that employee's authorized representative with the results of any testing conducted.

6.9.4.2 Attendant

The employer shall provide at least one attendant outside the permit space into which entry is authorized for the duration of entry operations. Attendants may be assigned to monitor more than one permit space provided the duties of this section can be effectively performed for each permit space that is monitored. Likewise, attendants may be stationed at any location outside the permit space to be monitored as long as the duties described in this section can be effectively performed for each permit space that is monitored.

If multiple spaces are to be monitored by a single attendant, include in the permit program the means and procedures to enable the attendant to respond to an emergency affecting one or more of the permit spaces being monitored without distraction from the attendant's responsibilities.

Designate the persons who are to have active roles (as, for example, authorized entrants, attendants, entry supervisors, or persons who test or monitor the atmosphere in a permit space) in entry operations, identify the duties of each such employee, and provide each such employee with the training required.

Employers should develop and implement procedures for summoning rescue and emergency services, for rescuing entrants from permit spaces, for providing necessary emergency services to rescued employees, and for preventing unauthorized personnel from attempting a rescue.

Employers should develop and implement a system for the preparation, issuance, use, and cancellation of entry permits as required by this section.

Employers should develop and implement procedures to coordinate entry operations when employees of more than one employer are working simultaneously as authorized entrants in a permit space, so that employees of one employer do not endanger the employees of any other employer.

Employers should develop and implement procedures (such as closing off a permit space and canceling the permit) necessary for concluding the entry after entry operations have been completed.

Employers should review entry operations when the employer has reason to believe that the measures taken under the permit space program may not protect employees and revise the program to correct deficiencies found to exist before subsequent entries are authorized. Examples of circumstances requiring the review of the permit space program are: any unauthorized entry of a permit space, the detection of a permit space hazard not covered by the permit, the detection of a condition prohibited by the permit, the occurrence of an injury or near-miss during entry, a change in the use or configuration of a permit space, and employee complaints about the effectiveness of the program.

Employers should review the permit space program, using the canceled permits retained within 1 year after each entry and revise the program as necessary, to ensure that employees participating in entry operations are protected from permit space hazards. Employers may perform a single annual review covering all entries performed during a 12-month period. If no entry is performed during a 12-month period, no review is necessary.

6.9.4.3 The Permit System

Before entry is authorized, the employer shall document the completion of measures required by this section by preparing an entry permit.

Before entry begins, the entry supervisor identified on the permit shall sign the entry permit to authorize entry.

The completed permit shall be made available at the time of entry to all authorized entrants or their authorized representatives, by posting it at the entry portal or by any other equally effective means, so that the entrants can confirm that pre-entry preparations have been completed.

The duration of the permit may not exceed the time required to complete the assigned task or job identified on the permit.

The entry supervisor shall terminate entry and cancel the entry permit when:

- The entry operations covered by the entry permit have been completed; or
- A condition that is not allowed under the entry permit arises in or near the permit space.

The employer shall retain each canceled entry permit for at least 1 year to facilitate the review of the permit-required confined space program required of this section. Any problems encountered during an entry operation shall be noted on the pertinent permit so that appropriate revisions to the permit space program can be made.

6.9.4.4 Entry Permit

The entry permit that documents compliance with this section and authorizes entry to a permit space shall identify:

- The permit space to be entered.
- The purpose of the entry.
- The date and the authorized duration of the entry permit.
- The authorized entrants within the permit space, by name or by such other means (for example, through the use of rosters or tracking systems) as will enable the attendant to determine quickly and accurately, for the duration of the permit, which authorized entrants are inside the permit space. This requirement may be met by inserting a reference on the entry permit as to the means used, such as a roster or tracking system, to keep track of the authorized entrants within the permit space.
- The personnel, by name, currently serving as attendants.
- The individual, by name, currently serving as entry supervisor, with a space for the signature or initials of the entry supervisor who originally authorized entry.
- The hazards of the permit space to be entered.
- The measures used to isolate the permit space and to eliminate or control permit space hazards before entry.

Those measures can include the lockout or tagging of equipment and procedures for purging, inerting, ventilating, and flushing permit spaces.

- The acceptable entry conditions.
- The results of initial and periodic tests performed, accompanied by the names or initials of the testers and by an indication of when the tests were performed.
- The rescue and emergency services that can be summoned and the means (such as the equipment to use and the numbers to call) for summoning those services;
- The communication procedures used by authorized entrants and attendants to maintain contact during the entry.
- Equipment, such as personal protective equipment, testing equipment, communications equipment, alarm systems, and rescue equipment, to be provided for compliance with this section.
- Any other information whose inclusion is necessary, given the circumstances of the particular confined space, in order to ensure employee safety; and any additional permits, such as for hot work, that have been issued to authorize work in the permit space.

6.9.4.5 *Training*

The employer shall provide training so that all employees whose work is regulated acquire the understanding, knowledge, and skills necessary for the safe performance of the duties.

Training shall be provided to each affected employee:

- Before the employee is first assigned duties under this section;
- Before there is a change in assigned duties;
- Whenever there is a change in permit space operations that presents a hazard about which an employee has not previously been trained; and
- Whenever the employer has reason to believe either that there are deviations from the permit space entry procedures or that there are inadequacies in the employee's knowledge or use of these procedures.

The training shall be designed to establish employee proficiency in the duties required by this section and shall introduce new or revised procedures, as necessary, for compliance with this section.

The employer shall certify the training. The certification shall contain each employee's name, the signatures or initials of the trainers, and the dates of training. The certification shall be available for inspection by employees and their authorized representatives.

6.9.4.6 Duties of Authorized Entrants

The employer shall ensure that all authorized entrants:

- Know the hazards that may be faced during entry, including information on the mode, signs or symptoms, and consequences of the exposure;
- Properly use equipment;
- Communicate with the attendant as necessary to enable the attendant to monitor entrant status and to enable the attendant to alert entrants of the need to evacuate the space;
- Alert the attendant whenever:
  - The entrant recognizes any warning sign or symptom of exposure to a dangerous situation, or
  - The entrant detects a prohibited condition; and
  - Exit from the permit space as quickly as possible whenever:
    - An order to evacuate is given by the attendant or the entry supervisor,
    - The entrant recognizes any warning sign or symptom of exposure to a dangerous situation,
    - The entrant detects a prohibited condition, or
    - An evacuation alarm is activated.

6.9.4.7 Duties of Attendants

The employer shall ensure that each attendant:

- Knows the hazards that may be faced during entry, including information on the mode, signs or symptoms, and consequences of the exposure;

- Is aware of possible behavioral effects of hazard exposure in authorized entrants;
- Continuously maintains an accurate count of authorized entrants in the permit space and ensures that the means used to identify authorized entrants under paragraph 6.4 of this section accurately identifies who is in the permit space;
- Remains outside the permit space during entry operations until relieved by another attendant. When the employer's permit entry program allows attendant entry for rescue, attendants may enter a permit space to attempt a rescue if they have been trained and equipped for rescue operations;
- Communicates with authorized entrants as necessary to monitor entrant status and to alert entrants of the need to evacuate the space;
- Monitors activities inside and outside the space to determine if it is safe for entrants to remain in the space and orders the authorized entrants to evacuate the permit space immediately under any of the following conditions:
  - If the attendant detects a prohibited condition;
  - If the attendant detects the behavioral effects of hazard exposure in an authorized entrant;
  - If the attendant detects a situation outside the space that could endanger the authorized entrants; or
  - If the attendant cannot effectively and safely perform all the duties required of this section;
- Summon rescue and other emergency services as soon as the attendant determines that authorized entrants may need assistance to escape from permit space hazards;
- Takes the following actions when unauthorized persons approach or enter a permit space while entry is underway:
  - Warn the unauthorized persons that they must stay away from the permit space;
  - Advise the unauthorized persons that they must exit immediately if they have entered the permit space; and

- Inform the authorized entrants and the entry supervisor if unauthorized persons have entered the permit space;
- Performs non-entry rescues as specified by the employer's rescue procedure; and
- Performs no duties that might interfere with the attendant's primary duty to monitor and protect the authorized entrants.

6.9.4.8 *Duties of Entry Supervisors*

The employer shall ensure that each entry supervisor:

- Knows the hazards that may be faced during entry, including information on the mode, signs or symptoms, and consequences of the exposure;
- Verifies, by checking that the appropriate entries have been made on the permit, that all tests specified by the permit have been conducted and that all procedures and equipment specified by the permit are in place before endorsing the permit and allowing entry to begin;
- Terminates the entry and cancels the permit;
- Verifies that rescue services are available and that the means for summoning them are operable;
- Removes unauthorized individuals who enter or who attempt to enter the permit space during entry operations; and
- Determines, whenever responsibility for a permit space entry operation is transferred and at intervals dictated by the hazards and operations performed within the space that entry operations remain consistent with terms of the entry permit and that acceptable entry conditions are maintained.

6.9.4.9 *Rescue and Emergency Services*

An employer who designates rescue and emergency services shall:

- Evaluate a prospective rescuer's ability to respond to a rescue summons in a timely manner, considering the hazard(s) identified; what will be considered timely will vary according to the specific hazards involved in each entry.

- Evaluate a prospective rescue service's ability, in terms of proficiency with rescue-related tasks and equipment, to function appropriately while rescuing entrants from the particular permit space or types of permit spaces identified;
- Select a rescue team or service from those evaluated that:
  - Has the capability to reach the victim(s) within a time frame that is appropriate for the permit space hazard(s) identified;
  - Is equipped for and proficient in performing the needed rescue services;
- Inform each rescue team or service of the hazards they may confront when called on to perform rescue at the site; and
- Provide the rescue team or service selected with access to all permit spaces from which rescue may be necessary so that the rescue service can develop appropriate rescue plans and practice rescue operations.

An employer whose employees have been designated to provide permit space rescue and emergency services shall take the following measures:

- Provide affected employees with the personal protective equipment (PPE) needed to conduct permit space rescues safely and train affected employees so they are proficient in the use of that PPE, at no cost to those employees;
- Train affected employees to perform assigned rescue duties. The employer must ensure that such employees successfully complete the training required to establish proficiency as an authorized entrant;
- Train affected employees in basic first-aid and cardiopulmonary resuscitation (CPR). The employer shall ensure that at least one member of the rescue team or service holding a current certification in first aid and CPR is available; and

- Ensure that affected employees practice making permit space rescues at least once every 12 months, by means of simulated rescue operations in which they remove dummies, manikins, or actual persons from the actual permit spaces or from representative permit spaces. Representative permit spaces shall, with respect to opening size, configuration, and accessibility, simulate the types of permit spaces from which a rescue is to be performed.

To facilitate non-entry rescue, retrieval systems or methods shall be used whenever an authorized entrant enters a permit space, unless the retrieval equipment would increase the overall risk of entry or would not contribute to the rescue of the entrant. Retrieval systems shall meet the following requirements.

- Each authorized entrant shall use a chest or full body harness, with a retrieval line attached at the center of the entrant's back near shoulder level, above the entrant's head, or at another point which the employer can establish presents a profile small enough for the successful removal of the entrant. Wristlets may be used in lieu of the chest or full body harness if the employer can demonstrate that the use of a chest or full body harness is infeasible or creates a greater hazard and that the use of wristlets is the safest and most effective alternative.
- The other end of the retrieval line shall be attached to a mechanical device or fixed point outside the permit space in such a manner that rescue can begin as soon as the rescuer becomes aware that rescue is necessary. A mechanical device shall be available to retrieve personnel from vertical type permit spaces more than 1.52 m deep.

If an injured entrant is exposed to a substance for which a material safety data sheet (MSDS) or other similar written information is required to be kept at the worksite, that MSDS or written information shall be made available to the medical facility treating the exposed entrant.

6.9.4.10 *Employee Participation*

Employers shall consult with affected employees and their authorized representatives on the development and implementation of all aspects of the permit space program.

Employers shall make available to affected employees and their authorized representatives all information required to be developed by this section.

6.9.4.11 *Procedures for Atmospheric Testing*

6.9.4.11.1 Purpose

Atmospheric testing is required for several purposes:

1. Evaluation testing. The atmosphere of a confined space should be analyzed using equipment of sufficient sensitivity and specificity to identify and evaluate any hazardous atmospheres that may exist or arise, so that appropriate permit entry procedures can be developed and acceptable entry conditions stipulated for that space. Evaluation and interpretation of these data, and development of the entry procedure, should be done by, or reviewed by, a technically qualified professional (e.g., a consultation service, or certified industrial hygienist, registered safety engineer, certified safety professional, certified marine chemist, etc.) based on evaluation of all serious hazards.
2. Verification testing. The atmosphere of a permit space which may contain a hazardous atmosphere should be tested for residues of all contaminants identified by evaluation testing using permit specified equipment to determine that residual concentrations at the time of testing and entry are within the range of acceptable entry conditions. Results of testing (i.e., actual concentration, etc.) should be recorded on the permit in the space provided adjacent to the stipulated acceptable entry condition.

6.9.4.11.2 Duration of Testing

Measurement of values for each atmospheric parameter should be made for at least the minimum response time of the test instrument specified by the manufacturer.

6.9.4.11.3 Testing Stratified Atmospheres

When monitoring for entries involving a descent into atmospheres that may be stratified, the atmospheric envelope should be tested at a distance of approximately 1.22 m in the direction of travel and to each side. If a sampling probe is used, the entrant's rate of progress should be slowed to accommodate the sampling speed and detector response.

6.9.4.11.4 Order of Testing

A test for oxygen is performed first because most combustible gas meters are oxygen dependent and will not provide reliable readings in an oxygen deficient atmosphere. Combustible gases are tested for next because the threat of fire or explosion is both more immediate and more life threatening, in most cases, than exposure to toxic gases and vapors. If tests for toxic gases and vapors are necessary, they are performed last.

6.10 Control of Hazardous Energy (Lockout/Tagout)

6.10.1 General Information

This standard covers the servicing and maintenance of machines and equipment in which the **unexpected** energization or startup of the machines or equipment, or release of stored energy could cause injury to employees. This standard establishes minimum performance requirements for the control of such hazardous energy.

This standard applies to the control of energy during servicing and/or maintenance of machines and equipment when:

- An employee is required to remove or bypass a guard or other safety device; or
- An employee is required to place any part of his or her body into an area on a machine or piece of equipment where work is actually performed upon the material being processed (point of operation) or where an associated danger zone exists during a machine operating cycle.

Note that minor tool changes and adjustments, and other minor servicing activities, which take place during normal production operations, are not covered by this standard if they are routine, repetitive, and integral to the use of the equipment for production, provided that the work is performed using alternative measures which provide effective protection.

This section requires employers to establish a program and utilize procedures for affixing appropriate lockout devices or tagout devices to energy isolating devices, and to otherwise disable machines or equipment to prevent unexpected energization, start up or release of stored energy in order to prevent injury to employees.

When other standards in this part require the use of lockout or tagout, they shall be used and supplemented by the procedural and training requirements of this section.

6.10.2 Definitions

Affected employee. An employee whose job requires him/her to operate or use a machine or equipment on which servicing or maintenance is being performed under lockout or tagout, or whose job requires him/her to work in an area in which such servicing or maintenance is being performed.

Authorized employee. A person who locks out or tags out machines or equipment in order to perform servicing or maintenance on that machine or equipment. An affected employee becomes an authorized employee when that employee's duties include performing servicing or maintenance covered under this section.

Capable of being locked out. An energy isolating device is capable of being locked out if it has a hasp or other means of attachment to which, or through which, a lock can be affixed, or it has a locking mechanism built into it. Other energy isolating devices are capable of being locked out, if lockout can be achieved without the need to dismantle, rebuild, or replace the energy isolating device or permanently alter its energy control capability.

Energized. Connected to an energy source or containing residual or stored energy.

Energy isolating device. A mechanical device that physically prevents the transmission or release of energy, including but not limited to the following: A manually operated electrical circuit breaker; a disconnect switch; a manually operated switch by which the conductors of a circuit can be disconnected from all ungrounded supply conductors, and, in addition, no pole can be operated independently; a line valve; a block; and any similar device used to block or isolate energy. Push buttons, selector switches and other control circuit type devices are not energy isolating devices.

Energy source. Any source of electrical, mechanical, hydraulic, pneumatic, chemical, thermal, or other energy.

Hot tap. A procedure used in the repair, maintenance and services activities which involves welding on a piece of equipment (pipelines, vessels or tanks) under pressure, in order to install connections or appurtenances. it is commonly used to replace or add sections of pipeline without the interruption of service for air, gas, water, steam, and petrochemical distribution systems.

Lockout. The placement of a lockout device on an energy isolating device, in accordance with an established procedure, ensuring that the energy isolating device and the equipment being controlled cannot be operated until the lockout device is removed.

Lockout device. A device that utilizes a positive means such as a lock, either key or combination type, to hold an energy isolating device in the safe position and prevent the energizing of a machine or equipment. Included are blank flanges and bolted slip blinds.

Normal production operations. The utilization of a machine or equipment to perform its intended production function.

Servicing and/or maintenance. Workplace activities such as constructing, installing, setting up, adjusting, inspecting, modifying, and maintaining and/or servicing machines or equipment. These activities include lubrication, cleaning or unjamming of machines or equipment and making adjustments or tool changes, where the employee may be exposed to the unexpected energization or startup of the equipment or release of hazardous energy.

Setting up. Any work performed to prepare a machine or equipment to perform its normal production operation.

Tagout. The placement of a tagout device on an energy isolating device, in accordance with an established procedure, to indicate that the energy isolating device and the equipment being controlled may not be operated until the tagout device is removed.

Tagout device. A prominent warning device, such as a tag and a means of attachment, which can be securely fastened to an energy isolating device in accordance with an established procedure, to indicate that the energy isolating device and the equipment being controlled may not be operated until the tagout device is removed.

Energy control program. The employer shall establish a program consisting of energy control procedures, employee training and periodic inspections to ensure that before any employee performs any servicing or maintenance on a machine or equipment where the unexpected energizing, startup or release of stored energy could occur and cause injury, the machine or equipment shall be isolated from the energy source and rendered inoperative.

6.10.3 Elements of Lockout/Tagout

If an energy isolating device is not capable of being locked out, the employer's energy control program under paragraph 3.1 of this section shall utilize a tagout system.

If an energy isolating device is capable of being locked out, the employer's energy control program shall utilize lockout, unless the employer can demonstrate that the utilization of a tagout system will provide full employee protection.

6.10.4 Full Employee Protection

When a tagout device is used on an energy isolating device which is capable of being locked out, the tagout device shall be attached at the same location that the lockout device would have been attached, and the employer shall demonstrate that the tagout program will provide a level of safety equivalent to that obtained by using a lockout program.

In demonstrating that a level of safety is achieved in the tagout program which is equivalent to the level of safety obtained by using a lockout program, the employer shall demonstrate full compliance with all tagout-related provisions of this standard together with such additional elements as are necessary to provide the equivalent safety available from the use of a lockout device. Additional means to be considered as part of the demonstration of full employee protection shall include the implementation of additional safety measures such as the removal of an isolating circuit element, blocking of a controlling switch, opening of an extra disconnecting device, or the removal of a valve handle to reduce the likelihood of inadvertent energization.

6.10.5 Energy Control Procedure

Procedures shall be developed, documented and utilized for the control of potentially hazardous energy when employees are engaged in the activities covered by this section.

The procedures shall clearly and specifically outline the scope, purpose, authorization, rules, and techniques to be utilized for the control of hazardous energy, and the means to enforce compliance including, but not limited to, the following:

- A specific statement of the intended use of the procedure;
- Specific procedural steps for shutting down, isolating, blocking and securing machines or equipment to control hazardous energy;
- Specific procedural steps for the placement, removal and transfer of lockout devices or tagout devices and the responsibility for them; and
- Specific requirements for testing a machine or equipment to determine and verify the effectiveness of lockout devices, tagout devices, and other energy control measures.

6.10.6 Protective Materials and Hardware

Locks, tags, chains, wedges, key blocks, adapter pins, self-locking fasteners, or other hardware shall be provided by the employer for isolating, securing or blocking of machines or equipment from energy sources.

Lockout devices and tagout devices shall be singularly identified; shall be the only devices(s) used for controlling energy; shall not be used for other purposes; and shall meet the following requirements.

Lockout and tagout devices shall be capable of withstanding the environment to which they are exposed for the maximum period of time that exposure is expected.

Tagout devices shall be constructed and printed so that exposure to weather conditions or wet and damp locations will not cause the tag to deteriorate or the message on the tag to become illegible.

Tags shall not deteriorate when used in corrosive environments such as areas where acid and alkali chemicals are handled and stored.

Lockout and tagout devices shall be standardized within the facility in at least one of the following criteria: color; shape; or size; and additionally, in the case of tagout devices, print and format shall be standardized.

Lockout devices shall be substantial enough to prevent removal without the use of excessive force or unusual techniques, such as with the use of bolt cutters or other metal cutting tools.

Tagout devices, including their means of attachment, shall be substantial enough to prevent inadvertent or accidental removal. Tagout device attachment means shall be of a non-reusable type, attachable by hand, self-locking, and non-releasable with a minimum unlocking strength of no less than 22.68 kilograms and having the general design and basic characteristics of being at least equivalent to a one-piece, all environment-tolerant nylon cable tie.

Lockout devices and tagout devices shall indicate the identity of the employee applying the device(s).

Tagout devices shall warn against hazardous conditions if the machine or equipment is energized and shall include a legend such as the following:

- Do Not Start.
- Do Not Open.
- Do Not Close.
- Do Not Energize.
- Do Not Operate.

6.10.7 Inspection

The employer shall conduct a periodic inspection of the energy control procedure at least annually to ensure that the procedure and the requirements of this standard are being followed.

The periodic inspection shall be performed by an authorized employee other than the ones(s) utilizing the energy control procedure being inspected.

The periodic inspection shall be conducted to correct any deviations or inadequacies identified.

Where lockout is used for energy control, the periodic inspection shall include a review, between the inspector and each authorized employee, of that employee's responsibilities under the energy control procedure being inspected.

Where tagout is used for energy control, the periodic inspection shall include a review, between the inspector and each authorized and affected employee, of that employee's responsibilities under the energy control procedure being inspected, and the elements set forth in this section.

The employer shall certify that the periodic inspections have been performed. The certification shall identify the machine or equipment on which the energy control procedure was being utilized, the date of the inspection, the employees included in the inspection, and the person performing the inspection.

6.10.8 Training and Communication

The employer shall provide training to ensure that the purpose and function of the energy control program are understood by employees and that the knowledge and skills required for the safe application, usage, and removal of the energy controls are acquired by employees. The training shall include the following:

- Each authorized employee shall receive training in the recognition of applicable hazardous energy sources, the type and magnitude of the energy available in the workplace, and the methods and means necessary for energy isolation and control.
- Each affected employee shall be instructed in the purpose and use of the energy control procedure.

All other employees whose work operations are or may be in an area where energy control procedures may be utilized, shall be instructed about the procedure, and about the prohibition relating to attempts to restart or reenergize machines or equipment which are locked out or tagged out.

When tagout systems are used, employees shall also be trained in the following limitations of tags:

- Tags are essentially warning devices affixed to energy isolating devices, and do not provide the physical restraint on those devices that is provided by a lock.
- When a tag is attached to an energy isolating means, it is not to be removed without authorization of the

authorized person responsible for it, and it is never to be bypassed, ignored, or otherwise defeated.

- Tags must be legible and understandable by all authorized employees, affected employees, and all other employees whose work operations are or may be in the area, in order to be effective.
- Tags and their means of attachment must be made of materials which will withstand the environmental conditions encountered in the workplace.
- Tags may evoke a false sense of security, and their meaning needs to be understood as part of the overall energy control program.
- Tags must be securely attached to energy isolating devices so that they cannot be inadvertently or accidentally detached during use.

Retraining shall be provided for all authorized and affected employees whenever there is a change in their job assignments, a change in machines, equipment or processes that present a new hazard, or when there is a change in the energy control procedures.

Additional retraining shall also be conducted whenever a periodic inspection is needed, or whenever the employer has reason to believe that there are deviations from or inadequacies in the employee's knowledge or use of the energy control procedures.

The retraining shall reestablish employee proficiency and introduce new or revised control methods and procedures, as necessary.

The employer shall certify that employee training has been accomplished and is being kept up to date. The certification shall contain each employee's name and dates of training.

6.10.9 Energy Isolation

Lockout or tagout shall be performed only by the authorized employees who are performing the servicing or maintenance.

Affected employees shall be notified by the employer or authorized employee of the application and removal of lockout devices or tagout devices. Notification shall be given before the controls are applied, and after they are removed from the machine or equipment.

The established procedures for the application of energy control (the lockout or tagout procedures) shall cover the following elements and actions and shall be done in the following sequence:

- Before an authorized or affected employee turns off a machine or equipment, the authorized employee shall have knowledge of the type and magnitude of the energy, the hazards of the energy to be controlled, and the method or means to control the energy.
- The machine or equipment shall be turned off or shut down using the procedures established for the machine or equipment. An orderly shutdown must be utilized to avoid any additional or increased hazard(s) to employees as a result of the equipment stoppage.
- All energy isolating devices that are needed to control the energy to the machine or equipment shall be physically located and operated in such a manner as to isolate the machine or equipment from the energy source(s).
- Lockout or tagout devices shall be affixed to each energy isolating device by authorized employees.
- Lockout devices, where used, shall be affixed in a manner to that will hold the energy isolating devices in a "safe" or "off" position.
- Tagout devices, where used, shall be affixed in such a manner as will clearly indicate that the operation or movement of energy isolating devices from the "safe" or "off" position is prohibited.
- Where tagout devices are used with energy isolating devices designed with the capability of being locked, the tag attachment shall be fastened at the same point at which the lock would have been attached.
- Where a tag cannot be affixed directly to the energy isolating device, the tag shall be located as close as safely possible to the device, in a position that will be immediately obvious to anyone attempting to operate the device.
- Following the application of lockout or tagout devices to energy isolating devices, all potentially hazardous stored or residual energy shall be relieved, disconnected, restrained, and otherwise rendered safe.

- If there is a possibility of reaccumulation of stored energy to a hazardous level, verification of isolation shall be continued until the servicing or maintenance is completed, or until the possibility of such accumulation no longer exists.
- Prior to starting work on machines or equipment that have been locked out or tagged out, the authorized employee shall verify that isolation and deenergization of the machine or equipment have been accomplished.

Before lockout or tagout devices are removed and energy is restored to the machine or equipment, procedures shall be followed and actions taken by the authorized employee(s) to ensure the following:

- The work area shall be inspected to ensure that nonessential items have been removed and to ensure that machine or equipment components are operationally intact.
- The work area shall be checked to ensure that all employees have been safely positioned or removed.
- After lockout or tagout devices have been removed and before a machine or equipment is started, affected employees shall be notified that the lockout or tagout device(s) have been removed.
- Each lockout or tagout device shall be removed from each energy isolating device by the employee who applied the device.

6.10.10 Group Lockout or Tagout

When servicing and/or maintenance is performed by a crew, craft, department or other group, they shall utilize a procedure which affords the employees a level of protection equivalent to that provided by the implementation of a personal lockout or tagout device.

Group lockout or tagout devices shall be used in accordance with the procedures including, but not necessarily limited to, the following specific requirements:

- Primary responsibility is vested in an authorized employee for a set number of employees working under the protection of a group lockout or tagout device (such as an operations lock).

- Provision for the authorized employee to ascertain the exposure status of individual group members with regard to the lockout or tagout of the machine or equipment.
- When more than one crew, craft, department, etc. is involved, assignment of overall job-associated lockout or tagout control responsibility to an authorized employee designated to coordinate affected work forces and ensure continuity of protection.
- Each authorized employee shall affix a personal lockout or tagout device to the group lockout device, group lockbox, or comparable mechanism when he or she begins work, and shall remove those devices when he or she stops working on the machine or equipment being serviced or maintained.

Specific procedures shall be utilized during shift or personnel changes to ensure the continuity of lockout or tagout protection, including provision for the orderly transfer of lockout or tagout device protection between off-going and oncoming employees, to minimize exposure to hazards from the unexpected energization or start-up of the machine or equipment, or the release of stored energy.

6.11 Medical Services and First Aid

The employer shall ensure the ready availability of medical personnel for advice and consultation on matters of plant health.

In the absence of an infirmary, clinic, or hospital in near proximity to the workplace which is used for the treatment of all injured employees, a person or persons shall be adequately trained to render first aid. Adequate first aid supplies shall be readily available.

Where the eyes or body of any person may be exposed to injurious corrosive materials, suitable facilities for quick drenching or flushing of the eyes and body shall be provided within the work area for immediate emergency use.

First aid supplies are required to be readily available. An example of the minimal contents of a generic first aid kit is described in American National Standard (ANSI) Z308.1-1998 "Minimum Requirements for Workplace First-aid Kits." The contents of the kit listed in the ANSI standard should be adequate for small worksites. When larger operations or multiple operations are being conducted at the same location, employers should determine the need for additional first aid kits at the worksite, additional types of first aid equipment and supplies and additional quantities and types of supplies and equipment in the first aid kits.

In a similar fashion, employers who have unique or changing first-aid needs in their workplace may need to enhance their first-aid kits.

If it is reasonably anticipated that employees will be exposed to blood or other potentially infectious materials while using first aid supplies, employers are required to provide appropriate personal protective equipment (PPE).

6.12 Fire Protection

6.12.1 Fire Brigades

It is recommended that pre-fire planning be conducted by the local fire department and/or the workplace fire brigade in order for them to be familiar with the workplace and process hazards. Involvement with the local fire department or fire prevention bureau is encouraged to facilitate coordination and cooperation between members of the fire brigade and those who might be called upon for assistance during a fire emergency.

Employees who cannot meet the physical capability requirement may still be members of the fire brigade as long as such employees do not perform interior structural fire fighting. It is suggested that fire brigade members who are unable to perform interior structural fire fighting be assigned less stressful and physically demanding fire brigade duties, e.g., certain types of training, recordkeeping, fire prevention inspection and maintenance, and fire pump operations.

Physically capable can also be determined by physical performance tests or by a physical examination when the examining physician is aware of the duties that the fire brigade member is expected to perform.

It is also recommended that fire brigade members participate in a physical fitness program. There are many benefits which can be attributed to being physically fit. It is believed that physical fitness may help to reduce the number of sprain and strain injuries as well as contributing to the improvement of the cardiovascular system.

At a minimum, hands-on training is required to be conducted annually for all fire brigade members. However, for those fire brigade members who are expected to perform interior structural fire fighting, some type of training or education session must be provided at least quarterly.

In addition to the required hands-on training, it is strongly recommended that fire brigade members receive other types of training and education such as: classroom instruction, review of emergency action procedures, pre-fire planning, review of special hazards in the workplace, and practice in the use of self-contained breathing apparatus.

It is not necessary for the employer to duplicate the same training or education that a fire brigade member receives as a member of a community volunteer fire department, rescue squad, or similar organization. However, such training or education must have been provided to the fire brigade member within the past year and it must be documented that the fire brigade member has received the training or education. For example: there is no need for a fire brigade member to receive another training class in the use of positive-pressure self-contained breathing apparatus if the fire brigade member has recently completed such training as a member of a community fire department. Instead, the fire brigade member should receive training or education covering other important equipment or duties of the fire brigade as they relate to the workplace hazards, facilities and processes.

It is generally recognized that the effectiveness of fire brigade training and education depends upon the expertise of those providing the training and education as well as the motivation of the fire brigade members. Fire brigade training instructors must receive a higher level of training and education than the fire brigade members they will be teaching. This includes being more knowledgeable about the functions to be performed by the fire brigade and the hazards involved.

The instructors should be qualified to train fire brigade members and demonstrate skills in communication, methods of teaching, and motivation. It is important for instructors and fire brigade members alike to be motivated toward the goals of the fire brigade and be aware of the importance of the service that they are providing for the protection of other employees and the workplace.

In order to be effective, fire brigades must have competent leadership and supervision. It is important for those who supervise the fire brigade during emergency situations, e.g., fire brigade chiefs, leaders, etc., to receive the necessary training and education for supervising fire brigade activities during these hazardous and stressful situations. These fire brigade members with leadership responsibilities should demonstrate skills in strategy and tactics, fire suppression and prevention techniques, leadership principles, pre-fire planning, and safety practices. It is again suggested that fire service training sources be consulted for determining the kinds of training and education which are necessary for those with fire brigade leadership responsibilities.

Fire brigade leaders and fire brigade instructors should receive more formalized training and education on a continuing basis by attending classes provided by such training sources as universities and university fire extension services.

The following recommendations should not be considered to be all of the necessary elements of a complete comprehensive training program, but the information may be helpful as a guide in developing a fire brigade training program.

All fire brigade members should be familiar with exit facilities and their location, emergency escape routes for handicapped workers, and the workplace "emergency action plan."

In addition, fire brigade members who are expected to control and extinguish fires in the incipient stage should, at a minimum, be trained in the use of fire extinguishers, standpipes, and other fire equipment they are assigned to use. They should also be aware of first aid medical procedures and procedures for dealing with special hazards to which they may be exposed. Training and education should include both classroom instruction and actual operation of the equipment under simulated emergency conditions. Hands-on type training must be conducted at least annually but some functions should be reviewed more often.

In addition to the above training, fire brigade members who are expected to perform emergency rescue and interior structural fire fighting should, at a minimum, be familiar with the proper techniques in rescue and fire suppression procedures. Training and education should include fire protection courses, classroom training, simulated fire situations including "wet drills" and, when feasible, extinguishment of actual mock fires. Frequency of training or education must be at least quarterly, but some drills or classroom training should be conducted as often as monthly or even weekly to maintain the proficiency of fire brigade members.

It is important that fire brigade members be informed about special hazards to which they may be exposed during fire and other emergencies. Such hazards as storage and use areas of flammable liquids and gases, toxic chemicals, water-reactive substances, etc., can pose difficult problems. There must be written procedures developed that describe the actions to be taken in situations involving special hazards. Fire brigade members must be trained in handling these special hazards as well as keeping abreast of any changes that occur in relation to these special hazards.

6.12.2 Fire Fighting Equipment

It is important that fire fighting equipment that is in damaged or unserviceable condition be removed from service and replaced. This will prevent fire brigade members from using unsafe equipment by mistake.

Fire fighting equipment, except portable fire extinguishers and respirators, must be inspected at least annually. Portable fire extinguishers and respirators are required to be inspected at least monthly.

It is not the intention of these standards to require employers to provide a full ensemble of protective clothing for every fire brigade member without consideration given to the types of hazardous environments to which the fire brigade member might be exposed. It is the intention of these standards to require adequate protection for those fire brigade members who might be exposed to fires in an advanced stage, smoke, toxic gases, and high temperatures. Therefore, the protective clothing requirements only apply to those fire brigade members who perform interior structural fire fighting operations.

Additionally, the protective clothing requirements do not apply to the protective clothing worn during outside fire fighting operations (brush and forest fires, crash crew operations) or other special fire fighting activities. It is important that the protective clothing to be worn during these types of fire fighting operations reflect the hazards which are expected to be encountered by fire brigade members.

Note that there is interdependence of protective clothing to cover one or more parts of the body. Therefore, an option is given so that fire brigade members may meet the foot and leg requirements by either wearing long fire-resistive coats in combination with fully extended boots, or by wearing shorter fire-resistive costs in combination with protective trousers and protective shoes or shorter boots.

Fire-resistive coats and protective trousers meeting all of the requirements contained in NFPA 1971-1975 "Protective Clothing for

Structural Fire Fighters," are acceptable as meeting the requirements of this standard.

The lining is required to be permanently attached to the outer shell. However, it is permissible to attach the lining to the outer shell material by stitching in one area such as at the neck. Fastener tape or snap fasteners may be used to secure the rest of the lining to the outer shell to facilitate cleaning. Reference to permanent lining does not refer to a winter liner which is a detachable extra lining used to give added protection to the wearer against the effects of cold weather and wind.

Hand protection may be met by protective gloves or a glove system. A glove system consists of a combination of different gloves. The usual components of a glove system consist of a pair of gloves, which provide thermal insulation to the hands, worn in combination with a second pair of gloves which provide protection against flame, cut, and puncture.

It is suggested that protective gloves provide dexterity and a sense of feel for objects. Criteria and test methods for dexterity are contained in the NIOSH publications, "The Development of Criteria for Firefighters' Gloves; Vol. I: Glove Requirements" and "Vol. II: Glove Criteria and Test Methods." These NIOSH publications also contain a permissible modified version of Federal Test Method 191, Method 5903, (paragraph .3 of Appendix E) for flame resistance when gloves, rather than glove material, are tested for flame resistance.

Head protective devices which meet the requirements contained in NFPA No. 1972 are acceptable as meeting the requirements of this standard for head protection.

Head protective devices are required to be provided with ear flaps so that the ear flaps will be available if needed. It is recommended that ear protection always be used while fighting interior structural fires.

It is recommended that a flame resistant protective head covering such as a hood or snood, which will not adversely affect the seal of a respirator facepiece, be worn during interior structural fire fighting operations to protect the sides of the face and hair.

Respiratory protection is required to be worn by fire brigade members while working inside buildings or confined spaces where toxic products of combustion or an oxygen deficiency is likely to be present; respirators are also to be worn during emergency situations involving toxic substances. When fire brigade members respond to emergency situations, they may be exposed to unknown contaminants in unknown concentrations. Therefore, it is imperative that fire brigade members wear proper respiratory protective devices during these situations. Additionally, there are many instances where toxic products of combustion are still present during mop-up and overhaul operations. Therefore, fire brigade members should continue to wear respirators during these types of operations.

A self-contained breathing apparatus is not required to be equipped with either a buddy-breathing device or a quick-disconnect valve. However, these accessories may be very useful and are acceptable as long as such accessories do not cause damage to the apparatus, restrict the air flow of the apparatus, or obstruct the normal operation of the apparatus.

Buddy-breathing devices are useful for emergency situations where a victim or another fire brigade member can share the same air supply with the wearer of the apparatus for emergency escape purposes.

The employer is encouraged to provide fire brigade members with an alternative means of respiratory protection to be used only for emergency escape purposes if the self-contained breathing apparatus becomes inoperative. Such alternative means of respiratory protection may be either a buddy-breathing device or an escape self-contained breathing apparatus (ESCBA). The ESCBA is a short-duration respiratory protective device which is approved for only emergency escape purposes. It is suggested that if ESCBA units are used, that they be of at least 5 minutes service life.

Quick-disconnect valves are devices which start the flow of air by insertion of the hose (which leads to the facepiece) into the regulator of self-contained breathing apparatus, and stop the flow of air by disconnecting the hose from the regulator. These devices are particularly useful for those positive-pressure self-contained breathing apparatus which do not have the capability of being switched from the demand to the positive-pressure mode.

The use of a self-contained breathing apparatus where the apparatus can be switched from a demand to a positive-pressure mode is acceptable as long as the apparatus is in the positive-pressure mode when performing interior structural fire fighting operations.

There are situations which require the use of respirators which have a duration of 2 hours or more. Presently, there are no approved positive-pressure apparatus with a rated service life of more than 2 hours. Consequently, negative-pressure self-contained breathing apparatus with a rated service life of more than 2 hours and which have a minimum protection factor of 5,000 as determined by an acceptable quantitative fit test performed on each individual, will be acceptable for use during situations which require long duration apparatus. Long duration apparatus may be needed in such instances as working in tunnels, subway systems, etc. Such negative-pressure breathing apparatus will continue to be acceptable for a maximum of 18 months after a positive-pressure apparatus with the same or longer rated service life of more than 2 hours is certified by NIOSH/MSHA. After this 18 month phase-in period, all self-contained breathing apparatus used for these long duration situations will have to be of the positive-pressure type.

Protection factor (sometimes called fit factor) is defined as the ratio of the contaminant concentrations outside of the respirator to the contaminant concentrations inside the facepiece of the respirator.

$$PF = \frac{\text{Concentration outside respirator}}{\text{Concentration inside facepiece}}$$

Protection factors are determined by quantitative fit tests. An acceptable quantitative fit test should include the following elements:

1. A fire brigade member who is physically and medically capable of wearing respirators, and who is trained in the use of respirators, dons a self-contained breathing apparatus equipped with a device that will monitor the concentration of a contaminant inside the facepiece.
2. The fire brigade member then performs a qualitative fit test to assure the best face to facepiece seal as possible. A qualitative fit test can consist of a negative-pressure test, positive-pressure test, isoamyl acetate vapor (banana oil) test, or an irritant smoke test.
3. The wearer should then perform physical activity which reflects the level of work activity which would be expected during fire fighting activities. The physical activity should include simulated fire-ground work activity or physical exercise such as running-in-place, a step test, etc.
4. Without readjusting the apparatus, the wearer is placed in a test atmosphere containing a non-toxic contaminant with a known, constant, concentration.

The protection factor is then determined by dividing the known concentration of the contaminant in the test atmosphere by the concentration of the contaminant inside the facepiece when the following exercises are performed:

1. Normal breathing with head motionless for one minute;
2. Deep breathing with head motionless for 30 seconds;
3. Turning head slowly from side to side while breathing normally, pausing for at least two breaths before changing direction. Continue for at least one minute;
4. Moving head slowly up and down while breathing normally, pausing for at least two breaths before changing direction. Continue for at least two minutes;
5. Reading from a prepared text, slowly and clearly, and loudly enough to be heard and understood. Continue for one minute; and
6. Normal breathing with head motionless for at least one minute.

The protection factor which is determined must be at least 5,000. The quantitative fit test should be conducted at least three times. It is acceptable to conduct all three tests on the same day. However, there should be at least one hour between tests to reflect the protection afforded by the apparatus during different times of the day.

The above elements are not meant to be a comprehensive, technical description of a quantitative fit test protocol. However, quantitative fit test procedures which include these elements are acceptable for determining protection factors. Procedures for a quantitative fit test are required to be available for inspection by the assistant secretary or authorized representative.

6.13 Handling Materials

Where mechanical handling equipment is used, sufficient safe clearances shall be allowed for aisles, at loading docks, through doorways and wherever turns or passage must be made. Aisles and passageways shall be kept clear and in good repair, with no obstruction across or in aisles that could create a hazard. Permanent aisles and passageways shall be appropriately marked.

Storage of material shall not create a hazard. Bags, containers, bundles, etc., stored in tiers shall be stacked, blocked, interlocked and limited in height so that they are stable and secure against sliding or collapse.

Storage areas shall be kept free from accumulation of materials that constitute hazards from tripping, fire, explosion, or pest harborage. Vegetation control will be exercised when necessary.

Clearance signs to warn of clearance limits shall be provided.

Derail and/or bumper blocks shall be provided on spur railroad tracks where a rolling car could contact other cars being worked, enter a building, work or traffic area.

Covers and/or guard- rails shall be provided to protect personnel from the hazards of open pits, tanks, vats, ditches, etc.

6.14 Slings

This section applies to slings used in conjunction with other material handling equipment for the movement of material by hoisting, in employments covered by this part. The types of slings covered are those made from alloy steel chain, wire rope, metal mesh, natural or synthetic fiber rope (conventional three strand construction), and synthetic web (nylon, polyester, and polypropylene). The following are important definitions.

Angle of loading is the inclination of a leg or branch of a sling measured from the horizontal or vertical plane; provided that an angle of loading of five degrees or less from the vertical may be considered a vertical angle of loading.

Basket hitch is a sling configuration whereby the sling is passed under the load and has both ends, end attachments, eyes or handles on the hook or a single master link.

Braided wire rope is a wire rope formed by plaiting component wire ropes.

Bridle wire rope sling is a sling composed of multiple wire rope legs with the top ends gathered in a fitting that goes over the lifting hook.

Cable laid endless sling-mechanical joint is a wire rope sling made endless by joining the ends of a single length of cable laid rope with one or more metallic fittings.

Cable laid grommet-hand tucked is an endless wire rope sling made from one length of rope wrapped six times around a core formed by hand tucking the ends of the rope inside the six wraps.

Cable laid rope is a wire rope composed of six wire ropes wrapped around a fiber or wire rope core.

Cable laid rope sling-mechanical joint is a wire rope sling made from a cable laid rope with eyes fabricated by pressing or swaging one or more metal sleeves over the rope junction.

Choker hitch is a sling configuration with one end of the sling passing under the load and through an end attachment, handle or eye on the other end of the sling.

Coating is an elastomer or other suitable material applied to a sling or to a sling component to impart desirable properties.

Cross rod is a wire used to join spirals of metal mesh to form a complete fabric.

Designated means selected or assigned by the employer or the employer's representative as being qualified to perform specific duties.

Equivalent entity is a person or organization (including an employer) which, by possession of equipment, technical knowledge and skills, can perform with equal competence the same repairs and tests as the person or organization with which it is equated.

Fabric (metal mesh) is the flexible portion of a metal mesh sling consisting of a series of transverse coils and cross rods.

Female handle (choker) is a handle with a handle eye and a slot of such dimension as to permit passage of a male handle thereby allowing the use of a metal mesh sling in a choker hitch.

Handle is a terminal fitting to which metal mesh fabric is attached.

Handle eye is an opening in a handle of a metal mesh sling shaped to accept a hook, shackle or other lifting device.

Hitch is a sling configuration whereby the sling is fastened to an object or load, either directly to it or around it.

Link is a single ring of a chain.

Male handle (triangle) is a handle with a handle eye.

Master coupling link is an alloy steel welded coupling link used as an intermediate link to join alloy steel chain to master links.

Master link or gathering ring is a forged or welded steel link used to support all members (legs) of an alloy steel chain sling or wire rope sling.

Mechanical coupling link is a nonwelded, mechanically closed steel link used to attach master links, hooks, etc., to alloy steel chain.

Proof load is the load applied in performance of a proof test.

Proof test is a nondestructive tension test performed by the sling manufacturer or an equivalent entity to verify construction and workmanship of a sling.

Rated capacity or working load limit is the maximum working load permitted by the provisions of this section.

Reach is the effective length of an alloy steel chain sling measured from the top bearing surface of the upper terminal component to the bottom bearing surface of the lower terminal component.

Selvage edge is the finished edge of synthetic webbing designed to prevent unraveling.

Sling is an assembly which connects the load to the material handling equipment.

Sling manufacturer is a person or organization that assembles sling components into their final form for sale to users.

Spiral is a single transverse coil that is the basic element from which metal mesh is fabricated.

Strand laid endless sling-mechanical joint is a wire rope sling made endless from one length of rope with the ends joined by one or more metallic fittings.

Strand laid grommet-hand tucked is an endless wire rope sling made from one length of strand wrapped six times around a core formed by hand tucking the ends of the strand inside the six wraps.

Strand laid rope is a wire rope made with strands (usually six or eight) wrapped around a fiber core, wire strand core, or independent wire rope core (IWRC).

Vertical hitch is a method of supporting a load by a single, vertical part or leg of the sling.

Whenever any sling is used, the following practices shall be observed:

- Slings that are damaged or defective shall not be used.
- Slings shall not be shortened with knots or bolts or other makeshift devices.
- Sling legs shall not be kinked.
- Slings shall not be loaded in excess of their rated capacities.
- Slings used in a basket hitch shall have the loads balanced to prevent slippage.
- Slings shall be securely attached to their loads.
- Slings shall be padded or protected from the sharp edges of their loads.

- Suspended loads shall be kept clear of all obstructions.
- All employees shall be kept clear of loads about to be lifted and of suspended loads.
- Hands or fingers shall not be placed between the sling and its load while the sling is being tightened around the load.
- Shock loading is prohibited.
- A sling shall not be pulled from under a load when the load is resting on the sling.

Each day before being used, the sling and all fastenings and attachments shall be inspected for damage or defects by a competent person designated by the employer. Additional inspections shall be performed during sling use, where service conditions warrant. Damaged or defective slings shall be immediately removed from service.

Alloy steel chain slings shall have permanently affixed durable identification stating size, grade, rated capacity, and reach. Table 6.10 in this section shows the rated capacity (working load limit), for alloy steel chain slings. Table 6.11 in this section lists the minimum allowable chain size at any point of link.

Hooks, rings, oblong links, pear shaped links, welded or mechanical coupling links or other attachments shall have a rated capacity at least equal to that of the alloy steel chain with which they are used or the sling shall not be used in excess of the rated capacity of the weakest component.

Makeshift links or fasteners formed from bolts or rods, or other such attachments, shall not be used.

A thorough periodic inspection of alloy steel chain slings in use shall be made on a regular basis, to be determined on the basis of (1) frequency of sling use; (2) severity of service conditions; (3) nature of lifts being made; and (4) experience gained on the service life of slings used in similar circumstances. Such inspections shall in no event be at intervals greater than once every 12 months.

The employer shall make and maintain a record of the most recent month in which each alloy steel chain sling was thoroughly inspected, and shall make such record available for examination.

The thorough inspection of alloy steel chain slings shall be performed by a competent person designated by the employer, and shall include a thorough inspection for wear, defective welds, deformation and increase in length. Where such defects or deterioration are present, the sling shall be immediately removed from service.

Table 6.10 Rated capacity (working load limit), for alloy steel chain slings rated capacity (working load limit), Newton meters [horizontal angles shown in parentheses].

Chain Size, cm	Single Branch Sling – 90º Loading	Double Sling Vertical Angle[1]			Triple and Quadruple Sling[3] Vertical Angle[1]		
		30º (60º)	45º (45º)	60º (30º)	30º (60º)	45º (45º)	60º (30º)
0.64	4,406	7,660	6,169	4,406	11,389	9,220	6,644
0.38	8,948	15,456	12,609	8,948	23,049	18,981	13,423
1.27	15,253	26,438	21,558	15,253	39,319	32,540	23,049
1.59	22,371	38,641	31,590	22,371	58,300	47,454	33,218
1.91	31,184	53,961	44,064	31,184	80,671	65,757	46,776
2.22	38,980	67,520	55,046	38,980	101,008	82,705	58,300
2.54	52,538	90,976	7,864	52,538	136,938	111,177	78,638
2.86	60,334	104,398	85,417	60,334	156,597	128,125	90,162
3.18	77,960	134,904	82,705	77,960	202,017	164,732	116,601
3.50	90,840	157,275	127,447	90,840	235,913	191,170	136,260
3.81	108,466	187,103	153,072	108,466	280,655	229,134	162,020
4.45	135,582	233,201	189,815	135,582	349,802	284,722	203,373

1. Rating of multileg slings adjusted for angle of loading measured as the included angle between the inclined leg and the vertical as shown in Figure N-184-5.

2. Rating of multileg slings adjusted for angle of loading between the inclined leg and the horizontal plane of the load, as shown in Figure N-184-5.

3. Quadruple sling rating is same as triple sling because normal lifting practice may not distribute load uniformly to all 4 legs.

Table 6.11 Minimum allowable chain size at any point of link.

Chain size,mm	Minimum Aallowable Chainsize, mm
6.4	5.6
3.8	7.5
12.7	9.9
15.9	12.3
19.1	15.1
22.2	18
25.4	20.7
28.6	23
31.8	25.4
35	27.8
38.1	30.8
44.5	35.7

Source: OSHA Standard 1910.184

The employer shall ensure that before use, each new, repaired, or reconditioned alloy steel chain sling, including all welded components in the sling assembly, shall be proof tested by the sling manufacturer or equivalent entity. The employer shall retain a certificate of the proof test and shall make it available for examination.

Alloy steel chain slings shall not be used with loads in excess of the rated capacities.

Alloy steel chain slings shall be permanently removed from service if they are heated above 537.8 degrees Celsius. When exposed to service temperatures in excess of 315.6 degrees Celsius, maximum working load limits permitted by the supplier shall be reduced in accordance with the chain or sling manufacturer's recommendations.

Worn or damaged alloy steel chain slings or attachments shall not be used until repaired. When welding or heat testing is performed, slings shall not be used unless repaired, reconditioned and proof tested by the sling manufacturer or an equivalent entity.

Mechanical coupling links or low carbon steel repair links shall not be used to repair broken lengths of chain.

Alloy steel chain slings with cracked or deformed master links, coupling links or other components shall be removed from service.

Wire rope slings shall not be used with loads in excess of the rated capacities shown in Tables N-184-3 through N-184-14. Slings not included in these tables shall be used only in accordance with the manufacturer's recommendations.

Cable laid and 6x19 and 6x37 slings shall have a minimum clear length of wire rope 10 times the component rope diameter between splices, sleeves or end fittings.

Braided slings shall have a minimum clear length of wire rope 40 times the component rope diameter between the loops or end fittings.

Cable laid grommets, strand laid grommets and endless slings shall have a minimum circumferential length of 96 times their body diameter.

Fiber core wire rope slings of all grades shall be permanently removed from service if they are exposed to temperatures in excess of 93.3 degrees Celsius. When nonfiber core wire rope slings of any grade are used at temperatures above 204.4 degrees Celsius or below minus 51.1 degrees Celsius, recommendations of the sling manufacturer regarding use at that temperature shall be followed.

Welding of end attachments, except covers to thimbles, shall be performed prior to the assembly of the sling.

All welded end attachments shall not be used unless proof tested by the manufacturer or equivalent entity at twice their rated capacity prior to initial use. The employer shall retain a certificate of the proof test, and make it available for examination.

6.15 Bibliography

American National Standard (ANSI) Z308.1-1998 "Minimum Requirements for Workplace First-aid Kits."

American Society for Testing and Materials (ASTM) D 120-87, Specification for Rubber Insulating Gloves.

ASTM D 178-93 (or D 178-88), Specification for Rubber Insulating Matting.

ASTM D 1048-93 (or D 1048-88a), Specification for Rubber Insulating Blankets.

ASTM D 1049-93 (or D 1049-88), Specification for Rubber Insulating Covers.

ASTM D 1050-90, Specification for Rubber Insulating Line Hose.

ASTM D 1051-87, Specification for Rubber Insulating Sleeves.

ANSI Safety Code for Building Construction, A10.2-1944.

Guide for the Selection of Chemical and Biological Decontamination Equipment for Emergency First Responders. National Institute of Justice Guide 103-00, Volume I, October 2001.

The Development of Criteria for Firefighters' Gloves; Vol. I: Glove Requirements.

The Development of Criteria for Firefighters' Gloves; Vol. II: Glove Criteria and Test Methods.

U.S. Department of Labor, Occupational Safety and Health Administration, Hazardous Waste Operations and Emergency Response (HAZWOPER) Standard (29 CFR 1910.120(q).

7

Recordkeeping, Training and Inspections, Accident Investigation and Reporting

7.1 Introduction

Occupational injury and illness records have several distinct functions or uses. One use is to provide information to employers whose employees are being injured or made ill by hazards in their workplace. The information in safety records makes employers more aware of the kinds of injuries and illnesses occurring in the workplace and the hazards that cause or contribute to them. When employers analyze and review the information in their records, they can identify and correct hazardous workplace conditions on their own. Injury and illness records are also an essential tool to help employers manage their company health and safety programs effectively.

Employees who have information about the occupational injuries and illnesses occurring in their workplace are also better informed about the hazards they face. They are therefore more likely to follow safe work practices and to report workplace hazards to their employers. When employees are aware of workplace hazards and participate in the identification and control of those hazards, the overall level of safety and health in the workplace improves.

This part provides procedures and recommended protocols for documentation, reporting and accident investigations.

7.2 Safety Recordkeeping Practices and Protocols

The extent of occupational illness and injury is much greater than government statistics indicate. The problem is recognized in the United States for example. Some U.S. researchers suspect that the number of injuries is twice as great as reported, and the number of illnesses may exceed the number reported by a factor of five. No one really knows. In most countries, accident statistics have only recently been recorded in the industry sector and there are no official published values. It's time that we learned the full extent and nature of occupational illness and injury in the workforce. The proper compensation for workers with workplace injuries and the direction of research on occupational disease prevention depend on accurate recordkeeping. Employers are required by law to record occupational injuries and illnesses. Records must be kept in the Daily Injury and Illness Log. An annual summary of that form is to be posted by the employer every February for the entire month. Records are to be kept for ten years.

Under-reporting of illness and injury may occur if employers aren't aware of the proper reporting procedures, if they misdiagnose conditions, if they fail to inform workers' compensation boards and insurance carriers when an illness/injury has occurred, if they make clerical errors in tabulating records, if they deliberately falsify records, etc.

Failure to comply with the recordkeeping regulations can bring stiff fines – up to $10,000 in for a willful violation. What's more, intentional falsification of records is a crime, grounds for a jail term.

7.2.1 Illnesses

It is required that employers record any work-related illness that is "diagnosed or recognized." A diagnosis is performed by a physician, registered nurse or someone with training and experience to make a diagnosis. It is required that employers record "recognized" (not diagnosed) illnesses because "employers, employees and others may be able to detect some illnesses such as skin diseases or disorders without the benefit of specialized medical training." Recordable illnesses include:

- skin diseases;
- lung diseases and other respiratory conditions;
- poisoning;
- disorders due to physical agents (such as heat or noise); and
- disorders associated with repeated trauma (such as carpal tunnel syndrome and tendonitis).

These cases do not require any lost time or medical treatment to be reported, only to be diagnosed or recognized.

7.2.2 Injuries

An injury which involves any of the following must be recorded:

- medical treatment (other than first aid);
- loss of consciousness;
- restriction of work motion;
- transfer to another job;
- termination of employment;
- electrical burns; and
- electrocution.

The distinction between an illness and an injury is not always obvious, but it is important, because all illnesses must be recorded, but some injuries do not need to be recorded. Whether a case is "an injury or an illness is determined by the nature of the original event or exposure which caused the case, not by the resulting condition." Injuries are caused by "instantaneous events." **A condition that is caused by anything other than an instantaneous event must be recorded as an illness.** "For example, a loss of hearing resulting from an explosion (an instantaneous event) is classified as an injury; the same condition resulting for exposure to industrial noise over a long period of time" must be classified as an illness, which means that it must be recorded even if it does not meet any of the criteria for a recordable injury.

7.2.3 Deaths

Any death of an employee or former employee that occurs as a result of an injury or illness that is recordable must also be recorded. If an employee has a recordable injury or illness and then is terminated or retires, and then dies as a result of the recordable injury or illness, the death must also be recorded.

7.2.4 What's Not Recordable

An injury that requires only first aid, that is, a minor injury that is resolved with one-time treatment (with follow-up for observation only, not additional treatment). A minor injury may be treated by a doctor or a nurse, but any injury that must be treated by a doctor or nurse or any injury that requires follow-up medical treatment is not minor, and must be recorded.

7.2.5 How to Analyze Injury and Illness Records

Local unions can analyze records to see if they are accurate and to pinpoint patterns of disease and injury in the workplace. Here's how:

1. Ask your employer in writing for the Incident Reporting daily log, not just the summary, for the past ten years. The right to review and copy these records is provided by the corresponding standard. The log and summary of all recordable occupational injuries and illnesses shall, upon request, be made available by the employer to any employee, former employee, and to their representatives for examination and copying in a reasonable manner and at reasonable times. The employee, former employee, and their representatives shall have access to the log for any establishment in which the employee is or has been employed.
2. Consult people knowledgeable about workers' compensation or insurance problems of the membership, since they know names of individual members who have filed claims which may not be recorded. The failure to list any cases known to the union is a tip-off that something is wrong.

3. Find out what current criteria the employer uses to record occupational illness and injury. This will also provide clues about how the employer interprets the guidelines when filling out the Incident Reporting log.
4. Obtain copies of employee exposure and medical records. Employers must provide this information to union representatives, free of charge, within 15 working days. This includes:
   - industrial hygiene reports;
   - biological monitoring;
   - medical histories and questionnaires*;
   - results of medical examinations & lab tests*;
   - medical opinions, diagnoses, & recommendations*; and
   - worker medical complaints*.

   *Requires prior written consent from affected employee(s).

7.2.6 Ensuring Accuracy in Recordkeeping

In order to avoid recordkeeping violations, agencies have adopted recognized procedures as a model for reporting injuries and illnesses. These procedures were developed by the U.S. company Chrysler Corporation in response to the record fines levied against them. Whether your employer adopts these guidelines or uses some other set of procedures, the union must actively police them to ensure accurate and complete reporting. The Chrysler procedures are reprinted below:

1. Each patient visiting the plant medical facility to report injuries or signs or symptoms of illnesses will be asked whether their medical condition is caused or aggravated by work. Responses will be recorded. Where the patient states that an injury or illness is work related, and that case otherwise meets the criteria for recording, that case will be entered on the log pending final determination of causality.
2. Injuries and illnesses for which workers' compensation claims are filed will be entered on the daily injury and illness log pending final determination. A list of claims paid voluntarily or by decisions will be maintained.

This list will be available for examination and copying by the union.

3. Sickness and accident claims for which occupational causation is claimed and which come to the attention of plant medical or safety departments will be investigated for occupational causation. Those instances which meet the recording criteria specified in the guidelines will be entered on the injury and illness log.
4. Instances where employees are granted medical restrictions or job transfers as a result of injuries or illnesses claimed to be of occupational origin will be entered on the log pending final determination. A list of such restrictions and transfers will be maintained. This list will be available to the union.
5. Where a case is removed from the log as a result of investigation, the reasons will be documented and the record of the investigation maintained. In general, investigations of illnesses will require on-site participation of medical personnel. The record of these investigations will be available to the union.
6. All personnel responsible for investigation of injury and illness reports and the maintenance of injury and illness records will be instructed. Special attention should be paid to diagnosis and recording criteria for repeated trauma disorders, strains and sprains, back injuries, and traumatic events.

7.3 Safety Training and Recordkeeping

7.3.1 Type of Training

Each organization should develop appropriate safety training programs. Programs should be tailored to the needs of the organization and based on a job hazards evaluation. At a minimum, all employees must receive hazard awareness training in addition to job specific training with emphasis on safe practices and familiarity with the safety standards. The following are some recommended programs that organizations may consider developing:

- Accident investigation training for supervisors and safety teams (2 hours) – Defines an "accident" in the workplace and explains why they should be investigated. Covers the rules of accident investigation and how to illicit information from victims and witnesses. Training should also cover the importance of investigating "near miss" incidents. Recommended for supervisors and safety committees.

- Back injury prevention (1 1/2 hours) – This session should cover statistics and causes of back pain and injury in the workplace. It also should discuss contributing factors such as medical conditions, lifestyle, posture and body mechanics. It should emphasize employee and employer responsibilities to help prevent back injuries in the workplace.
- Bloodborne pathogens awareness training (1 hour awareness, 2 hour in-depth) – Such a course will inform employees of the hazards due to occupational exposure to bloodborne pathogens (Contained in blood and other potentially infectious materials). Participants should be informed of the employer's exposure control plan and its contents. Simple awareness training is usually adequate for non-medical personnel.
- Confined space entry training (1 hour unless performing rescue) – The course should explain what confined spaces are and requirements for working in and around these spaces. This training can be performed according to the employer's confined space entry program and can also include a simulated rescue utilizing the employer's rescue equipment.
- General safety awareness training (1 1/2 hours) – This session should touch on all requirements regarding employer responsibilities for worker safety. It is recommended that the session include information on general duty clause, hazard communication, fire extinguishers in the workplace, first aid and bloodborne pathogens, lock out, tag out, personal protective equipment, employer safety programs, confined space entry, hearing conservation programs, lift truck requirements, respiratory protection, citation trends, and common shop violations.
- Hearing conservation training (1 hour) – This session should cover requirements for employers who have employees exposed to noise in the workplace. It should

examine time-weighted-averages and decibel level exposures. It should also cover the employers' responsibility for having a written hearing conservation program and the components required in it. It should also discuss requirements for annual audiometric testing and follow-up, training, access to information and recordkeeping for employees included in the program.

- Heat stress in the workplace (1/2 to 1 hour) – This session should cover risk factors (human and environmental) associated with heat stress injuries/illnesses. It should define signs and symptoms of heat-related illnesses including heat rash, heat cramps, heat syncope, heat exhaustion and heat stroke. It should also cover first aid procedures related to each as well as engineering and work practice controls to avoid heat stress.
- Ladder safety (1/2 hour) – It is recommended that a basic session detailing requirements for ladders and proper usage in the workplace including proper selection, inspection, set-up and use be implemented.
- Lock out, tag out (1 hour awareness, 2 to 3 hour in-depth) – This session should inform employees of safety requirements. This course should cover the proper measures for preventing the release or escape of hazardous energy (electrical, mechanical, thermal and other potential sources). Elements of this program should discuss required documented energy control procedures (written plans), employee training requirements and periodic inspections of the use of these procedures. Awareness training should be done on employees working on or around equipment that has been locked out and in-depth training should be done for those employees who actually perform the lock-out function.
- Office safety (1/2 to 1 hour) – Office personnel are not immune to workplace injuries. This session should detail workplace hazards in an office setting including chairs and desks, stacking and storage, stairs, filing cabinets, office machines, slips trips and falls, proper office layout, lighting, housekeeping, ergonomics, material handling, electrical/heat-generating appliances, chemical safety and lock-out, tag-out.

- Log training (recordkeeping requirements – 2 to 3 hours) – This session should cover requirements of record keeping standard (Part 4) including how to fill out incident reporting forms. Such a program will help those required to keep these forms to know what constitutes a recordable injury and/or illness and the importance of not over or under-recording.
- Safety committees (2 hours) – This program should detail how to put together an effective safety committee, how an effective committee operates and the purpose they exist. It should cover proper make-up of a committee and discuss why safety committees sometimes fail to meet their objectives. Good, practical advice on how to have a successful and productive safety committee should be offered.
- Safety training for supervisors and managers (1 to 2 hours) – This session should approach safety in the workplace from the manager or supervisor's standpoint. This program will help those in authority prioritize safe work practices and convey the importance of safety to their employees.

The employer may find the need to develop other programs. Not all programs should be mandatory – rather those which address site-specific hazards, job functions and working conditions. In addition to the recommended programs above, facilities should consider developing written safety programs for some or all of the topics listed below and then familiarizing workers with them in safety meetings and training sessions:

- Asbestos Safety Program
- Assured Equipment Grounding Program
- Benzene Safety Program
- Chemical Hygiene Plan
- Chemical Labeling Program
- Compressed Gas Safety Program
- Contractor Safety Verification Program
- Crane Safety Program
- Crisis Management Plan
- Drug Free Workplace Program – State and Federal
- Electrical Safety Hazards Program

- Eye/Face Protection Program
- Eyewash Station Program
- Facility Evacuation Plan
- Facility Smoking Program
- Facility Physical Security Plan
- Fall Protection Program
- Fire Extinguisher Program
- Fire Prevention Program
- First Aid Response Program
- Flammable and Combustible Liquids Safety Program
- Forklift Safety Program
- Hand and Power Tool Safety Program
- Hazard Communication Program
- Industrial Housekeeping Program
- Industrial Burn Safety
- Laboratory Safety Program
- Laser Safety Program
- Light Duty - Return To Work Program
- Loading Dock Safety Program
- Machine Guarding Program
- Marking Industrial Hazards Program
- Medical Services and First Aid Program
- MSDS Development Program
- New Products, Procedures, Equipment Program
- Overall Company Safety Program
- Portable Ladder Safety Program
- Power Press Safety Program
- PPE and Job Hazard Analysis Program
- Preparing for Workplace Emergencies
- Radiation Safety for X-ray Units
- Respiratory Protection Program
- Safety Committee Program
- Sand Abrasive Blasting Safety Program
- Scaffolding Safety Program
- Sling Safety Program
- Slips, Trips, & Falls Safety Program
- Spill Prevention, Control and Countermeasure
- Stairway and Ladder Safety Program
- Supervisor Safety Program
- Trenching and Excavations Program
- Violence in the Workplace Program

- Welding Safety Program
- Working in Hot Conditions Program
- Workplace Back Safety
- Bomb Threat Response Program
- Aerial Lift and Scissors Lift Safety Program
- Electrical Safety for Unqualified Employees Program.

7.3.2 Recordkeeping

Employers should maintain records of employee training for all employees for the life of employment and up to five years beyond retirement or termination. The records shall include name of employee, name of training session, dates of training, and number of contact hours.

The records should be made available to inspectors upon request. Maintaining accurate records will protect both employees and employers from liabilities and will help to identify areas where safer working environments can be created.

It is recommended that each organization adopt the use of continuing education units (CEUs) to keep track of overall training records for employees. The continuing education unit (CEU) is an internationally recognized method of quantifying the time spent in the classroom during professional development and training activities. Ten hours of instruction = 1.0 CEU. One hour of instruction = 0.1 CEU.

7.4 OHSAS 18001 (Occupation Health and Safety Assessment Series)

The following section is general information on the OHSAS 18000 series. OHSAS 18000 is a voluntary program which is recommended that organizations consider adopting to improve safety performance.

OHSAS 18000 is an international occupational health and safety management system specification. It comprises two parts, 18001 and 18002, and embraces BS8800 (British Standard) and a number of other publications.

OHSAS 18001 was created via a concerted effort from a number of the worlds leading national standards bodies, certification bodies, and specialist consultancies. OHSAS helps in a variety of respects. It helps: minimize risk to employees/etc; improve an existing

OH&S (Occupational Health and Safety) management system; demonstrate diligence; gain assurance; etc. Central to OHSAS requirements are strong policies and procedures.

Occupational Health & Safety is a subject that must be addressed by all organizations large and small. The organization's management system should identify all legislative requirements, identify the hazards and control the risks of the organization.

Progressive businesses will aim to go beyond compulsory measures and promote continuous improvement on health and safety matters.

Managing the health and safety of an organization can be approached using a structured management system and it can be integrated into current systems, to reduce the burden of bureaucracy.

A formal H&S management system will provide the following benefits:

- A system for continually identifying legal and other requirements.
- A clear management structure delegating authority and responsibility.
- A clear set of objectives for improvement, with measurable results.
- A structured approach to risk assessment within the organization.
- A planned and documented approach to health and safety.
- The monitoring of health and safety management issues, auditing of performance and review of policies and objectives.

Time spent on improving an organization's health and safety could provide a financial return in terms of:

- Reduced accidents and occupational ill health.
- Reduced stress and greater productivity.
- An improvement in underwriting risk.
- A reduction in the likelihood of paying regulatory compliance fees.

For many years, there has been demand for a certification scheme for occupational health and safety, which intensified with the publication of BS 8800 in 1996. However, while BS 8800 offers guidance on implementing an occupational health & safety management system, it is not and never was intended for certification purposes. The pressure was for a certification scheme that could offer independent verification that an organization has taken all reasonable measures to minimize risks and prevent accidents.

The situation prompted many certification bodies to develop their own specifications based on BS 8800. The inevitable irregularities between the specifications made this an undesirable way forward. In response, a committee was formed in November 1998 chaired by the British Standards Institution, and consisted of the major certification bodies and other national standard organizations known to be active in health and safety, with the goal of creating a single specification. This resulted in the occupational health and safety assessment series OHSAS 18001, which unified the existing schemes. Guidance to this specification can be found in OHSAS 18002.

The structure of OHSAS 18001 is:

- General requirements
- OH&S policy
- Hazard identification, risk assessment and determining controls
- Legal and other requirements
- Objectives and programs
- Resources, roles, responsibility, accountability and authority
- Competence, training and awareness
- Communication, participation and consultation
- Documentation
- Control of documents
- Operational control
- Emergency preparedness and response
- Performance measurement and monitoring
- Evaluation of compliance
- Incident investigation, nonconformity, corrective action and preventive action
- Control of records
- Internal audit
- Management review.

The certification process is similar to that of ISO 9001 or ISO 14001. Once the certification body receives a completed application form, a stage one assessment is undertaken on site, to determine the state of the policy, procedures and work instructions. If the readiness is satisfactory, then a date is set for the stage two assessment, which will assess the level of implementation. Once satisfactory, a certificate is issued. Annual surveillance ensures continued conformance. Information on the certification process can be obtained from the internet.

There are currently no costs published for certification; however it is anticipated that initially costs are high. As in the case of ISO 14001 (Environmental Management Series) companies may initially opt to sidestep certification and formally adopt the principles of OHSAS 18001. Many companies opt to formally adopt the management system and establish their own system following the OHSAS structure. Analogously, this approach has worked well in the environmental sector as there are many more companies throughout the world that have self declared ISO 14001 programs than certified programs.

7.5 Bibliography

International Standards Organization, ISO 9001.

International Standards Organization, ISO 14001.

U.S. Department of Labor, Occupation Health and Safety Assessment Series, OHSAS 18001.

U.S. Department of Labor, Occupation Health and Safety Assessment Series, OHSAS 18002.

U.S. Department of Labor, Occupational Safety and Health Administration, 300 Log records.

8

Risk and Vulnerability Assessments

8.1 Risk Management

8.1.1 General

What constitutes a work-related injury? Any illness or injury incurred in the course of employment. This includes industrial accidents, of course, but the injury need not result from one specific event. Repetitive stress injuries, such as back strain or carpal tunnel syndrome, are legitimate worker injuries.

In order to reduce accidents it is important to both identify and prevent accidents at the start, but also to immediately report incidents and to learn from them.

Men are more likely to be hurt on the job than women, filing two thirds of "lost time" claims. Men are also more susceptible to traumatic and permanent injuries. Women, by contrast, are more likely to file mental stress and cumulative-injury claims.

Fueled by statistics like these, on-site safety training or safety-management sessions with managers and employees should be viewed as critical. Companies might promote this as a bonus policy feature, and if you're a small company with few safety-management resources, this can be a huge help in reducing injuries and increasing efficiency. From a practical standpoint, companies may view this as loss-control services. If you're a smaller employer, chances are you don't maintain an internal risk management department. There are, however, many independent risk managers and safety consultants who can work with you to implement safety policies — good accident-reporting practices, safety training for employees, and meeting requirements for your industry can help you keep loss time down by having the right procedures (and the right safety record) in place.

Safety should not be something you talk about once a month. Safety should be in everything you do. Employers need to establish a safety culture – practices and policies that can reduce workers injuries from a humanitarian and a financial standpoint. First and foremost in safety management, is a safety committee, a group of managers and workers that discuss aspects of the workplace and what can be done to improve upon safety and health conditions.

The safety committee should be weighted to the side of the workers. They're involved in the day-to-day processes, and can bring ideas to the table on how to make them safer. Involving employees in that kind of forum is extremely important, because they see themselves as being involved in the process and can take that perspective back to the rest of the employees.

8.1.2 Steps to Risk Assessment

A risk assessment is an important step in protecting your workers and your business, as well as complying with the law. It helps you focus on the risks that really matter in your workplace – the ones with the potential to cause real harm. In many instances, straightforward measures can readily control risks, for example ensuring spillages are cleaned up promptly so people do not slip, or cupboard drawers are kept closed to ensure people do not trip. For most, that means simple, cheap and effective measures to ensure your most valuable asset - your workforce - is protected.

A risk assessment is simply a careful examination of what, in your work, could cause harm to people, so that you can weigh up whether you have taken enough precautions or should do more to prevent harm. Workers and others have a right to be protected from harm caused by a failure to take reasonable control measures.

Accidents and ill health can ruin lives and affect your business too if output is lost, machinery is damaged, insurance costs increase or you have to go to court. You are legally required to assess the risks in your workplace so that you put in place a plan to control the risks.

To assess the risks in your workplace follow the five steps below:

1. Identify the hazards;
2. Decide who might be harmed and how;
3. Evaluate the risks and decide on precaution;
4. Record your findings and implement them; and
5. Review your assessment and update if necessary.

8.1.2.1 Step 1: Identify the Hazards

First you need to work out how people could be harmed. When you work in a place everyday it is easy to overlook some hazards, so here are some tips to help you identify the ones that matter:

- Walk around your workplace and look at what could reasonably be expected to cause harm.
- Ask your employees or their representatives what they think. They may have noticed things that are not immediately obvious to you.
- If you are a member of a trade association, contact them. Many produce very helpful guidance.
- Check manufacturers' instructions or data sheets for chemicals and equipment as they can be very helpful in spelling out the hazards and putting them in their true perspective.
- Have a look back at your accident and ill-health records – these often help to identify the less obvious hazards.
- Remember to think about long-term hazards to health (e.g., high levels of noise or exposure to harmful substances) as well as safety hazards.

8.1.2.2 Step 2: Decide Who Might Be Harmed and How

For each hazard you need to be clear about who might be harmed; it will help you identify the best way of managing the risk. That doesn't mean listing everyone by name, but rather identifying groups of people (e.g. 'people working in the storeroom' or 'passers-by').

Remember:

- some workers have particular requirements, e.g. new and young workers, new or expectant mothers and people with disabilities may be at particular risk. Extra thought will be needed for some hazards;
- cleaners, visitors, contractors, maintenance workers etc, who may not be in the workplace all the time;
- members of the public, if they could be hurt by your activities;
- if you share your workplace, you will need to think about how your work affects others present, as well as how their work affects your staff – talk to them; and
- ask your staff if they can think of anyone you may have missed.

In each case, identify how they might be harmed, i.e. what type of injury or ill health might occur. For example, 'shelf stackers may suffer back injury from repeated lifting of boxes.'

8.1.2.3 Step 3: Evaluate the Risks and Decide on Precautions

Having spotted the hazards, you then have to decide what to do about them. Standards require you to do everything 'reasonably practicable' to protect people from harm. You can work this out for yourself, but the easiest way is to compare what you are doing with good practice.

Initially, look at what you're already doing, think about what controls you have in place and how the work is organized. Then compare this with the good practice and see if there's more you should be doing to bring yourself up to standard. In asking yourself this, consider:

- Can I get rid of the hazard altogether?
- If not, how can I control the risks so that harm is unlikely?

When controlling risks, apply the principles below, if possible in the following order:

- try a less risky option (e.g., switch to using a less hazardous chemical);

- prevent access to the hazard (e.g., by guarding);
- organize work to reduce exposure to the hazard (e.g., put barriers between pedestrians and traffic);
- issue personal protective equipment (e.g., clothing, footwear, goggles etc); and then
- provide welfare facilities (e.g., first aid and washing facilities for removal of contamination).

Improving health and safety need not cost a lot. For instance, placing a mirror on a dangerous blind corner to help prevent vehicular accidents is a low-cost precaution considering the risks. Failure to take simple precautions can cost you a lot more if an accident does happen.

Involve staff, so that you can be sure that what you propose to do will work in practice and won't introduce any new hazards.

8.1.2.4 Step 4: Record Your Findings and Implement Them

Putting the results of your risk assessment into practice will make a difference when looking after people and your business.

Writing down the results of your risk assessment, and sharing them with your staff, encourages you to do this. If you have fewer than five employees you do not have to write anything down, though it is useful so that you can review it at a later date if, for example, something changes.

When writing down your results, keep it simple, for example 'Tripping over rubbish: bins provided, staff instructed, weekly housekeeping checks', or 'Fume from welding: local exhaust ventilation used and regularly checked.'

We do not expect a risk assessment to be perfect – but it must be suitable and sufficient. You need to be able to show that:

- A proper check was made;
- You asked who might be affected;
- You dealt with all the obvious significant hazards, taking into account the number of people who could be involved;
- The precautions are reasonable, and the remaining risk is low; and
- You involved your staff or their representatives in the process.

If, like many businesses, you find that there are quite a lot of improvements that you could make, big and small, don't try to do everything at once. Make a plan of action to deal with the most important things first. Health and safety inspectors acknowledge the efforts of businesses that are clearly trying to make improvements.

A good plan of action often includes a mixture of different things such as:

- a few cheap or easy improvements that can be done quickly, perhaps as a temporary solution until more reliable controls are in place;
- long-term solutions to those risks most likely to cause accidents or ill health;
- long-term solutions to those risks with the worst potential consequences;
- arrangements for training employees on the main risks that remain and how they are to be controlled;
- regular checks to make sure that the control measures stay in place; and
- clear responsibilities – who will lead on what action, and by when.

Remember, prioritize and tackle the most important things first. As you complete each action, tick it off your plan.

8.1.2.5 Step 5: Review Your Risk Assessment and Update If Necessary

Few workplaces stay the same. Sooner or later, you will bring in new equipment, substances and procedures that could lead to new hazards. It makes sense therefore, to review what you are doing on an ongoing basis. Every year or so, formally review where you are to make sure you are still improving, or at least not sliding back.

Look at your risk assessment again. Have there been any changes? Are there improvements you still need to make? Have your workers spotted a problem? Have you learned anything from accidents or near misses? Make sure your risk assessment stays up to date.

When you are running a business, it's all too easy to forget about reviewing your risk assessment – until something has gone wrong and it's too late. Why not set a review date for this risk assessment now? Write it down and note it in your diary as an annual event.

During the year, if there is a significant change, don't wait: check your risk assessment and where necessary, amend it. If possible, it is best to think about the risk assessment when you're planning your change – that way you give yourself more flexibility.

8.1.2.6 *Sensible Risk Management Tips*

Risk management should be about practical steps to protect people from real harm and suffering – not bureaucratic back covering. If you believe some of the stories you hear, health and safety is all about stopping any activity that might possibly lead to harm. This is not the proper vision of sensible health and safety – rather the focus should be on saving lives, not stopping them. Organizations must seek a balance between the unachievable aim of absolute safety and the kind of poor management of risk that damages lives and the economy.

Don't overcomplicate the process. In many organizations, the risks are well known and the necessary control measures are easy to apply. You probably already know whether, for example, you have employees who move heavy loads and could harm their backs, or where people are most likely to slip or trip. If so, check that you have taken reasonable precautions to avoid injury.

If you run a small organization and you are confident you understand what's involved, you can do the assessment yourself. You don't have to be a health and safety expert.

If you work in a larger organization, you could ask a health and safety advisor to help you. If you are not confident, get help from someone who is competent. In all cases, you should make sure that you involve your staff or their representatives in the process. They will have useful information about how the work is done that will make your assessment of the risk more thorough and effective. But remember, you are responsible for seeing that the assessment is carried out properly.

When thinking about your risk assessment, remember:

- a hazard is anything that may cause harm, such as chemicals, electricity, working from ladders, an open drawer etc; and
- the risk is the chance, high or low, that somebody could be harmed by these and other hazards, together with an indication of how serious the harm could be.

Some frequently asked questions:

1. What if the work I do varies a lot, or I (or my employees) move from one site to another?
    a. Identify the hazards you can reasonably expect and assess the risks from them. This general assessment should stand you in good stead for the majority of your work. Where you do take on work or a new site that is different, cover any new or different hazards with a specific assessment. You do not have to start from scratch each time.
2. What if I share a workplace?
    a. Tell the other employers and self-employed people there about any risks your work could cause them, and what precautions you are taking. Also, think about the risks to your own workforce from those who share your workplace.
3. Do my employees have responsibilities?
    a. Yes. Employees have legal responsibilities to co-operate with their employer's efforts to improve health and safety (e.g., they must wear protective equipment when it is provided), and to look out for each other.
4. What if one of my employee's circumstances change?
    a. You'll need to look again at the risk assessment. You are required to carry out a specific risk assessment for new or expectant mothers, as some tasks (heavy lifting or work with chemicals for example) may not be appropriate. If an employee develops a disability then you are required to make reasonable adjustments. People returning to work following major surgery may also have particular requirements. If you put your mind to it, you can almost always find a way forward that works for you and your employees.
5. What if I have already assessed some of the risks?
    a. If, for example, you use hazardous chemicals and you have already assessed the risks to health and the precautions you can consider them 'checked' and move on.

Safety risk assessments, to the maximum extent feasible:

1. Are scientifically objective.
2. Are unbiased.
3. Include all relevant data available.
4. Employ default or conservative assumptions only if situation-specific information is not reasonably available. The basis of these assumptions must be clearly identified.
5. Distinguish clearly as to what risks would be affected by the decision and what risks would not.
6. Are reasonably detailed and accurate.
7. Relate to current risk or the risk resulting from not adopting the proposal being considered.
8. Allow for unknown and/or unquantifiable risks.

The principles to be applied when preparing safety risk assessments are:

1. Each risk assessment should first analyze the two elements of risk: severity of the hazard and likelihood of occurrence. Risk assessment is then performed by comparing the combined effect of their characteristics to acceptable criteria as determined in the plan.
2. A risk assessment may be qualitative and/or quantitative. To the maximum extent practicable, these risk assessments will be quantitative.
3. The selection of a risk assessment methodology should be flexible.
4. Basic assumptions should be documented or, if only bounds can be estimated reliably, the range encompassed should be described.
5. Significant risk assessment assumptions, inferences, or models should:
   a. Describe any model used in the risk assessment and make explicit the assumptions incorporated in the model.
   b. Identify any policy or value judgments.
   c. Explain the basis for choices.

   d. Indicate the extent that the model and the assumptions incorporated have been validated by or conflict with empirical data.
6. All safety risk assessments should include or summarize the information gathered. This record should be maintained by the organization performing the assessment.

8.1.3 Safety Risk Management Committees

A safety risk management committee can provide a valuable service to the organization for safety risk management planning. It can meet periodically to exchange risk management ideas and information. The committee can provide advice and counsel to the managers when requested.

The safety risk management committee provides a communication and support team to supplement the overall risk analysis capability and efficiency of key managers. The committee supports safety risk management activities. It provides advice and guidance, upon request from responsible program offices, to help them fulfill their authority and responsibility to incorporate safety risk management as a decision-making tool. It serves as an internal vehicle for risk management process communication, for coordination of risk analysis methods, and for use of common practices where appropriate. This includes, but is not limited to:

1. Continuing the internal exchange of risk management information.
2. Fostering the exchange of risk management ideas and information to avoid duplication of effort.
3. Providing risk analysis/management advice and guidance.
4. Identifying and recommending needed enhancements to risk analysis/management capabilities and/or efficiencies upon request.
5. Maintaining a risk management resources directory that includes:
   a. Risk methodologies productively employed,
   b. Specific internal risk analysis/management expertise by methodology or tool and organizational contact point(s), and
   c. A central contact point for resource identification assistance.

6. Encouraging the establishment of a directory of safety information resources via the internet.
7. Assisting in the identification of suitable risk analysis tools and initiate appropriate training in the use of these tools.

8.2 Crisis Management

Crisis prevention is very important for small public entities, companies and nonprofit organizations. Small organizations often have fewer resources to draw on when a crisis erupts, and insurance and other risk financing tools may not be an available due to the organization's meager financial resources. But every organization, from the smallest to the largest can and should take steps to prevent the preventable and prepare for the unavoidable. The key is to select the strategies that appeal to your organization and best suit your situation. Once you've undertaken some activities, your organization will be that much more fortified to withstand a crisis.

Crisis management requires your investment of time and common sense, rather than a large budget. You do what you can, as you can, keeping the final goal of preserving your vital mission at the forefront. Consider your organization as a physician considers a patient. Check your nonprofit's vital signs — the ones that enable your organization to fulfill its mission by meeting critical community and citizen/customer/client needs — to establish a baseline for future diagnosis. When you detect an aberration, determine the source, identify treatment methods, apply the methods and evaluate the results. Regular checkups should be scheduled to monitor the organization's vital signs (e.g., funding stream, cash flow, employee and volunteer turnover rates, past incidents, and losses and lawsuits). This process will make a critical difference in your organization's future health.

If your organization is healthy, determine what you can do to keep it that way for a long term. However, if you find weaknesses, you'll need to assess the symptoms, make a diagnosis and begin a treatment plan to cure its ailments so that it can thrive or, at the very least, ease the symptoms to enable it to survive. Weaknesses might exist in financial management, human resources, fund raising or volunteer management. The symptoms could manifest as uneven cash flow; a sharp increase in formal grievances; a complaint from a major constituency, investor or donor; or a steep reduction in volunteer hours. The financial treatment plan could involve purposefully delaying the launch of a new program to coincide with your funding cycle or applying for a line of credit. The human resources treatment plan might be to revise the staff handbook and re-train supervisors on your agency's policies and procedures. The volunteer management plan might involve identifying ways to involve volunteers more deeply in projects that are vital to the realization of your mission, or providing a wider range of opportunities for your volunteer work force.

If possible, you'll want to make gradual changes, starting with the ones that will bring your organization the most benefits. You don't want the treatment to cause more damage than the disease. Instead of forcing a mountain of new policies and procedures on a staff accustomed to an informal work environment, which could send key staff running for the want ads, begin by looking at what works and what's missing.

As a leader who wants a healthier entity, company or nonprofit, visualize what the organization would look like if it were strong and fortified to survive crises. Then divide and conquer the tasks required to reach your goal. How much less stressful this will be on you and your staff — and more do-able than thinking: OK, today I'm writing a crisis management plan and putting safety initiatives into practice and — and — and — or the place will fall apart tomorrow. The critical difference between success and failure is how you approach crisis management. Ease into making your organization healthier and more likely to avoid a preventable crisis and survive the one fated to occur, or you and your staff risk burnout without completing the job.

Another strategy for getting "buy-in" from personnel is to involve a diverse group of people in your crisis planning activities. If there are nay-sayers in your agency, why not get them involved in the process of identifying the crisis risks you want to focus on and determining the strategies you'll use to address these risks? If they are part of the problem-solving process, you can turn them into ambassadors for the strategies developed by the group, even when these strategies include new policies and procedures.

First, create a comprehensive directory of the organization's staff, board and key volunteers. Include home addresses, phone/fax/wireless/beeper numbers, as well as emergency contact information. Distribute the list to employees and keep copies off site, as well as in your offices. Update and redistribute the directory annually (perhaps the first business day in January or the first day of your fiscal year) or more often if you have a high turnover rate.

Next, maintain a backup of your computer file server, key databases, and financial files. Update the backup weekly (at least) and store a copy off site or on site in a fireproof safe.

Then, conduct an inventory of your nonprofit's assets. Include equipment, furniture, databases, records and anything else you need to fulfill your mission. Your inventory should include brand names, model numbers, location, purchase price and other key details necessary for insurance claims and replacement. Store a copy of the inventory on site (preferably in a fireproof safe) and off site.

Finally, identify an attorney licensed in your state who you can call upon from time to time for advice and assistance. (If your nonprofit can't afford to pay a monthly retainer or an hourly rate, consider soliciting bids from prominent law firms, emphasizing the charitable work of your agency. Don't be surprised if you receive proposals offering pro bono or dramatically discounted legal services).

For many small organizations it makes sense to convene one group to develop a crisis management plan for the organization and a second group that will serve as the nonprofit's "crisis response team." The crisis-planning group should be diverse and include people in the organization who fully understand the risks your operation faces. For example, if you own a building, including the person responsible for maintenance on your crisis planning team can help ensure that building-related hazards that could cause a crisis will be identified and addressed. Including the director of volunteers might enable the team to spot a volunteer-related hazard looming on the horizon.

A crisis response team is the group of people who coordinate your nonprofit's reaction and response to a crisis. The composition of an organization's crisis response team will vary based on a wide range of factors, including: the size of the organization, the nature of the products or services provided, the likely sources of crisis in the organization, and the organization's prior experience responding to a crisis. For example, in an organization with more than 50 staff, the crisis response team may include a handful of key department heads plus the CEO. In an organization with fewer than 10 paid staff, the crisis response team may include one or more board members, a couple of staff and outside professional advisors.

The composition of a crisis response team will vary with respect to the products and services that you offer. A team at an environmental advocacy group will differ from the team that responds to a crisis at a daycare center. In the former, the team may include an experienced lobbyist and an environmental scientist. In the latter, the team may include the organization's retained counsel, an expert on child-abuse prevention or playground safety, and parents of enrolled children.

The likely sources/causes of crisis in the organization should also be considered in forming a crisis response team. Is the organization more likely to face a crisis stemming from allegations of client/staff mistreatment or a crisis caused by inadequate financial resources? The ranking of crisis risks will suggest areas of expertise and training that may be required during a crisis and individuals with special talents or expertise may be identified as necessary members of the crisis response team. If your public entity, company or nonprofit has successfully weathered a crisis in the past, you'll want to include the people who were effective in addressing that situation on your crisis response team.

8.3 Vulnerability Assessments

A vulnerability assessment is the process of identifying, quantifying, and prioritizing (or ranking) the vulnerabilities in a system or organization. Examples of systems for which vulnerability assessments are performed include, but are not limited to, nuclear power plants, hydroelectric power plants, information technology systems, energy supply systems, water supply systems, transportation systems, and communication systems. Vulnerability assessments can be conducted for small businesses to large regional infrastructures.

A vulnerability assessment has many things in common with a risk assessment. Assessments are typically performed according to the following steps:

1. Cataloging assets and capabilities (resources) in a system.
2. Assigning quantifiable value (or at least rank order) and importance to those resources.
3. Identifying the vulnerabilities or potential threats to each resource.
4. Mitigating or eliminating the most serious vulnerabilities for the most valuable resources.

Classical risk analysis is principally concerned with investigating the risks surrounding physical plant (or some other object), its design, and operations. Such analyses tend to focus on causes and the direct consequences for the studied object. Vulnerability analyses, in contrast, focuses both on consequences for the object itself and on primary and secondary consequences for the surrounding environment. It also concerns itself with the possibilities of reducing such consequences and of improving the capacity to manage future incidents. In general, a vulnerability analysis serves to categorize key assets and drive the risk management process.

Energy utilities should routinely perform vulnerability assessments to better understand threats and vulnerabilities, determine acceptable levels of risk, and stimulate action to mitigate identified vulnerabilities. According to the U.S. Department of Energy, the direct benefits of performing a vulnerability assessment include:

- **Build and broaden awareness.** The assessment process directs senior management's attention to security. Security issues, risks, vulnerabilities, mitigation options, and best practices are brought to the surface. Awareness is one of the least expensive and most effective methods for improving the organization's overall security posture.
- **Establish or evaluate against a baseline**. If a baseline has been previously established, an assessment is an opportunity for a checkup to gauge the improvement or deterioration of an organization's security posture. If no previous baseline has been performed (or the work was not uniform or comprehensive), an assessment is an opportunity to integrate and unify previous efforts, define common metrics, and establish a definitive baseline. The baseline also can be compared against best practices to provide perspective on an organization's security posture.
- **Identify vulnerabilities and develop responses.** Generating lists of vulnerabilities and potential responses is usually a core activity and outcome of an assessment. Sometimes, due to budget, time, complexity, and risk considerations, the response selected for many of the vulnerabilities may be non-action, but after completing the assessment process, these decisions will be conscious ones, with a documented decision process

and item-by-item rationale available for revisiting issues at scheduled intervals. This information can help drive or motivate the development of a risk management process.

- **Categorize key assets and drive the risk management process.** An assessment can be a vehicle for reaching corporate-wide consensus on a hierarchy of key assets. This ranking, combined with threat, vulnerability, and risk analysis, is at the heart of any risk management process. For many organizations, the Y2K threat was the first time a company-wide inventory and ranking of key assets was attempted. An assessment allows an organization to revisit that list from a broader and more comprehensive perspective.
- **Develop and build internal skills and expertise.** A security assessment, when not implemented in an "audit" mode, can serve as an excellent opportunity to build security skills and expertise within an organization. A well-structured assessment can have elements that serve as a forum for cross-cutting groups to come together and share issues, experiences, and expertise. External assessors can be instructed to emphasize "teaching and collaborating" rather than "evaluating" (the traditional role). Whatever an organization's current level of sophistication, a long-term goal should be to move that organization towards a capability for self-assessment.
- **Promote action.** Although disparate security efforts may be underway in an organization, an assessment can crystallize and focus management attention and resources on solving specific and systemic security problems. Often, the people in the trenches are well aware of security issues (and even potential solutions) but are unable to convert their awareness to action. An assessment provides an outlet for their concerns and the potential to surface these issues at appropriate levels (legal, financial, executive) and achieve action. A well-designed and executed assessment not only identifies vulnerabilities and makes recommendations, it also gains executive buy-in, identifies key players, and establishes a set of cross-cutting groups that can convert those recommendations into action.

- **Kick off an ongoing security effort.** An assessment can be used as a catalyst to involve people throughout the organization in security issues, build cross-cutting teams, establish permanent forums and councils, and harness the momentum generated by the assessment to build an ongoing institutional security effort. The assessment can lead to the creation of either an actual or a virtual (matrixed) security organization.

The assessment methodology consists of 10 elements:

1. Network architecture;
2. Threat environment;
3. Penetration testing;
4. Physical security;
5. Physical asset analysis;
6. Operations security;
7. Policies and procedures;
8. Impact analysis;
9. Infrastructure interdependencies; and
10. Risk characterization.

8.3.1 Network Architecture

This element provides an analysis of the information assurance features of the information network(s) associated with the organization's critical information systems. Information examined should include network topology and connectivity (including subnets), principal information assets, interface and communication protocols, function and linkage of major software and hardware components (especially those associated with information security such as intrusion detectors), and policies and procedures that govern security features of the network. Procedures for information assurance in the system, including authentication of access and management of access authorization, should be reviewed. The assessment should identify any obvious concerns related to architectural vulnerabilities, as well as operating procedures.

Existing security plans should be evaluated, and the results of any prior testing should beanalyzed. Results from the network architecture assessment should include potential recommendations for changes in the information architecture, functional areas and categories where testing is needed, and suggestions regarding system design that would enable more effective information and information system protection.

Three techniques are used in conducting the network architecture assessment:

1. Analysis of network and system documentation during and after the site visit;
2. Interviews with facility staff, managers, and chief information officer; and
3. Tours and physical inspections of key facilities.

8.3.2 Threat Environment

Development of a clear understanding of the threat environment is a fundamental element of risk management. When combined with an appreciation of the value of the information assets and systems, and the impact of unauthorized access and subsequent malicious activity, an understanding of threats provides a basis for better defining the level of investment needed to prevent such access.

The threat of a terrorist attack to the electric power infrastructure is real and could come from several areas, including physical, cyber, and interdependency. In addition, threats could come from individuals or organizations motivated by financial gain or persons who derive pleasure from such penetration (e.g., recreational hackers, disgruntled employees). Other possible sources of threats are those who want to accomplish extremist goals (e.g., environmental terrorists, antinuclear advocates) or embarrass one or more organizations.

This element should include a characterization of these and other threats, identification of trends in these threats, and ways in which vulnerabilities are exploited. To the extent possible, characterization of the threat environment should be localized, that is, within the organization's service area.

8.3.3 Penetration Testing

The purpose of network penetration testing is to utilize active scanning and penetration tools to identify vulnerabilities that a determined adversary could easily exploit. Penetration testing can be customized to meet the specific needs and concerns of the utility. In general, penetration testing should include a test plan and details on the rules of engagement (ROE). It should also include a general characterization of the access points to the critical information systems and communication interface connections, modem network connections, access points to principal network routers, and other external connections. Finally, penetration testing should include identified vulnerabilities and, in particular, whether access could be gained to the control network or specific subsystems or devices that have a critical role in assuring continuity of service.

Penetration testing consists of an overall process for establishing the ground rules or ROE for the test; establishing a white cell for continuous communication; developing a format or methodology for the test; conducting the test; and generating a final report that details methods, findings, and recommendations.

Penetration testing methodology consists of three phases: reconnaissance, scenario development, and exploitation. A one-time penetration test can provide the utility with valuable feedback; however, it is far more effective if performed on a regular basis. Repeated testing is recommended because new threats develop continuously, and the networks, computers, and architecture of the utility are likely to change over time.

8.3.4 Physical Security

The purpose of a physical security assessment is to examine and evaluate the systems in place (or being planned) and to identify potential improvements in this area for the sites evaluated.

Physical security systems include access controls, barriers, locks and keys, badges and passes, intrusion detection devices and associated alarm reporting and display, closed-circuit television (assessment and surveillance), communications equipment (telephone, two-way radio, intercom, cellular), lighting (interior and exterior), power sources (line, battery, generator), inventory control, postings (signs), security system wiring, and protective force. Physical security systems are reviewed for design, installation, operation, maintenance, and testing.

The physical security assessment should focus on those sites directly related to the critical facilities, including information systems and assets required for operation. Typically included are facilities that house critical equipment or information assets or networks dedicated to the operation of electric or gas transmission, storage, or delivery systems. Other facilities can be included on the basis of criteria specified by the organization being assessed. Appropriate levels of physical security are contingent upon the value of company assets, the potential threats to these assets, and the cost associated with protecting the assets. Once the cost of implementing/maintaining physical security programs is known, it can be compared to the value of the company assets, thus providing the necessary information for risk management decisions. The focus of the physical security assessment task is determined by prioritizing the company assets; that is, the most critical assets receive the majority of the assessment activity.

At the start of the assessment, survey personnel should develop a prioritized listing of company assets. This list should be discussed with company personnel to identify areas of security strengths and weaknesses. During these initial interviews, assessment areas that would provide the most benefit to the company should be identified; once known, they should become the major focus of the assessment activities.

The physical security assessment of each focus area usually consists of the following:

- Physical security program (general);
- Physical security program (planning);
- Barriers;
- Access controls/badges;
- Locks/keys;
- Intrusion detection systems;
- Communications equipment; and
- Protective force/local law enforcement agency.

The key to reviewing the above topics is not to just identify if they exist, but to determine the appropriate level that is necessary and consistent with the value of the asset being protected. The physical security assessment worksheets provide guidance on appropriate levels of protection.

Once the focus and content of the assessment task have been identified, the approach to conducting the assessment can be either at the "implementation level" or at the "organizational level." The approach taken depends on the maturity of the security program.

For example, a company with a solid security infrastructure (staffing, plans/procedures, funding) should receive a cursory review of these items. However, facilities where the security programs are being implemented should receive a detailed review. The security staff can act upon deficiencies found at the facilities, once reported.

For companies with an insufficient security organization, the majority of time spent on the assessment should take place at the organizational level to identify the appropriate staffing/funding necessary to implement security programs to protect company assets. Research into specific facility deficiencies should be limited to finding just enough examples to support any staffing/funding recommendations.

8.3.5 Physical Asset Analysis

The purpose of the physical asset analysis is to examine the systems and physical operational assets to ascertain whether vulnerabilities exist. Included in this element is an examination of asset utilization, system redundancies, and emergency operating procedures. Consideration should also be given to the topology and operating practices for electric and gas transmission, processing, storage, and delivery, looking specifically for those elements that either singly or in concert with other factors provide a high potential for disrupting service. This portion of the assessment determines company and industry trends regarding these physical assets. Historic trends, such as asset utilization, maintenance, new infrastructure investments, spare parts, SCADA linkages, and field personnel are part of the scoping element.

Key output from analysis should be graphs that show trends. The historic data analysis should be supplemented with on-site interviews and visits. Items to focus on during a site visit include the following:

- Trends in maintenance expenditures and field staffing;
- Trends in infrastructure investments;
- Historic infrastructure outages;
- Critical system components and potential system bottlenecks;
- Overall system operation controls;
- Use and dependency of SCADA systems;

- Linkages of operation staff with physical and IT security;
- Adequate policies and procedures;
- Communications with other regional utilities;
- Communications with external infrastructure providers; and
- Adequate organizational structure.

8.3.6 Operations Security

Operations security (OPSEC) is the systematic process of denying potential adversaries (including competitors or their agents) information about capabilities and intentions of the host organization. OPSEC involves identifying, controlling, and protecting generally non-sensitive activities concerning planning and execution of sensitive activities. The OPSEC assessment reviews the processes and practices employed for denying adversary access to sensitive and non-sensitive information that might inappropriately aid or abet an individual's or organization's disproportionate influence over system operation (e.g., electric markets, grid operations). This assessment should include a review of security training and awareness programs, discussions with key staff, and tours of appropriate principal facilities. Information that might be available through public access should also be reviewed.

8.3.7 Policies and Procedures

The policies and procedures by which security is administered:

1. provide the basis for identifying and resolving issues;
2. establish the standards of reference for policy implementation; and
3. define and communicate roles, responsibilities, authorities, and accountabilities for all individuals and organizations that interface with critical systems. They are the backbone for decisions and day-to-day security operations.

Security policies and procedures become particularly important at times when multiple parties must interact to effect a desired level of security and when substantial legal ramifications could result from policy violations.Policies and procedures should be reviewed to determine whether they

1. address the key factors affecting security;
2. enable effective compliance, implementation, and enforcement;
3. reference or conform to established standards;
4. provide clear and comprehensive guidance; and
5. effectively address the risks.

The objective of the policies and procedures assessment task is to develop a comprehensive understanding of how a facility protects its critical assets through the development and implementation of policies and procedures. Understanding and assessing this area provide a means of identifying strengths and areas for improvements that can be achieved through:

- Modification of current policies and procedures;
- Implementation of current policies and procedures;
- Development and implementation of new policies and procedures;
- Assurance of compliance with policies and procedures; and
- Cancellation of policies and procedures that are no longer relevant, or are inappropriate, for the facility's current strategy and operations

8.3.8 Impact Analysis

A detailed analysis should be conducted to determine the influence that exploitation of unauthorized access to critical facilities or information systems might have on an organization's operations (e.g., market and/or physical operations). In general, such an analysis would require thorough understanding of:

1. the applications and their information processing;
2. decisions influenced by this information;
3. independent checks and balances that might exist regarding information upon which decisions are made;

4. factors that might mitigate the impact of unauthorized access; and
5. secondary impacts of such access (e.g., potential destabilization of organizations serving the grid, particularly those affecting reliability or safety). Similarly, the physical chain of events following disruption, including the primary, secondary, and tertiary impacts of disruption, should be examined.

The purpose of the impact analysis is to help estimate the impact that outages could have on a utility. Outages in electric power, natural gas, and oil can have significant financial and external consequences to a utility. The impact analysis provides an introduction to risk characterization by providing quantitative estimates of these impacts so that the utility can implement a risk management program and weigh the risks and costs of various mitigation measures.

8.3.9 Infrastructure Interdependencies

The term "infrastructure interdependencies" refers to the physical and electronic (cyber) linkages within and among our nation's critical infrastructures — energy (electric power, oil, natural gas), telecommunications, transportation, water supply systems, banking and finance, emergency services, and government services. This task identifies the direct infrastructure linkages between and among the infrastructures that support critical facilities as recognized by the organization. Performance of this task requires a detailed understanding of an organization's functions, internal infrastructures, and how these link to external infrastructures.

The purpose of the infrastructure interdependencies assessment is to examine and evaluate the infrastructures (internal and external) that support critical facility functions, along with their associated interdependencies and vulnerabilities.

8.3.10 Risk Characterization

Risk characterization provides a framework for prioritizing recommendations across all task areas. The recommendations for each task area are judged against a set of criteria to help prioritize the recommendations and assist the organization in determining the appropriate course of action. It provides a framework for assessing vulnerabilities, threats, and potential impacts (determined in the other tasks). In addition, the existing risk analysis and management process at the organization should be reviewed and, if appropriate, utilized for prioritizing recommendations. The degree to which corporate risk management includes security factors is also evaluated.

8.3.11 Post-Assessment

The post-assessment phase involves prioritizing assessment recommendations, developing an action plan, capturing lessons learned and best practices, and conducting training. The risk characterization element results provide the basis for the post-assessment by providing prioritized lists of recommendations that are ranked by key criteria. The company should take the prioritized lists and validate the recommendations and costs. Recommendations that are low cost or result in cost savings should be singled out for special attention. Other recommendations, however, might require formidable financial resources for implementation and require knowledge of the current company financial situation and posture toward risk.

Each company should carefully evaluate the costs and benefits of each recommendation.Recommendations compared in this section include making trade-offs in improvements in each of the other element areas. For example, which physical security measures should be selected versus changes in policies and procedures and network architecture? These are difficult decisions to make and a risk management framework combined with a diverse group of company decision makers should be a part of this decision making process.

The next step is to develop an action plan that includes timelines, staffing assignments, and budgets to implement the proposed recommendations. Lessons learned should be captured along the way to improve the overall process in the future. Training and other technical support activities, such as workshops, are also appropriate throughout the process.

Appendix A

Chemical Exposure Tables

Table 1 OSHA Permissible Exposure Limits (PEL).

Chemical Name	ppm (a)(1)	mg/m³ (b) (1)	Skin designation
Acetaldehyde	200	360	
Acetic acid	10	25	
Acetic anhydride	5	20	
Acetone	1000	2400	
Acetonitrile	40	70	
2-Acetylaminofluorene			
Acetylene tetrabromide	1	14	
Acrolein	0.1	0.25	
Acrylamide		0.3	X
Acrylonitrile			
Aldrin		0.25	X
Allyl alcohol	2	5	X
Allyl chloride	1	3	
Allyl glycidyl ether	(C)10	(C)45	
Allyl propyl disulfide	2	12	

Table 1 (cont.) OSHA Permissible Exposure Limits (PEL).

Chemical Name	ppm (a)(1)	mg/m³ (b) (1)	Skin designation
alpha-Alumina			
Total dust		15	
Respirable fraction		5	
Aluminum Metal (as Al).			
Total dust		15	
Respirable fraction		5	
4-Aminodiphenyl			
2-Aminopyridine	0.5	2	
Ammonia	50	35	
Ammonium sulfamate			
Total dust		15	
Respirable fraction		5	
n-Amyl acetate	100	525	
sec-Amyl acetate	125	650	
Aniline and homologs	5	19	X
Anisidine (o-,p-isomers)		0.5	X
Antimony and compounds (as Sb)		0.5	
Naphthylthiourea		0.3	
Arsenic, inorganic (as As)			
Arsenic, organic compounds (as As)		0.5	
Arsine	0.05	0.2	
Asbestos		(4)	
Azinphos-methyl		0.2	X
Barium, soluble compounds (as Ba)		0.5	
Barium sulfate			

Table 1 (cont.) OSHA Permissible Exposure Limits (PEL).

Chemical Name	**ppm (a)(1)**	**mg/m³ (b) (1)**	**Skin designation**
Total dust		15	
Respirable fraction		5	
Benomyl			
Total dust		15	
Respirable fraction		5	
Benzoyl peroxide		5	
Benzyl chloride	1	5	
Beryllium and beryllium compounds (as Be)		(2)	
Bismuth telluride			
Undoped			
Total dust		15	
Respirable fraction		5	
Boron oxide			
Total dust		15	
Boron trifluoride	(C)1	(C)3	
Bromine	0.1	0.7	
Bromoform	0.5	5	X
Butadiene	1 ppm/ 5 ppm STEL		
Methyl ethyl ketone	200	590	
2-Butoxyethanol	50	240	X
n-Butyl-acetate	150	710	
sec-Butyl acetate	200	950	
tert-Butyl-acetate	200	950	
n-Butyl alcohol	100	300	
sec-Butyl alcohol	150	450	
tert-Butyl alcohol	100	300	
Butylamine	(C)5	(C)15	X
tert-Butyl chromate (as CrO(3)			

Table 1 (cont.) OSHA Permissible Exposure Limits (PEL).

Chemical Name	ppm (a)(1)	mg/m³ (b) (1)	Skin designation
n-Butyl glycidyl ether (BGE)	50	270	
Butyl mercaptan	10	35	
p-tert-Butyltoluene	10	60	
Cadmium (as Cd)			
Calcium Carbonate			
Total dust		15	
Respirable fraction		5	
Calcium hydroxide			
Total dust		15	
Respirable fraction		5	
Calcium oxide		5	
Calcium silicate			
Total dust		15	
Respirable fraction		5	
Calcium sulfate			
Total dust		15	
Respirable fraction		5	
Camphor, synthetic		2	
Carbaryl (Sevin)		5	
Carbon black		3.5	
Carbon dioxide	5000	9000	
Carbon disulfide		(2)	
Carbon monoxide	50	55	
Carbon tetrachloride		(2)	
Cellulose			
Total dust		15	
Respirable fraction		5	
Chlordane		0.5	X
Chlorinated camphene		0.5	X
Chlorinated diphenyl oxide		0.5	

Table 1 (cont.) OSHA Permissible Exposure Limits (PEL).

Chemical Name	**ppm (a)(1)**	**mg/m³ (b) (1)**	**Skin designation**
Chlorine	(C)1	(C)3	
Chlorine dioxide	0.1	0.3	
Chlorine trifluoride	(C)0.1	(C)0.4	
Chloroacetaldehyde	(C)1	(C)3	
a-Chloroacetophenone (Phenacyl chloride)	0.05	0.3	
Chlorobenzene	75	350	
o-Chlorobenzylidene malononitrile	0.05	0.4	
Chlorobromomethane	200	1050	
Chlorodiphenyl (42% Chlorine)(PCB)		1	X
Chlorodiphenyl (54% Chlorine)(PCB)		0.5	X
Chloroform (Trichloromethane)	(C)50	(C)240	
1-Chloro-1-nitropropane	20	100	
Chloropicrin	0.1	0.7	
beta-Chloroprene	25	90	X
2-Chloro-6 (trichloromethyl) pyridine			
Total dust		15	
Respirable fraction		5	
Chromic acid and chromates (as CrO(3))		(2)	
Chromium (II) compounds (as Cr)		0.5	
Chromium (III) compounds (as Cr)		0.5	
Chromium metal and insol. salts (as Cr)		1	
Clopidol			
Total dust		15	

Table 1 (cont.) OSHA Permissible Exposure Limits (PEL).

Chemical Name	ppm (a)(1)	mg/m³ (b) (1)	Skin designation
Respirable fraction		5	
Coal dust (less than 5% SiO(2)), respirable fraction		(3)	
Coal dust (greater than or equal to 5% SiO(2)), respirable fraction		(3)	
Coal tar pitch volatiles (benzene soluble fraction), anthracene, BaP, phenanthrene, acridine, chrysene, pyrene		0.2	
Cobalt metal, dust, and fume (as Co)		0.1	
Copper			
Fume (as Cu)		0.1	
Dusts and mists (as Cu)		1	
Crag herbicide (Sesone)			
Total dust		15	
Respirable fraction		5	
Cresol, all isomers	5	22	X
Crotonaldehyde	2	6	
Cumene	50	245	X
Cyanides (as CN)		5	X
Cyclohexane	300	1050	
Cyclohexanol	50	200	
Cyclohexanone	50	200	
Cyclohexene	300	1015	
Cyclopentadiene	75	200	
2,4-D (Dichlorophenoxyacetic acid)		10	
Decaborane	0.05	0.3	X
Demeton (Systox)		0.1	X

Table 1 (cont.) OSHA Permissible Exposure Limits (PEL).

Chemical Name	ppm (a)(1)	mg/m³ (b) (1)	Skin designation
Diacetone alcohol (4-Hydroxy-4-methyl-2-pentanone)	50	240	
Diazomethane	0.2	0.4	
Diborane	0.1	0.1	
Dibutyl phosphate	1	5	
Dibutyl phthalate		5	
o-Dichlorobenzene	(C)50	(C)300	
p-Dichlorobenzene	75	450	
Dichlorodifluoromethane	1000	4950	
1,3-Dichloro-5,5-dimethyl hydantoin		0.2	
Dichlorodiphenyltri-chloroethane (DDT)		1	X
1,1-Dichloroethane	100	400	
1,2-Dichloroethylene	200	790	
Dichloroethyl ether	(C)15	(C)90	X
Dichloromonofluoro-methane	1000	4200	
1,1-Dichloro-1-nitroethane	(C)10	(C)60	
Dichlorotetrafluoro-ethane	1000	7000	
Dichlorvos (DDVP)		1	X
Dicyclopentadienyl iron			
Total dust		15	
Respirable fraction		5	
Dieldrin		0.25	X
Diethylamine	25	75	
2-Diethylaminoethanol	10	50	X
Difluorodibromomethane	100	860	
Diglycidyl ether (DGE)	(C)0.5	(C)2.8	
Diisobutyl ketone	50	290	
Diisopropylamine	5	20	X

Table 1 (cont.) OSHA Permissible Exposure Limits (PEL).

Chemical Name	**ppm (a)(1)**	**mg/m³ (b) (1)**	**Skin designation**
Dimethyl acetamide	10	35	X
Dimethylamine	10	18	
Dimethylaniline (N,N-Dimethylaniline)	5	25	X
Dimethyl-1,2-dibromo-2, 2-dichloroethyl phosphate		3	
Dimethylformamide	10	30	X
1,1-Dimethylhydrazine	0.5	1	X
Dimethylphthalate	–	5	
Dimethyl sulfate	1	5	X
Dinitrobenzene (all isomers)		1	X
(ortho)			
(meta)			
(para)			
Dinitro-o-cresol		0.2	X
Dinitrotoluene		1.5	X
Dioxane (Diethylene dioxide)	100	360	X
Diphenyl (Biphenyl)	0.2	1	
Dipropylene glycol methyl ether	100	600	X
Di-sec octyl phthalate (Di-(2-ethylhexyl) phthalate)		5	
Emery			
Total dust		15	
Respirable fraction		5	
Endrin		0.1	X
Epichlorohydrin	5	19	X
EPN		0.5	X
Ethanolamine	3	6	
2-Ethoxyethanol (Cellosolve)	200	740	X
2-Ethoxyethyl acetate (Cellosolve acetate)	100	540	X

Table 1 (cont.) OSHA Permissible Exposure Limits (PEL).

Chemical Name	ppm (a)(1)	mg/m³ (b) (1)	Skin designation
Ethyl acetate	400	1400	
Ethyl acrylate	25	100	X
Ethyl alcohol (Ethanol)	1000	1900	
Ethylamine	10	18	
Ethyl amyl ketone (5-Methyl-3-heptanone)	25	130	
Ethyl benzene	100	435	
Ethyl bromide	200	890	
Ethyl butyl ketone (3-Heptanone)	50	230	
Ethyl chloride	1000	2600	
Ethyl ether	400	1200	
Ethyl formate	100	300	
Ethyl mercaptan	(C)10	(C)25	
Ethyl silicate	100	850	
Ethylene chlorohydrin	5	16	X
Ethylenediamine	10	25	
Ethylene dibromide		(2)	
Ethylene dichloride (1, 2-Dichloroethane)		(2)	
Ethylene glycol dinitrate	(C)0.2	(C)1	X
N-Ethylmorpholine	20	94	X
Ferbam			
Total dust		15	
Ferrovanadium dust		1	
Fluorides (as F)		2.5	
Fluorine	0.1	0.2	
Fluorotrichloromethane (Trichlorofluoromethane)	1000	5600	
Formic acid	5	9	
Furfural	5	20	X
Furfuryl alcohol	50	200	

Table 1 (cont.) OSHA Permissible Exposure Limits (PEL).

Chemical Name	**ppm (a)(1)**	**mg/m³ (b) (1)**	**Skin designation**
Grain dust (oat, wheat, barley)		10	
Glycerin (mist)			
Total dust		15	
Respirable fraction		5	
Glycidol	50	150	
Graphite, natural			
respirable dust		(3)	
Graphite, synthetic			
Total dust		15	
Respirable Fraction		5	
Gypsum			
Total dust		15	
Respirable fraction		5	
Hafnium		0.5	
Heptachlor		0.5	X
Heptane (n-Heptane)	500	2000	
Hexachloroethane	1	10	X
Hexachloronaphthalene		0.2	X
n-Hexane	500	1800	
2-Hexanone (Methyl n-butyl ketone)	100	410	
Hexone (Methyl isobutyl ketone)	100	410	
sec-Hexyl acetate	50	300	
Hydrazine	1	1.3	X
Hydrogen bromide	3	10	
Hydrogen chloride	(C)5	(C)7	
Hydrogen cyanide	10	11	X
Hydrogen fluoride (as F)		(2)	
Hydrogen peroxide	1	1.4	
Hydrogen selenide (as Se)	0.05	0.2	

Table 1 (cont.) OSHA Permissible Exposure Limits (PEL).

Chemical Name	ppm (a)(1)	mg/m^3 (b) (1)	Skin designation
Hydrogen sulfide		(2)	
Hydroquinone		2	
Iodine	(C)0.1	(C)1	
Iron oxide fume		10	
Isomyl acetate	100	525	
Isomyl alcohol (primary and secondary)	100	360	
Isobutyl acetate	150	700	
Isobutyl alcohol	100	300	
Isophorone	25	140	
Isopropyl acetate	250	950	
Isopropyl alcohol	400	980	
Isopropylamine	5	12	
Isopropyl ether	500	2100	
Isopropyl glycidyl ether (IGE)	50	240	
Kaolin			
Total dust		15	
Respirable fraction		5	
Ketene	0.5	0.9	
Limestone			
Total dust		15	
Respirable fraction		5	
Lindane		0.5	X
Lithium hydride		0.025	
L.P.G. (Liquified petroleum gas)	1000	1800	
Magnesite			
Total dust		15	
Respirable fraction		5	
Magnesium oxide fume			
Total Particulate		15	

Table 1 (cont.) OSHA Permissible Exposure Limits (PEL).

Chemical Name	ppm (a)(1)	mg/m^3 (b) (1)	Skin designation
Malathion			
Total dust		15	X
Maleic anhydride	0.25	1	
Manganese compounds (as Mn)		(C)5	
Manganese fume (as Mn)		(C)5	
Marble			
Total dust		15	
Respirable fraction		5	
Mercury (aryl and inorganic)(as Hg)		(2)	
Mercury (organo) alkyl compounds (as Hg)		(2)	
Mercury (vapor) (as Hg)		(2)	
Mesityl oxide	25	100	
Methoxychlor			
Total dust		15	
2-Methoxyethanol (Methyl cellosolve)	25	80	X
2-Methoxyethyl acetate (Methyl cellosolve acetate)	25	120	X
Methyl acetate	200	610	
Methyl acetylene (Propyne)	1000	1650	
Methyl acetylene propadiene mixture (MAPP)	1000	1800	
Methyl acrylate	10	35	X
Methylal (Dimethoxymethane)	1000	3100	
Methyl alcohol	200	260	
Methylamine	10	12	
Methyl n-amyl ketone	100	465	
Methyl bromide	(C)20	(C)80	X
Methyl chloride		(2)	

Table 1 (cont.) OSHA Permissible Exposure Limits (PEL).

Chemical Name	ppm (a)(1)	mg/m³ (b) (1)	Skin designation
Methyl chloroform (1,1,1-Trichloro-ethane)	350	1900	
Methylcyclohexane	500	2000	
Methylcyclohexanol	100	470	
o-Methylcyclohexanone	100	460	X
Methylene chloride		(2)	
Methyl formate	100	250	
Methyl hydrazine (Monomethyl hydrazine)	(C)0.2	(C)0.35	X
Methyl iodide	5	28	X
Methyl isoamyl ketone	100	475	
Methyl isobutyl carbinol	25	100	X
Methyl isocyanate	0.02	0.05	X
Methyl mercaptan	(C)10	(C)20	
Methyl methacrylate	100	410	
alpha-Methyl styrene	(C)100	(C)480	
Methylene bisphenyl isocyanate (MDI)	(C)0.02	(C)0.2	
Molybdenum (as Mo)			
Soluble compounds		5	
Total dust		15	
Monomethyl aniline	2	9	X
Morpholine	20	70	X
Naphtha (Coal tar)	100	400	
Naphthalene	10	50	
Nickel carbonyl (as Ni)	0.001	0.007	
Nickel, metal and insoluble compounds (as Ni)		1	
Nickel, soluble compounds (as Ni)		1	
Nicotine		0.5	X
Nitric acid	2	5	

Table 1 (cont.) OSHA Permissible Exposure Limits (PEL).

Chemical Name	ppm (a)(1)	mg/m³ (b) (1)	Skin designation
Nitric oxide	25	30	
p-Nitroaniline	1	6	X
Nitrobenzene	1	5	X
p-Nitrochlorobenzene		1	X
Nitroethane	100	310	
Nitrogen dioxide	(C)5	(C)9	
Nitrogen trifluoride	10	29	
Nitroglycerin	(C)0.2	(C)2	X
Nitromethane	100	250	
1-Nitropropane	25	90	
2-Nitropropane	25	90	
Nitrotoluene (all isomers)	5	30	X
o-isomer			
m-isomer			
p-isomer			
Octachloronaphthalene		0.1	X
Octane	500	2350	
Oil mist, mineral		5	
Osmium tetroxide (as Os)		0.002	
Oxalic acid		1	
Oxygen difluoride	0.05	0.1	
Ozone	0.1	0.2	
Paraquat, respirable dust		0.5	X
Parathion		0.1	X
Particulates not otherwise regulated (PNOR)(f)			
Total dust		15	
Respirable fraction		5	
Pentaborane	0.005	0.01	
Pentachloronaphthalene		0.5	X
Pentachlorophenol		0.5	X

Table 1 (cont.) OSHA Permissible Exposure Limits (PEL).

Chemical Name	ppm (a)(1)	mg/m³ (b) (1)	Skin designation
Pentaerythritol			
Total dust		15	
Respirable fraction		5	
Pentane	1000	2950	
2-Pentanone (Methyl propyl ketone)	200	700	
Perchloroethylene (Tetrachloroethylene)		(2)	
Perchloromethyl mercaptan	0.1	0.8	
Perchloryl fluoride	3	13.5	
Petroleum distillates (Naphtha)(Rubber Solvent)	500	2000	
Phenol	5	19	X
p-Phenylene diamine		0.1	X
Phenyl ether, vapor	1	7	
Phenyl ether-biphenyl mixture, vapor	1	7	
Phenyl glycidyl ether (PGE)	10	60	
Phenylhydrazine	5	22	X
Phosdrin (Mevinphos)		0.1	X
Phosgene (Carbonyl chloride)	0.1	0.4	
Phosphine	0.3	0.4	
Phosphoric acid		1	
Phosphorus (yellow)		0.1	
Phosphorus pentachloride		1	
Phosphorus pentasulfide		1	
Phosphorus trichloride	0.5	3	
Phthalic anhydride	2	12	
Picloram			
Total dust		15	
Respirable fraction		5	

Table 1 (cont.) OSHA Permissible Exposure Limits (PEL).

Chemical Name	ppm (a)(1)	mg/m^3 (b) (1)	Skin designation
Picric acid		0.1	X
Pindone (2-Pivalyl-1, 3-indandione)		0.1	
Plaster of paris			
Total dust		15	
Respirable fraction		5	
Platinum (as Pt)			
Soluble Salts		0.002	
Portland cement			
Total dust		15	
Respirable fraction		5	
Propane	1000	1800	
n-Propyl acetate	200	840	
n-Propyl alcohol	200	500	
n-Propyl nitrate	25	110	
Propylene dichloride	75	350	
Propylene imine	2	5	X
Propylene oxide	100	240	
Pyrethrum		5	
Pyridine	5	15	
Quinone	0.1	0.4	
Rhodium (as Rh), metal fume and insoluble compounds		0.1	
Rhodium (as Rh), soluble compounds		0.001	
Ronnel		15	
Rotenone		5	
Rouge			
Total dust		15	
Respirable fraction		5	
Selenium compounds (as Se)		0.2	
Selenium hexafluoride (as Se)	0.05	0.4	

Table 1 (cont.) OSHA Permissible Exposure Limits (PEL).

Chemical Name	**ppm (a)(1)**	**mg/m³ (b) (1)**	**Skin designation**
Silica, amorphous, precipitated and gel		(3)	
Silica, amorphous, diatomaceous earth, containing less than 1% crystalline silica		(3)	
Silica, crystalline cristobalite, respirable dust		(3)	
Silica, crystalline quartz, respirable dust		(3)	
Silica, crystalline tripoli (as quartz), respirable dust		(3)	
Silica, crystalline tridymite, respirable dust		(3)	
Silica, fused, respirable dust		(3)	
Silicates (less than 1% crystalline silica)			
Mica (respirable dust)		(3)	
Soapstone, total dust		(3)	
Soapstone, respirable dust		(3)	
Talc (containing no asbestos), respirable dust		(3)	
Silicon			
Total dust		15	
Respirable fraction		5	
Silicon carbide			
Total dust		15	
Respirable fraction		5	
Silver, metal and soluble compounds (as Ag)		0.01	
Sodium fluoroacetate		0.05	X
Sodium hydroxide		2	
Starch			
Total dust		15	

Table 1 (cont.) OSHA Permissible Exposure Limits (PEL).

Chemical Name	ppm (a)(1)	mg/m³ (b) (1)	Skin designation
Respirable fraction		5	
Stibine	0.1	0.5	
Stoddard solvent	500	2900	
Strychnine		0.15	
Styrene		(2)	
Sucrose			
Total dust		15	
Respirable fraction		5	
Sulfur dioxide	5	13	
Sulfur hexafluoride	1000	6000	
Sulfuric acid		1	
Sulfur monochloride	1	6	
Sulfur pentafluoride	0.025	0.25	
Sulfuryl fluoride	5	20	
2,4,5-T (2,4,5-tri-chlorophenoxyacetic acid)		10	
Tantalum, metal and oxide dust.		5	
TEDP (Sulfotep)		0.2	X
Tellurium and compounds (as Te)		0.1	
Tellurium hexafluoride (as Te)	0.02	0.2	
Temephos			
Total dust		15	
Respirable fraction		5	
TEPP (Tetraethyl pyrophosphaate)		0.05	X
Terphenylis	(C)1	(C)9	
1,1,1,2-Tetrachloro-2, 2-difluoroethane	500	4170	
1,1,2,2-Tetrachloro-1, 2-difluoroethane	500	4170	

Table 1 (cont.) OSHA Permissible Exposure Limits (PEL).

Chemical Name	ppm (a)(1)	mg/m³ (b) (1)	Skin designation
1,1,2,2-Tetrachloro-ethane	5	35	X
Tetrachloronaphthalene.		2	X
Tetraethyl lead (as Pb)		0.075	X
Tetrahydrofuran	200	590	
Tetramethyl lead, (as Pb)		0.075	X
Tetramethyl succinonitrile	0.5	3	X
Tetranitromethane	1	8	
Tetryl (2,4,6-Trinitro-phenylmethyl-nitramine)		1.5	X
Thallium, soluble compounds (as Tl)		0.1	X
4,4′-Thiobis(6-tert, Butyl-m-cresol)			
Total dust		15	
Respirable fraction		5	
Thiram		5	
Tin, inorganic compounds (except oxides) (as Sn)		2	
Tin, organic compounds (as Sn)		0.1	
Titanium dioxide			
Total dust		15	
Toluene		(2)	
Toluene-2,4-diisocyanate (TDI)	(C)0.02	(C)0.14	
o-Toluidine	5	22	X
Tributyl phosphate		5	
1,1,2-Trichloroethane	10	45	X
Trichloroethylene		(2)	
Trichloronaphthalene		5	X
1,2,3-Trichloropropane	50	300	
1,1,2-Trichloro-1,2, 2-trifluoroethane	1000	7600	

Table 1 (cont.) OSHA Permissible Exposure Limits (PEL).

Chemical Name	**ppm (a)(1)**	**mg/m³ (b) (1)**	**Skin designation**
Triethylamine	25	100	
Trifluorobromomethane	1000	6100	
2,4,6-Trinitrotoluene (TNT)		1.5	X
Triorthocresyl phosphate		0.1	
Triphenyl phosphate		3	
Turpentine	100	560	
Uranium (as U)			
Soluble compounds		0.05	
Insoluble compounds		0.25	
Vanadium			
Respirable dust (as V2O5)		(C)0.5	
Fume (as V2O5)		(C)0.1	
Vegetable oil mist			
Total dust		15	
Respirable fraction		5	
Vinyl toluene	100	480	
Warfarin		0.1	
Xylenes (o-, m-, p-isomers)	100	435	
Xylidine	5	25	X
Yttrium		1	
Zinc chloride fume		1	
Zinc oxide fume		5	
Zinc oxide			
Total dust		15	
Respirable fraction		5	
Zinc stearate			
Total dust		15	

Table 1 (cont.) OSHA Permissible Exposure Limits (PEL).

Chemical Name	**ppm (a)(1)**	**mg/m³ (b) (1)**	**Skin designation**
Respirable fraction		5	
Zirconium compounds (as Zr)		5	

29 CFR Part Number: 1910 Part Title: Occupational Safety and Health Standards Subpart: Z Subpart Title: Toxic and Hazardous Substances, Standard Number: 1910.1000 TABLE Z-1, Title: TABLE Z-1 Limits for Air Contaminants. Available HTTP://www.osha.gov/pls/oshaweb/owadisp.show_document?p_table=STANDARDS&p_id=9992

Table 2 OSHA Permissible Exposure Limits (PEL) Time Weighted Averages.

Substance	8-hour TWA	Acceptable ceiling concentration	Acceptable maximum peak above the acceptable ceiling concentration for an 8-hr shift	Maximum duration
Benzene(a)	10 ppm	25 ppm	50 ppm	10 minutes
Beryllium and beryllium compounds	2 ug/m(3)	5 ug/m(3)	25 ug/m(3)	30 minutes
Cadmium fume(b)	0.1 mg/m(3)	0.3 mg/m(3)		
Cadmium dust(b)	0.2 mg/m(3)	0.6 mg/m(3)		
Carbon disulfide	20 ppm	30 ppm	100 ppm	30 minutes
Carbon tetrachloride	10 ppm	25 ppm	200 ppm	5 min. in any 3 hrs
Chromic acid and chromates		1 mg/10 m(3)		
Ethylene dibromide	20 ppm	30 ppm	50 ppm	5 minutes
Ethylene dichloride	50 ppm	100 ppm	200 ppm	5 min. in any 3 hrs
Fluoride as dust	2.5 mg/m(3)			
Formaldehyde				
Hydrogen fluoride	3 ppm			

Hydrogen sulfide		20 ppm	50 ppm	10 mins. once only if no other meas. exp. occurs
Mercury		1 mg/10m(3)		
Methyl chloride	100 ppm	200 ppm	300 ppm	5 mins. in any 3 hrs
Methylene Chloride				
Organo (alkyl) mercury	0.01mg/m(3)	0.04 mg/m(3)		
Styrene	100 ppm	200 ppm	600 ppm	5 mins. in any 3 hrs
Tetrachloroethylene	100 ppm	200 ppm	300 ppm	5 mins. in any 3 hrs
Toluene	200 ppm	300 ppm	500 ppm	10 minutes
Trichloroethylene	100 ppm	200 ppm	300 ppm	5 mins. in any 2 hrs

29 CFR Part Number: 1910 Part Title: Occupational Safety and Health Standards Subpart: Z Subpart Title: Toxic and Hazardous Substances Standard Number: 1910.1000 TABLE Z-2 Title: TABLE Z-2. Available HTTP://www.osha.gov/pls/oshaweb/owadisp.show_document?p_table= STANDARDS&p_id=9993

Table 3 OSHA IDLH.

Chemical Name	IDLH	STEL/ Ceiling
	ppm	ppm
Acetaldehyde	2000++	25+
Acetic Acid	50	15
Acetic Anhydride	200	
Acetone	2500	1000
Acetonitrile	500	60
Acetylene tetrabromide	8	
Acrolein	2	0.3
Acrylamide	60 mg/m^3++	
Aldrin	25 mg/m^3++	
Allyl alcohol	20	4
Allyl chloride	250	2
Allyl glycidyl ether	50	10+
Allyl propyl disulfide		3
2-Aminopyridine	5	
Ammonia	300	35
Ammonium sulfamate (td)	1500 mg/m^3	
n-Amyl acetate	1000	
sec-Amyl acetate	1000	
Aniline and homologs	100 ++	
Anisidine o-, p-isomers	50 mg/m^3	
Antimony & compounds	50 mg/m^3	
ANTU	100 mg/m^3	
Arsine	3 ++	
Azinphos-methyl	10 mg/m^3	
Benzoyl peroxide	1500 mg/m^3	
Benzyl chloride	10	
Boron oxide (td)	2000 mg/m^3	
Boron trifluoride	25	1+
Bromine	3	0.2

Table 3 (cont.) OSHA IDLH.

Chemical Name	IDLH	STEL/ Ceiling
	ppm	ppm
Bromoform	850	
1,3 Butadiene	2000++	
2-Butanone (MEK)	3000	300
2-Butoxyethanol	700	
n-Butyl-acetate	1700	200
sec-Butyl acetate	1700	
tert-Butly acetate	1500	
n-Butyl alcohol	1400	50+
sec-Butyl alcohol	2000	
tert-Butyl alcohol	1600	
Butylamine	300	5+
tert-Butyl chromate	15 mg/m^3	0.1 mg/m^3+
n-Butyl glycidyl ether	250	
Butyl mercaptan	500	
p-tert-Butyl-1-toluene	100	
Calcium oxide	25 mg/m^3	
Camphor, synthetic	200 mg/m^3	19 mg/m^3+
Carbaryl	100 mg/m^3	
Carbon black	1750 mg/m^3	
Carbon dioxide	40,000	30,000
Carbon Disulfide	500	
Carbon monoxide	1200	
Chlorodane	100 mg/m^3++	
Chlorinated camphene	200 mg/m^3CA	
Chlorinated diphenyl oxide	5 mg/m^3	
Chlorine	10	1
Chlorine dioxide	5	0.3
Chlorine trifluoride	20	0.1+
Chloroacetaldehyde	45	1+

Table 3 (cont.) OSHA IDLH.

Chemical Name	**IDLH**	**STEL/ Ceiling**
	ppm	**ppm**
a-chloroacetophenone	15 mg/m^3	
Chlorobenzene	1000	
o-Chlorobenzylidene malononitrile	2 mg/m^3	0.05+
Chlorobromomethane	2000	
Chlorodiphenyl (42% Cl)	5 mg/m^3++	
Chlorobiphenyl (54% Cl)	5 mg/m^3++	
1-Chloro-1-nitro- propane	100	
Chloropicrin	2	
beta-Chloroprene	300	
2-Chloro-6-trichloro- methyl) pyridine (td)		20 mg/m^3
Chromium (II) compounds	250 mg/m^3	
Chromium (III) compounds	25 mg/m^3	
Chromium metal and insoluble salts	250 mg/m^3	
Coal tar pitch volatiles	80 mg/m^3++	
Cobalt metal, dust, and fume	20 mg/m^3	
Copper fume	100 mg/m^3	
Copper dusts and mists	100 mg/m^3	
Cotton dust	100 mg/m^3	
Crag herbicide (td)	500 mg/m^3	
Cresol, all isomers	250	
Crotonaldehyde	50	
Cumene	900	
Cyclohexane	1300	
Cyclohexanol	400	
Cyclohexanone	700	
Cyclohexene	2000	
Cyclopentadiene	750	

Table 3 (cont.) OSHA IDLH.

Chemical Name	**IDLH**	**STEL/ Ceiling**
	ppm	**ppm**
2,4-D	100 mg/m^3	
Decaborane	15 mg/m^3	0.15
Demeton	10 V	
Diacetone alcohol	1800	
Diazomethane	2	
Diborane	15	
Dibutyl phosphate	30	2
Dibutyl phthalate	4000 mg/m^3	
o-Dichlorobenzene	200	
p-Dichlorobenzene	150++	
1,1Dichloroethane	3000	
1,2-Dichloroethylene	1000	
Dichloroethyl ether	100++	15
Dichloromono- fluomethane	5000	
1,1 Dichloro-1- nitroethane	25	10 +
Dichlorotetra- fluoroethane	15,000	
Dichlorvos (DDVP)	100 mg/m^3	
Dieldrin	50 mg/m^3++	
Diethylamine	200	15
2-Diethylaminoethanol	100	
Difluorodibromomethane	2000	
Diglycidyl ether (DGE)	10++	0.5+
Diisobutyl ketone	500	
Diisopropylamine	200	
Dimethyl acetamide	300	
Dimethylamine	500	15
Dimethylaniline (N,N-Dimethylaniline)	100	10
Dimethyl-1,2-dibromo- 2, 2-dichloroethyl phosphate	200 mg/m^3	

Table 3 (cont.) OSHA IDLH.

Chemical Name	**IDLH**	**STEL/ Ceiling**
	ppm	**ppm**
Dimethylformamide	500	
1,1-Dimethylhydrazine	15++	
Dimethylphthalate	2000 mg/m^3	
Dimethyl sulfate	7++	
Dinitrobenzene (all isomers)	50 mg/m^3	
Dinitro-o-cresol	5 mg/m^3	
Dinitrotoluene	50 mg/m^3++	
Dioxane (Diethylene dioxide)	500++	
Diphenyl (Biphenyl)	100 mg/m^3	
Dipropylene glycol methyl ether	600	150
Di-sec octyl phthalate (Di-(2-ethylhexyl) phthalate)	10 mg/m^3+	
Endrin	2 mg/m^3	
Epichlorohydrin	75++	
EPN	5 mg/m^3	
Ethanolamine	30	6
2-Ethoxyethanol (Cellosolve)	500	
2-Ethoxyethyl acetate (cellosolve acetate)	500	
Ethyl acetate	2000	
Ethyl acrylate	300	15
Ethyl alcohol (Ethanol)	3300	
Ethylamine	600	15
Ethyl benzene	800	125
Ethyl bromide	2000	
Ethyl butyl ketone (3- Heptanone)	1000	
Ethyl chloride	3800	
Ethylene chlorohydrin	7	
Ethylene glycol dinitrate	75 mg/m^3	0.2+
Ethyl ether	1900	500

Table 3 (cont.) OSHA IDLH.

Chemical Name	IDLH	STEL/ Ceiling
	ppm	ppm
Ethyl formate	1500	
Ethyl mercaptan	500	10+
Ethyl silicate	700	
Ethylenediamine	1000	
N-Ethylmorpholine	100	
Ferbam (td)	800 mg/m^3	
Ferrovanadium dust	500 mg/m^3	3 mg/m^3+
Fluorine	25	2
Fluorotrichloromethane (Trichlorofluoromethane)	2000	1000+
Formic acid	30	10
Furfural	100	
Furfuryl alcohol	75	15
Glycidol	150	
Hafnium	50 mg/m^3	
Heptachlor	35 mg/m^3++	
Heptane (n-Heptane)	750	500
Hexachloroethane	300++	
Hexachloronaph- thalene	2 mg/m^3	
n-Hexane	1100	
2-Hexanone (Methyl n-butyl ketone)	1600	
Hexone (Methyl isobutyl ketone)	500	75
sec-Hexyl acetate	500	
Hydrazine	50	
Hydrogen bromide	30	3+
Hydrogen chloride	50	5+
Hydrogen cyanide	50	4.7+
Hydrogen peroxide	75	
Hydrogen selenide (as Se)	1	

Table 3 (cont.) OSHA IDLH.

Chemical Name	IDLH	STEL/ Ceiling
	ppm	ppm
Hydroquinone	50 mg/m^3	
Iodine	2	0.1+
Iron oxide fume	2500 mg/m^3	
Isoamyl acetate	1000	
Isoamyl alcohol (primary and secondary)	500	125
Isobutyl acetate	1300	
Isobutyl alcohol	1600	
Isophorone	200	5+
Isopropyl acetate	1800	310
Isopropyl alcohol	2000	500
Isopropylamine	750	10
Isopropyl ether	1400	310
Isopropyl glycidyl ether (IGE)	400	75
Ketene	5	1.5
Lindane	50 mg/m^3	
Lithium hydride	0.5 mg/m^3	
L.P.G. (Liquefied petroleum gas)	2000	
Magnesium oxide fume (total particulate)	750 mg/m^3	
Malathion (td)	250 mg/m^3	
Maleic anhydride	10 mg/m^3	
Manganese compounds (as Mn)	500 mg/m^3	5 mg/m^3+
Manganese fume (as Mn)	500 mg/m^3	5 mg/m^3+
Mesityl oxide	1400	25
Methoxychlor (td)	5000 mg/m^3+	
2-Methoxyethanol (Methyl cellosolve)	200	
2-Methoxyethyl acetate (Methyl cellosolve acetate)	200	
Methyl acetate	3100	250

Table 3 (cont.) OSHA IDLH.

Chemical Name	IDLH	STEL/ Ceiling
	ppm	ppm
Methyl acetylene (propyne)	1700	
Methyl acetylene- propadiene mixture (MAPP)	3400	1250
Methyl acrylate	250	
Methylal (Dimetoxy- methane)	2200	
Methyl alcohol	6000	250
Methylamine	100	15
Methyl n-amyl ketone	800	
Methyl bromide	250++	20+
Methyl chloroform (1,1,1- Trichloroethane)	700	450
Methylcyclohexane	1200	
Methylcyclohexanol	500	
o-Methylcyclohexanone	600	75
Methylene bisphenyl isocyanate (MDI)	75 mg/m^3	0.02+
Methyl formate	4500	150
Methyl hydrazine (Mono- methyl hydrazine)	20++	0.2+
Methyl iodide	100++	
Methyl isobutyl carbinol	400	40
Methyl isocyanate	3	
Methyl mercaptan	150	10+
Methyl methacrylate	1000	
alpha-Methyl styrene	700	100+
Molybdenum (as Mo) Soluble compounds	1000 mg/m^3	
Molybdenum (as Mo) Insoluble compounds (td)	5000 mg/m^3	
Monomethyl aniline	100	
Morpholine	1400	

Table 3 (cont.) OSHA IDLH.

Chemical Name	**IDLH**	**STEL/ Ceiling**
	ppm	**ppm**
Naphtha (Coal tar)	1000	
Naphthalene	250	15
Nickel carbonyl (as Ni)	2++	
Nickel, metal and insoluble compounds (as Ni)	10 mg/m^3++	
Nicotine	5 mg/m^3	
Nitric acid	25	4
Nitric oxide	100	
p-Nitroaniline	300 mg/m^3	
Nitrobenzene	200	
p-Nitrochlorobenzene	100 mg/m^3+	
Nitroethane	1000	
Nitrogen dioxide	20	5
Nitrogen trifluoride	1000	
Nitroglycerin	75 mg/m^3	0.2+
Nitromethane	750	
1-Nitropropane	1000	
2-Nitropropane	100	
Nitrotoluene (all isomers)	200	
Octachloronaphthalene	Unknown	0.3 mg/m^3+
Octane	1000	375
Oil mist, mineral	2500 mg/m^3	10 mg/m^3
Osmium tetroxide (as Os)	1 mg/m^3	0.0006
Oxalic acid	500 mg/m^3	2 mg/m^3
Oxygen difluoride	0.5	0.05+
Ozone	5	0.1+
Paraquat (rd)	1 mg/m^3	
Parathion	10 mg/m^3	
Pentaborane	1	0.015
Pentachloronaphthalene	Unknown	

Table 3 (cont.) OSHA IDLH.

Chemical Name	**IDLH**	**STEL/ Ceiling**
	ppm	**ppm**
Pentachlorophenol	2.5 mg/m^3	
Pentane	1500	750
2-Pentanone (Methyl propyl ketone)	1500	250
Perchloromethyl mercaptan	10	
Perchloryl fluoride	100	25+
Petroleum distillates (Naphtha) (Rubber Solvent)	1100	1480
Phenol	250	
p-Phenylene diamine	25 mg/m^3	
Phenyl ether, vapor	100	2
Phenyl ether-biphenyl mixture, vapor	10	
Phenyl glycidyl ether (PGE)	100++	
Phenylhydrazine	15++	
Phosdrin (mevinphos)	4	0.03
Phosgene (Carbonyl chloride)	2	
Phosphine	50	1
Phosphoric acid	1000 mg/m^3	
Phosphorus (yellow)	5 mg/m^3	
Phosphorus pentachloride	70 mg/m^3	
Phosphorus pentasulfide	250 mg/m^3	3 mg/m^3
Phosphorus trichloride	25	0.5
Phthalic anhydride	60 mg/m^3	
Picric acid	75 mg/m^3	
Pindone (2-Pivalyl-1, 3-indandione)	100 mg/m^3	
Platinum (as Pt) Soluble salts	4 mg/m^3	
Portland cement (td)	5000 mg/m^3	
Propane	2100	
n-Propyl acetate	1700	250

Table 3 (cont.) OSHA IDLH.

Chemical Name	**IDLH**	**STEL/ Ceiling**
	ppm	**ppm**
n-Propyl alcohol	800	250
Propylene dichloride	400++	110
Propylene imine	100++	
Propylene oxide	400++	
n-Propyl nitrate	500	40
Pyrethrum	5000 mg/m^3	
Pyridine	1000	
Quinone	100 mg/m^3	
Rhodium, metal fume and insoluble compounds	100 mg/m^3	
Rhodium, soluble compounds	2 mg/m^3	
Ronnel	300 mg/m^3	
Rotenone	2500 mg/m^3	
Selenium compounds (as Se)	1 mg/m^3	
Selenium hexafluoride (as Se)	2	
Silver, metal and soluble compounds (as Ag)	10 mg/m^3	
Sodium fluoro- acetate	2.5 mg/m^3	
Sodium hydroxide	10 mg/m^3	2 mg/m^3+
Stibine	5	
Stoddard solvent	20,000 mg/m^3	
Strychnine	3 mg/m^3	
Sulfur dioxide	100	5
Sulfuric acid	15 mg/m^3	3 mg/m^3+
Sulfur monochloride	5	1+
Sulfur pentafluoride	1	0.01+
Sulfuryl fluoride	200	10
2,4,5-T (2,4,5-trichloro-phenoxyacetic acid)	250 mg/m^3	
Tantalum, metal and oxide dust	2500 mg/m^3	
TEDP (Sulfotep)	10 mg/m^3	

Table 3 (cont.) OSHA IDLH.

Chemical Name	**IDLH**	**STEL/ Ceiling**
	ppm	**ppm**
Tellurium and com- pounds (as Te)	25 mg/m^3	
Tellurium hexafluoride (as Te)	1	
TEPP (Tetraethyl pyrophosphate)	5 mg/m^3	
Terphenyls	500 mg/m^3	1
1,1,1,2-Tetrachloro-2, 2-difluoroethane	2000	
1,1,2,2-Tetrachloro-1, 2-difluoroethane	2000	
1,1,2,2-Tetrachlo- roethane	100++	
Tetrachloronaph- thalene	Unknown	
Tetraethyl lead (as Pb)	40 mg/m^3	
Tetrahydrofuran	2000	250
Tetramethyl lead (as Pb)	40 mg/m^3	
Tetramethyl succinonitrile	5	
Tetranitromethane	4	
Tetryl (2,4,6- Trinitro- phenylmethylni- tramine)	750 mg/m^3	
Thallium, soluble compounds (as Tl)	15 mg/m^3	
Thiram	100 mg/m^3	
Tin, inorganic compounds (except oxides) (as Sn)	100 mg/m^3	
Tin, organic com- pounds (as Sn)	25 mg/m^3	
Titanium dioxide (td)	5000 mg/m^3+	
Toluene-2,4- diisocyanate (TDI)	2.5++	0.02+
o-Toluidine	50++	
Tributyl phosphate	30	
1,1,2-Trichloroethane	100++	
Trichloronaphthalene	Unknown	
1,2,3-Trichloropropane	100++	

Table 3 (cont.) OSHA IDLH.

Chemical Name	**IDLH**	**STEL/ Ceiling**
	ppm	**ppm**
1,1,2-Trichloro-1,2,2- trifluoroethane	2000	1250
Triethylamine	200	3
Trifluorobromo- methane	40,000	
2,4,6-Trinitrotoluene (TNT)	500 mg/m^3	
Triorthocresyl phosphate	40 mg/m^3	
Triphenyl phosphate	1000 mg/m^3	
Turpentine	800	
Uranium (as U) Soluble compounds	10 mg/m^3++	
Uranium (as U) Insoluble compounds	10 mg/m^3++	
Vanadium (rd) (as V2O5)	35 mg/m^3	0.5 mg/m^3
Vanadium (fume) (as V2O5)	35 mg/m^3	0.1 mg/m^3
Vinyl toluene	400	100
Warfarin	100 mg/m^3	
Xylenes (o-, m-, p-isomers)	900	150
Xylidine	50	
Yttrium	500 mg/m^3	
Zinc chloride fume	50 mg/m^3	
Zinc oxide fume	500 mg/m^3	
Zinc oxide (td)	500 mg/m^3	
Zinc oxide (rf)	500 mg/m^3	
Zirconium compounds (as Zr)	50 mg/m^3	10 mg/m^3

"+" indicates a Ceiling Value.

"++" indicates that the chemical is believed, by NIOSH, to be a potential Carcinogen.

**Reference: http://www.labsafety.com/refinfo/ezfacts/ezf232.htm

Table 4 NIOSH Recommended Exposure Limits (REL).

Chemical Name	NIOSH REL TWA (ppm)	NIOSH REL TWA (mg/m^3)	NIOSH Ceiling REL (ppm)	NIOSH Ceiling REL (mg/m^3)	NIOSH REL STEL (ppm)	NIOSH REL STEL (mg/m^3)
AA	2 (skin)	5 (skin)			4 (skin)	10 (skin)
2-AAF						
AAF						
Abate®		10 (total) 5 (resp)				
Acetaldehyde						
Acetic acid	10	25			15	37
Acetic anhydride	5	20				
Acetone	250	590				
Acetone cyanohydrin			1 (15min)	4 (15 min)		
Acetonitrile	20	34				
2-Acetylaminofluorene						
Acetylene			2500	2662		
Acetylene tetrabromide						
Acetylsalicyclic acid		5				
Acrolein	0.1	0.25			0.3	0.8
Acrylamide		0.03 (skin)				

Table 4 (cont.) NIOSH Recommended Exposure Limits (REL).

Chemical Name	NIOSH REL TWA (ppm)	NIOSH REL TWA (mg/m^3)	NIOSH Ceiling REL (ppm)	NIOSH Ceiling REL (mg/m^3)	NIOSH REL STEL (ppm)	NIOSH REL STEL (mg/m^3)
Acrylic acid	2 (skin)	6 (skin)				
Acrylonitrile	1 (15min/ skin)		10 (15min/ skin)			
3-(alpha-Acetonyl) -benzyl-4-hydroxycoumarin		0.1				
2-Acetoxybenzoic acid		5				
o-Acetoxybenzoic acid		5				
1-Acetoxyethylene			4 (15min)	15 (15min)		
2-Acetylaminofluorene						
N-Acetyl-2-aminofluorene						
Acetylene black		3.5 (0.1 mg PAHs/m^3)				
Acetylene dichloride	200	790				
cis-Acetylene dichloride	200	790				
trans-Acetylene dichloride	200	790				
Acetylene tetrabromide						
Acetylene tetrachloride	1 (skin)	7 (skin)				
Acetyl mercaptan	1 (skin)	4 (skin)				

Acetyl oxide			5	20		
2-Acetyl propane	200	705				
Acetylsalicyclic acid		5				
Acheson graphite						
Acraldehyde	0.1	0.25			0.3	0.8
Acridine		0.1 (cyclohexane-extractable)				
Acroleic acid	2 (skin)	6 (skin)				
Acrolein	0.1	0.25			0.3	0.8
Acrylaldehyde	0.1	0.25			0.3	0.8
Acrylamide		0.03 (skin)				
Acrylamide monomer		0.03 (skin)				
Acrylic acid	2 (skin)	6 (skin)				
Acrylic aldehyde	0.1	0.25			0.3	0.8
Acrylic amide		0.03 (skin)				
Acrylonitrile	1 (15min/ skin)		10 (15min/ skin)			
Acrylonitrile monomer	1 (15min/ skin)		10 (15min/ skin)			
Actinolite						

Table 4 (cont.) NIOSH Recommended Exposure Limits (REL).

Chemical Name	NIOSH REL TWA (ppm)	NIOSH REL TWA (mg/m^3)	NIOSH Ceiling REL (ppm)	NIOSH Ceiling REL (mg/m^3)	NIOSH REL STEL (ppm)	NIOSH REL STEL (mg/m^3)
Actinolite asbestos						
Adiponitrile	4	18				
AGE	5 (skin)	22 (skin)			10 (skin)	44 (skin)
Alcohol	1000	1900				
Aldrin		0.25(skin)				
Aliphatic petroleum naphtha		350 (15min)		1800 (15min)		
Allyl alcohol	2 (skin)	5 (skin)			4 (skin)	10 (skin)
Allyl aldehyde	0.1	0.25			0.3	0.8
Allyl chloride	1	3			2	6
Allylene	1000	1650				
Allyl glycidyl ether	5 (skin)	22 (skin)			10 (skin)	44 (skin)
Allylic alcohol	2 (skin)	5 (skin)			4 (skin)	10 (skin)
1-Allyloxy-2, 3-epoxypropane	5 (skin)	22 (skin)			10 (skin)	44 (skin)
Allyl propyl disulfide	2	12			3	18
Allyl trichloride	10 (skin)	60 (skin)				
alpha-Alumina						

Alumina						
Aluminum		10 (total) 5 (resp)				
Aluminum metal		10 (total) 5 (resp)				
Aluminum oxide						
Aluminum oxide						
Aluminum powder		10 (total) 5 (resp)				
Aluminum (pyro powders and welding fumes, as Al)		5				
Aluminum (soluble salts and alkyls, as Al)		2				
Aluminum trioxide						
Amidocyanogen		2				
4-Aminoaniline		0.1 (skin)				
ortho-Aminoanisole		0.5 (skin)				
para-Aminoanisole		0.5 (skin)				
Aminobenzene						
4-Aminobiphenyl						
p-Aminobiphenyl						
1-Aminobutane			5 (skin)	15 (skin)		

Table 4 (cont.) NIOSH Recommended Exposure Limits (REL).

Chemical Name	NIOSH REL TWA (ppm)	NIOSH REL TWA (mg/m^3)	NIOSH Ceiling REL (ppm)	NIOSH Ceiling REL (mg/m^3)	NIOSH REL STEL (ppm)	NIOSH REL STEL (mg/m^3)
Aminocaproic lactam	0.22 (vapor)	1 (dust) 1 (vapor)			0.66 (vapor)	3 (dust) 3 (vapor)
Aminocyclohexane	10	40				
Aminodimethylbenzene	2 (skin)	10 (skin)				
4-Amino-6-(1, 1-dimethyl-ethyl) -3-(methylthio) -1, 2, 4-triazin-5(4H) -one		5				
4-Aminodiphenyl						
p-Aminodiphenyl						
1-Aminobutane			5 (skin)	15 (skin)		
Aminocaproic lactam	0.22 (vapor)	1 (dust) 1 (vapor)			0.66 (vapor)	3 (dust) 3 (vapor)
Aminocyclohexane	10	40				
Aminodimethylbenzene	2 (skin)	10 (skin)				
4-Amino-6-(1, 1-dimethyl-ethyl) -3-(methylthio) -1, 2, 4-triazin-5(4H) -one		5				
4-Aminodiphenyl						
p-Aminodiphenyl						

Aminoethane	10	18				
2-Aminoethanol	3	8			6	15
beta-Aminoethyl alcohol	3	8			6	15
bis(2-Aminoethyl) amine	1 (skin)	4 (skin)				
Aminoethylene						
N-(2-Aminoethyl)-1, 2-ethanediamine	1 (skin)	4 (skin)				
Aminohexahydrobenzene	10	40				
Aminomethane	10	12				
1, 3-bis(Aminomethyl) benzene				0.1 (skin)		
3-Amino-1-methylbenzene						
1-Aminonaphthalene						
2-Aminonaphthalene						
para-Aminonitrobenzene		3 (skin)				
1-Aminophenylmethane						
2-Aminopropane						
2-Aminopyridine	0.5	2				
alpha-Aminopyridine	0.5	2				
2-Aminotoluene						

Table 4 (cont.) NIOSH Recommended Exposure Limits (REL).

Chemical Name	NIOSH REL TWA (ppm)	NIOSH REL TWA (mg/m^3)	NIOSH Ceiling REL (ppm)	NIOSH Ceiling REL (mg/m^3)	NIOSH REL STEL (ppm)	NIOSH REL STEL (mg/m^3)
4-Aminotoluene						
m-Aminotoluene						
o-Aminotoluene						
2-Amino-1, 3, 4-triazole		0.2				
3-Amino-1, 2, 4-triazole		0.2				
3-Aminotriazole		0.2				
Aminotriazole		0.2				
4-Amino-3, 5, 6-trichloro-2-picolinic acid						
4-Amino-3, 5, 6-trichloropic-olinic acid						
Aminoxylene	2 (skin)	10 (skin)				
Amitrole		0.2				
Ammate herbicide		10 (total) 5 (resp)				
Ammonia	25	18			35	27
Ammonium amidosulfonate		10 (total) 5 (resp)				
Ammonium chloride		10				20

Ammonium chloride fume		10				20
Ammonium muriate fume		10				20
Ammonium sulfamate		10 (total) 5 (resp)				
Amosite (cummingtonite-grunerite)						
AMS	50	240			100	485
AMS		10 (total) 5 (resp)				
n-Amyl acetate	100	525				
sec-Amyl acetate	125	650				
Amyl acetic ester	100	525				
Amyl acetic ether	100	525				
Amyl aldehyde	50	175				
Amyl ethyl ketone	25	130				
Amyl hydrosulfide			0.5 (15min)	2.1 (15min)		
Amyl mercaptan			0.5 (15min)	2.1 (15min)		
Amyl methyl ketone	100	465				
n-Amyl methyl ketone	100	465				
Amyl sulfhydrate			0.5 (15min)	2.1 (15min)		
AN	1 (15min/ skin)		10 (15min/ skin)			
Anhydrous ammonia	25	18			35	27

Table 4 (cont.) NIOSH Recommended Exposure Limits (REL).

Chemical Name	NIOSH REL TWA (ppm)	NIOSH REL TWA (mg/m^3)	NIOSH Ceiling REL (ppm)	NIOSH Ceiling REL (mg/m^3)	NIOSH REL STEL (ppm)	NIOSH REL STEL (mg/m^3)
Anhydrous borax		1				
Anhydrous calcium sulfate		10 (total) 5 (resp)				
Anhydrous gypsum		10 (total) 5 (resp)				
Anhydrous hydrogen bromide			3	10		
Anhydrous hydrogen chloride			5	7		
Anhydrous hydrogen fluoride	3 (15min)	2.5 (15min)	6 (15min)	5 (15min)		
Anhydrous sulfate of lime		10 (total) 5 (resp)				
Aniline (and homologs)						
Aniline oil						
Anilinobenzene		10				
2-Anilinonaphthalene						
2-Anisidine		0.5 (skin)				
4-Anisidine		0.5 (skin)				
o-Anisidine		0.5 (skin)				
p-Anisidine		0.5 (skin)				

Anol	50 (skin)	200 (skin)				
Anone	25 (skin)	100 (skin)				
Antabuse®		2				
Anthophyllite						
Anthophyllite asbestos						
Anthracene & benzo(a) pyrene) .		0.1				
Anthracite coal dust						
Antimony		0.5				
Antimony hydride	0.1	0.5				
Antimony metal		0.5				
Antimony powder		0.5				
Antimony trihydride	0.1	0.5				
ANTU		0.3				
Aprocarb®		0.5				
Aqua ammonia	25	18			35	27
Aqua fortis	2	5			4	10
Aqueous acrylic acid (technical grade is 94%)	2 (skin)	6 (skin)				
Aqueous ammonia	25	18			35	27

Table 4 (cont.) NIOSH Recommended Exposure Limits (REL).

Chemical Name	NIOSH REL TWA (ppm)	NIOSH REL TWA (mg/m^3)	NIOSH Ceiling REL (ppm)	NIOSH Ceiling REL (mg/m^3)	NIOSH REL STEL (ppm)	NIOSH REL STEL (mg/m^3)
Aqueous hydrogen bromide (i.e., Hydrobromic acid)			3	10		
Aqueous hydrogen chloride (i.e., Hydrochloric acid, Muriatic acid)			5	7		
Aqueous hydrogen fluoride (i.e., Hydrofluoric acid)	3 (15min)	2.5 (15min)	6 (15min)	5 (15min)		
Aroclor® 1242		0.001				
Aroclor® 1254		0.001				
Arsenic hydride				0.002 (15min)		
Arsenic (inorganic compounds, as As)				0.002 (15min)		
Arsenic metal: Arsenia				0.002 (15min)		
Arsenic, organic compounds (as As)						
Arsenic trihydride				0.002 (15min)		

Arseniuretted hydrogen				0.002 (15min)		
Arsenous hydride				0.002 (15min)		
Arsine				0.002 (15min)		
Artificial barite		10 (total) 5 (resp)				
Arificial graphite						
Asbestos						
Asphalt: Asphaltum				5 (15min)		
Asphalt fumes				5 mg/m^3 (15min)		
Aspirin				5		
Asymmetrical trimethylbenzene	25	125				
Asymmetrical dichloroethane	100	400				
ATCP						
Atrazine		5				
Aurum paradoxum		0.1				
Azabenzene	5	15				

Table 4 (cont.) NIOSH Recommended Exposure Limits (REL).

Chemical Name	**NIOSH REL TWA (ppm)**	**NIOSH REL TWA (mg/m³)**	**NIOSH Ceiling REL (ppm)**	**NIOSH Ceiling REL (mg/m³)**	**NIOSH REL STEL (ppm)**	**NIOSH REL STEL (mg/m³)**
Azide			0.1 (as HN3 on skin)	0.3 (as NaN3 on skin)		
Azimethylene	0.2	0.4				
Azine	5	15				
Azinphos-methyl		0.2 (skin)				
Azirane						
Aziridine						
Azium				0.1 (as HN3 on skin)	0.3 (as NaN3 on skin)	
Azomethylene	0.2	0.4				
Azophos®		0.2 (skin)				
Barium chloride (as Ba)		0.5				
Barium nitrate (as Ba)		0.5				
Barium sulfate		10 (total) 5 (resp)				
Benomyl		15 (total) 5 (resp)				
Benzene	0.1				1	

Benzenethiol			.1 (15min)	0.5 (15min)		
Benzidine						
Benzoyl peroxide		5				
Benzyl chloride			1 (15min)	5 (15min)		
Beryllium & beryllium compounds (as Be)		not to exceed 0.0005				
Bismuth telluride, doped with Selenium sulfide (as Bi2Te3)		5				
Bismuth telluride, undoped		10 (total) 5 (resp)				
Borates, tetra, sodium salts (Anhydrous)		1				
Borates, tetra, sodium salts (Decahydrate)		5				
Borates, tetra, sodium salts (Pentahydrate)		1				
Boron oxide		10				
Boron tribromide			1	10		
Boron trifluoride			1	3		
Bromacil	1	10				
Bromine	0.1	0.7			0.3	2
Bromine pentafluoride	0.1	0.7				

Table 4 (cont.) NIOSH Recommended Exposure Limits (REL).

Chemical Name	NIOSH REL TWA (ppm)	NIOSH REL TWA (mg/m^3)	NIOSH Ceiling REL (ppm)	NIOSH Ceiling REL (mg/m^3)	NIOSH REL STEL (ppm)	NIOSH REL STEL (mg/m^3)
Bromoform	0.5 (skin)	5 (skin)				
1, 3-Butadiene						
n-Butane	800	1900				
2-Butanone	200	590			300	885
2-Butoxyethanol	5 (skin)	24 (skin)				
2-Butoxyethanol acetate	5	33				
n-Butyl acetate	150	710			200	950
sec-Butyl acetate	200	950				
tert-Butyl acetate	200	950				
Butyl acrylate	10	55				
n-Butyl alcohol			50 (skin)	150 (skin)		
sec-Butyl alcohol	100	305			150	455
tert-Butyl alcohol	100	300			150	450
n-Butylamine			5 (skin)	15 (skin)		
tert-Butyl chromate		0.001 mg $Cr(VI)/m^3$				
n-Butyl glycidyl ether			5.6 (15min)	30 (15min)		

n-Butyl lactate	5	25				
n-Butyl mercaptan			0.5 (15min)	1.8 (15min)		
o-sec-Butylphenol	5 (skin)	30 (skin)				
p-tert-Butyltoluene	10	60			20	120
n-Butyronitrile	8	22				
Cadmium dust (as Cd)						
Cadmium fume (as Cd)						
Calcium arsenate (as As)				0.0002 (15min)		
Calcium carbonate		10 (total) 5 (resp)				
Calcium cyanamide		0.5				
Calcium hydroxide		5				
Calcium oxide		2				
Calcium silicate		10 (total) 5 (resp)				
Calcium sulfate		10 (total) 5 (resp)				
Camphor (synthetic)		2				
Caprolactam	0.22 (vapor)	1 (dust) 1 (vapor)			0.66 (vapor)	3 (dust) 3 (vapor)
Captan		5				
Carbaryl		5				
Carbofuran		0.1				

Table 4 (cont.) NIOSH Recommended Exposure Limits (REL).

Chemical Name	NIOSH REL TWA (ppm)	NIOSH REL TWA (mg/m^3)	NIOSH Ceiling REL (ppm)	NIOSH Ceiling REL (mg/m^3)	NIOSH REL STEL (ppm)	NIOSH REL STEL (mg/m^3)
Carbon black		3.5 (0.1 mg PAHs/m^3… carbon black in the presence of PAHs)				
Carbon dioxide	5000	9000			30000	54000
Carbon disulfide	1 (skin)	3 (skin)			10 (skin)	30 (skin)
Carbon monoxide	35	40	200	229		
Carbon tetrabromide	0.1	1.4			0.3	4
Carbon tetrachloride					2 (60min)	12.6 (60min)
Carbonyl fluoride	2	5			5	15
Catechol	5 (skin)	20 (skin)				
Cellulose		10 (total) 5 (resp)				
Cesium hydroxide		2				
Chlordane		0.5 (skin)				
Chlorinated camphene						
Chlorinated diphenyl oxide		0.5				

Chlorine			0.5 (15min)	1.45 (15min)		
Chlorine dioxide	0.1	0.3			0.3	0.9
Chlorine trifluoride			0.1	0.4		
Chloroacetaldehyde			1	3		
alpha-Chloroacetophenone	0.05	0.3				
Chloroacetyl chloride	0.05	0.2				
Chlorobenzene						
o-Chlorobenzylidene malononitrile			0.05 (skin)	0.4 (skin)		
Chlorobromomethane	200	1050				
Chlorodifluoromethane	1000	3500			1250	4375
Chlorodiphenyl (42% chlorine)		0.001				
Chlorodiphenyl (54% chlorine)		0.001				
Chloroform					2 (60min)	9.78 (60min)
bis-Chloromethyl ether						
Chloromethyl methyl ether						
1-Chloro-1-nitropropane	2	10				
Chloropentafluoroethane	1000	6320				

Table 4 (cont.) NIOSH Recommended Exposure Limits (REL).

Chemical Name	NIOSH REL TWA (ppm)	NIOSH REL TWA (mg/m^3)	NIOSH Ceiling REL (ppm)	NIOSH Ceiling REL (mg/m^3)	NIOSH REL STEL (ppm)	NIOSH REL STEL (mg/m^3)
Chloropicrin	0.1	0.7				
beta-Chloroprene			1 (15min)	3.6 (15min)		
o-Chlorostyrene	50	285			75	428
o-Chlorotoluene	50	250			75	375
2-Chloro-6-trichloromethyl pyridine		10 (total) 5 (resp)				20 (total)
Chlorpyrifos		0.2 (skin)			0.6 (skin)	
Chromic acid and chromates		0.001				
Chromium(II) compounds (as Cr)		0.5				
Chromium(III) compounds (as Cr)		0.5				
Chromium metal		0.5				
Chromyl chloride		0.001 mg Cr (VI)/m^3				
Clopidol		10 (total) 5 (resp)				20 (total)

Coal dust		2.4 (resp, <5% SiO_2) 10 (resp, greater than or equal to 5% SiO_2)				
Coal tar pitch volatiles		0.1 (cyclohexane-extractable fraction)				
Cobalt carbonyl (as Co)		0.1				
Cobalt hydrocarbonyl (as Co)		0.1				
Cobalt metal dust and fume (as Co)		0.05				
Coke oven emissions		0.2 (benzene-soluble fraction)				
Copper (dusts and mists, as Cu)		1				
Copper fume (as Cu)		0.1				
Cotton dust (raw)		<0.2				
Crag® herbicide		10 (total) 5 (resp)				
m-Cresol	2.3	10				
o-Cresol	2.3	10				
p-Cresol	2.3	10				

Table 4 (cont.) NIOSH Recommended Exposure Limits (REL).

Chemical Name	NIOSH REL TWA (ppm)	NIOSH REL TWA (mg/m^3)	NIOSH Ceiling REL (ppm)	NIOSH Ceiling REL (mg/m^3)	NIOSH REL STEL (ppm)	NIOSH REL STEL (mg/m^3)
Crotonaldehyde	2	6				
Crufomate		5				20
Cumene	50 (skin)	245 (skin)				
Cyanamide		2				
Cyanogen	10	20				
Cyanogen chloride			0.3	0.6		
Cyclohexane	300	1050				
Cyclohexanethiol			0.5 (15min)	2.4 (15min)		
Cyclohexanol	50 (skin)	200 (skin)				
Cyclohexanone	25 (skin)	100 (skin)				
Cyclohexene	300	1015				
Cyclohexylamine	10	40				
Cyclonite		1.5 (skin)				3 (skin)
Cyclopentadiene	75	200				
Cyclopentane	600	1720				
Cyhexatin		5				
2, 4-D		10				

DDT		0.5				
Decaborane	0.05 (skin)	0.3 (skin)			0.15 (skin)	0.9 (skin)
1-Decanethiol			0.5 (15min)	3.6 (15min)		
Demeton		0.1 (skin)				
Diacetone alcohol	50	240				
2, 4-Diaminoanisole (and its salts)	(minimize occupa-tional exposure)					
o-Dianisidine						
Diazinon®		0.1 (skin)				
Diazomethane	0.2	0.4				
Diborane	0.1	0.1				
1, 2-Dibromo-3-chloropropane						
2-N-Dibutylaminoethanol	2 (skin)	14 (skin)				
2, 6-Di-tert-butyl-p-cresol		10				
Dibutyl phosphate	1	5			2	10
Dibutyl phthalate		5				
Dichloroacetylene			0.1	0.4		
o-Dichlorobenzene			50	300		
p-Dichlorobenzene						

Table 4 (cont.) NIOSH Recommended Exposure Limits (REL).

Chemical Name	NIOSH REL TWA (ppm)	NIOSH REL TWA (mg/m³)	NIOSH Ceiling REL (ppm)	NIOSH Ceiling REL (mg/m³)	NIOSH REL STEL (ppm)	NIOSH REL STEL (mg/m³)
3, 3′-Dichlorobenzidine (and its salts)						
Dichlorodifluoromethane	1000	4950				
1, 3-Dichloro-5, 5-dimethylhydantoin		0.2				0.4
1, 1-Dichloroethane	100	400				
1, 2-Dichloroethylene	200	790				
Dichloroethyl ether	5 (skin)	30 (skin)			10 (skin)	60 (skin)
Dichloromonofluoromethane	10	40				
1, 1-Dichloro-1-nitroethane	2	10				
1, 3-Dichloropropene	1 (skin)	5 (skin)				
2, 2-Dichloropropionic acid	1	6				
Dichlorotetrafluoroethane	1000	7000				
Dichlorvos		1(skin)				
Dicrotophos		0.25 (skin)				
Dicyclopentadiene	5	30				
Dicyclopentadienyl iron		10 (total) 5 (resp)				

Dieldrin		0.25 (skin)				
Diesel exhaust						
Diethanolamine	3	15				
Diethylamine	10	30			25	75
2-Diethylaminoethanol	10 (skin)	50 (skin)				
Diethylenetriamine	1 (skin)	4 (skin)				
Diethyl ketone	200	705				
Diethyl phthalate		5				
Difluorodibromomethane	100	860				
Diglycidyl ether	0.1	0.5				
Diisobutyl ketone	25	150				
Diisopropylamine	5 (skin)	20 (skin)				
Dimethyl acetamide	10 (skin)	35 (skin)				
Dimethylamine	10	18				
4-Dimethylaminoazobenzene						
bis(2-(Dimethylamino) ethyl) ether						
Dimethylaminopropionitrile						
N, N-Dimethylaniline	5 (skin)	25 (skin)			10 (skin)	50 (skin)
Dimethyl carbamoyl chloride						

Table 4 (cont.) NIOSH Recommended Exposure Limits (REL).

Chemical Name	NIOSH REL TWA (ppm)	NIOSH REL TWA (mg/m^3)	NIOSH Ceiling REL (ppm)	NIOSH Ceiling REL (mg/m^3)	NIOSH REL STEL (ppm)	NIOSH REL STEL (mg/m^3)
Dimethyl-1, 2-dibromo-2, 2-dichlorethyl phosphate		3 (skin)				
Dimethylformamide	10 (skin)	30 (skin)				
1, 1-Dimethylhydrazine			0.06 (2hr)	0.15 (2hr)		
Dimethylphthalate		5				
Dimethyl sulfate	0.1 (skin)	0.5 (skin)				
Dinitolmide		5				
m-Dinitrobenzene		1 (skin)				
o-Dinitrobenzene		1 (skin)				
p-Dinitrobenzene		1 (skin)				
Dinitro-o-cresol		0.2 (skin)				
Dinitrotoluene		1.5 (skin)				
Di-sec octyl phthalate		5				10
Dioxane			1 (30min)	3.6 (30min)		
Dioxathion		0.2 (skin)				
Diphenyl	0.2	1				
Diphenylamine		10				

Dipropylene glycol methyl ether	100 (skin)	600 (skin)			150 (skin)	900 (skin)
Dipropyl ketone	50	235				
Diquat (Diquat dibromide)		0.5				
Disulfiram		2				
Disulfoton		0.1 (skin)				
Diuron		10				
Divinyl benzene	10	50				
1-Dodecanethiol			0.5(15min)	4.1 (15min)		
Emery						
Endosulfan		0.1 (skin)				
Endrin		0.1 (skin)				
Enflurane			2 (60min)	15.1 (60min)		
Epichlorohydrin	5 (skin)	19 (skin)				
EPN		0.5 (skin)				
Ethanolamine	3	8			6	15
Ethion		0.4 (skin)				
2-Ethoxyethanol	0.5 (skin)	1.8 (skin)				
2-Ethoxyethyl acetate	0.5 (skin)	2.7 (skin)				
Ethyl acetate	400	1400				
Ethyl acrylate						

Table 4 (cont.) NIOSH Recommended Exposure Limits (REL).

Chemical Name	**NIOSH REL TWA (ppm)**	**NIOSH REL TWA (mg/m³)**	**NIOSH Ceiling REL (ppm)**	**NIOSH Ceiling REL (mg/m³)**	**NIOSH REL STEL (ppm)**	**NIOSH REL STEL (mg/m³)**
Ethyl alcohol	1000	1900				
Ethylamine	10	18				
Ethyl benzene	100	435			125	545
Ethyl bromide						
Ethyl butyl ketone	50	230				
Ethyl chloride	(handle with caution in the workplace)					
Ethylene chlorohydrin			1 (skin)	3 (skin)		
Ethylenediamine	10	25				
Ethylene dibromide	0.045 (15min)		0.13 (15min)			
Ethylene dichloride	1	4			2	8
Ethylene glycol						
Ethylene glycol dinitrate						0.1 (skin)
Ethyleneimine	<0.1 (10min)	0.18 (10min)	5 (10min)	9 (10min)		

Ethylene oxide	<0.1 (10min)	0.18 (10min)	5 (10min)	9 (10min)		
Ethylene thiourea						
Ethyl ether						
Ethyl formate	100	300				
Ethylidene norbornene			5	25		
Ethyl mercaptan			0.5 (15min)	1.3 (15min)		
N-Ethylmorpholine	5 (skin)	23 (skin)				
Ethyl silicate	10	85				
Fenamiphos		0.1 (skin)				
Fensulfothion		0.1				
Fenthion						
Ferbam		10				
Ferrovanadium dust		1				3
Fibrous glass dust		5 (total) 3 fibers/ cm^3 (fiber diameter < or = to 3.5um)				
Fluorine	0.1	0.2				
Fluorotrichloromethane			1000	5600		
Fluoroxene			2 (60min)	10.3 (60min)		

Table 4 (cont.) NIOSH Recommended Exposure Limits (REL).

Chemical Name	NIOSH REL TWA (ppm)	NIOSH REL TWA (mg/m^3)	NIOSH Ceiling REL (ppm)	NIOSH Ceiling REL (mg/m^3)	NIOSH REL STEL (ppm)	NIOSH REL STEL (mg/m^3)
Fonofos		0.1 (skin)				
Formaldehyde	0.016 (15min)		0.1 (15min)			
Formalin (as formaldehyde)	0.016 (15min)		0.1 (15min)			
Formamide	10 (skin)	15 (skin)				
Formic acid	5	9				
Furfural						
Furfuryl alcohol	10 (skin)	40 (skin)			15 (skin)	60 (skin)
Gasoline						
Germanium tetrahydride	0.2	0.6				
Glutaraldehyde			0.2	0.8		
Glycerin (mist)						
Glycidol	25	75				
Glycolonitrile			2 (15min)	5 (15min)		
Grain dust (oat, wheat, barley)		4				
Graphite (natural)		2.5 (resp)				

Graphite (synthetic)		15 (total) 5 (resp)				
Gypsum		10 (total) 5 (resp)				
Hafnium		0.5				
Halothane			2 (60min)	16.2 (60min)		
Heptachlor		0.5 (skin)				
n-Heptane	85 (15min)	350 (15min)	440 (15min)	1800 (15min)		
1-Heptanethiol			0.5 (15min)	2.7 (15min)		
Hexachlorobutadiene	0.02 (skin)	0.24 (skin)				
Hexachlorocyclopentadiene	0.01	0.1				
Hexachloroethane	1 (skin)	10 (skin)				
Hexachloronaphthalene		0.2 (skin)				
1-Hexadecanethiol			0.5 (15min)	5.3 (15min)		
Hexafluoroacetone	0.1 (skin)	0.7 (skin)				
Hexamethylene diisocyanate	0.005 (10min)	0.035 (10min)	0.020 (10min)	0.140 (10min)		
Hexamethyl phosphoramide						
n-Hexane	50	180				
Hexane isomers (excluding n-Hexane)	100 (15min)	350 (15min)	510 (15min)	1800 (15min)		
n-Hexanethiol			0.5 (15min)	2.7 (15min)		
2-Hexanone	1	4				

Table 4 (cont.) NIOSH Recommended Exposure Limits (REL).

Chemical Name	NIOSH REL TWA (ppm)	NIOSH REL TWA (mg/m^3)	NIOSH Ceiling REL (ppm)	NIOSH Ceiling REL (mg/m^3)	NIOSH REL STEL (ppm)	NIOSH REL STEL (mg/m^3)
Hexone	50	205			75	300
sec-Hexyl acetate	50	300				
Hexylene glycol			25	125		
Hydrazine			0.03 (2hour)	0.04 (2hour)		
Hydrogenated terphenyls	0.5	5				
Hydrogen bromide			3	10		
Hydrogen chloride			5	7		
Hydrogen cyanide					4.7 (skin)	5 (skin)
Hydrogen fluoride	3(15min)	2.5(15min)	6(15min)	5 (15min)		
Hydrogen peroxide	1	1.4				
Hydrogen selenide	0.05	0.2				
Hydrogen sulfide			10 (10min)	15 (10min)		
Hydroquinone				2 (15min)		
2-Hydroxypropyl acrylate	0.5 (skin)	3 (skin)				
Indene	10	45				
Indium		0.1				

Iodine			0.1	1		
Iodoform	0.6	10				
Iron oxide dust and fume (as Fe)		5				
Iron pentacarbonyl (as Fe)	0.1	0.23			0.2	0.45
Iron salts (soluble, as Fe)		1				
Isoamyl acetate	100	525				
Isoamyl alcohol (secondary)	100	360			125	450
Isoamyl alcohol (primary)	100	360			125	450
Isobutane	800	1900				
Isobutyl acetate	150	700				
Isobutyl alcohol	50	150				
Isobutyronitrile	8	22				
Isooctyl alcohol	50 (skin)	270 (skin)				
Isophorone	4	23				
Isophorone diisocyanate	0.005 (skin)	0.045 (skin)			0.02 (skin)	0.180 (skin)
2-Isopropoxyethanol						
Isopropyl acetate						
Isopropyl alcohol	400	980			500	1225
Isopropylamine						

Table 4 (cont.) NIOSH Recommended Exposure Limits (REL).

Chemical Name	**NIOSH REL TWA (ppm)**	**NIOSH REL TWA (mg/m³)**	**NIOSH Ceiling REL (ppm)**	**NIOSH Ceiling REL (mg/m³)**	**NIOSH REL STEL (ppm)**	**NIOSH REL STEL (mg/m³)**
N-Isopropylaniline	2 (skin)	10 (skin)				
Isopropyl ether	500	2100				
Isopropyl glycidyl ether			50 (15min)	240 (15min)		
Kaolin		10 (total) 5 (resp)				
Kepone		0.001				
Kerosene		100				
Ketene	0.5	0.9			1.5	3
Lead		0.050				
Limestone		10 (total) 5 (resp)				
Lindane		0.5 (skin)				
Lithium hydride		0.025				
L.P.G.	1000	1800				
Magnesite		10 (total) 5 (resp)				
Magnesium oxide fume						
Malathion		10 (skin)				
Maleic anhydride	0.25	1				
Malonaldehyde						

Malononitrile	3	8				
Manganese compounds and fume (as Mn)		1				3
Manganese cyclopentadienyl tricarbonyl (as Mn)		0.1 (skin)				
Manganese tetroxide (as Mn)						
Marble		10 (total) 5 (resp)				
Mercury compounds [except (organo) alkyls] (as Hg)		0.05 (vapor/ skin)		0.1 (skin)		
Mercury (organo) alkyl compounds (as Hg)		0.01 (skin)				0.03 (skin)
Mesityl oxide	10	40				
Methacrylic acid	20 (skin)	70 (skin)				
Methomyl		2.5				
Methoxychlor						
Methoxyflurane			2 (60min)	13.5 (60min)		
4-Methoxyphenol		5				
Methyl acetate	200	610			250	760
Methyl acetylene	1000	1650				
Methyl acetylene-propadiene mixture	1000	1800			1250	2250
Methyl acrylate	10 (skin)	35 (skin)				

Table 4 (cont.) NIOSH Recommended Exposure Limits (REL).

Chemical Name	NIOSH REL TWA (ppm)	NIOSH REL TWA (mg/m³)	NIOSH Ceiling REL (ppm)	NIOSH Ceiling REL (mg/m³)	NIOSH REL STEL (ppm)	NIOSH REL STEL (mg/m³)
Methylacrylonitrile	1 (skin)	3 (skin)				
Methylal	1000	3100				
Methyl alcohol	200 (skin)	260 (skin)			250 (skin)	325 (skin)
Methylamine	10	12				
Methyl bromide						
Methyl Cellosolve®	0.1 (skin)	0.3 (skin)				
Methyl Cellosolve® acetate	0.1 (skin)	0.5 (skin)				
Methyl chloride						
Methyl chloroform			350 (15min)	1900 (15min)		
Methyl-2-cyanoacrylate	2	8			4	16
Methylcyclohexane	400	1600				
Methylcyclohexanol	50	235				
o-Methylcyclohexanone	50 (skin)	230 (skin)			75 (skin)	345 (skin)
Methyl cyclopentadienyl manganese tricarbonyl (as Mn)		0.2 (skin)				
Methyl demeton		0.5 (skin)				

Methylene bisphenyl isocyanate	0.005 (10min)	0.05 (10min)	0.02 (10min)	0.20 (10min)		
Methylene chloride						
4, 4′-Methylenebis (2-chloroaniline)		0.003 (skin)				
Methylene bis (4-cyclohexylisocyanate)			0.01	0.11		
4, 4′-Methylenedianiline						
Methyl ethyl ketone peroxide			0.2	1.5		
Methyl formate	100	250			150	375
5-Methyl-3-heptanone	25	130				
Methyl hydrazine			0.04 (2hour)	0.08 (2hour)		
Methyl iodide	2 (skin)	10 (skin)				
Methyl isoamyl ketone	50	240				
Methyl isobutyl carbinol	25 (skin)	100 (skin)			40 (skin)	165 (skin)
Methyl isocyanate	0.02 (skin)	0.05 (skin)				
Methyl isopropyl ketone	200	705				
Methyl mercaptan			0.5 (15min)	1 (15min)		
Methyl methacrylate	100	410				
Methyl (n-amyl) ketone	100	465				

Table 4 (cont.) NIOSH Recommended Exposure Limits (REL).

Chemical Name	NIOSH REL TWA (ppm)	NIOSH REL TWA (mg/m^3)	NIOSH Ceiling REL (ppm)	NIOSH Ceiling REL (mg/m^3)	NIOSH REL STEL (ppm)	NIOSH REL STEL (mg/m^3)
Methyl parathion		0.2 (skin)				
Methyl silicate	1	6				
alpha-Methyl styrene	50	240			100	485
Metribuzin		5				
Mica (containing less than 1% quartz)		3 (resp)				
Mineral wool fiber		3 fibers/cm^3 (fiber diameter < or = to 3.5um) 5 mg/m^3 (total)				
Molybdenum						
Molybdenum (soluble compounds, as Mo)						
Monocrotophos		0.25				
Monomethyl aniline	0.5 (skin)	2 (skin)				
Morpholine	20 (skin)	70 (skin)			30 (skin)	105 (skin)
Naphtha (coal tar)	100	400				
Naphthalene	10	50			15	75

Naphthalene diisocyanate	0.005	0.040	0.020 (10min)	0.170 (10min)		
alpha-Naphthylamine						
beta-Naphthylamine						
Niax® Catalyst ESN						
Nickel carbonyl	0.001	0.007				
Nickel metal and other compounds (as Ni)		0.015				
Nicotine		0.5 (skin)				
Nitric acid	2	5			4	10
Nitric oxide	25	30				
p-Nitroaniline		3 (skin)				
Nitrobenzene	1 (skin)	5 (skin)				
4-Nitrobiphenyl						
p-Nitrochlorobenzene						
Nitroethane	100	310				
Nitrogen dioxide					1	1.8
Nitrogen trifluoride	10	29				
Nitroglycerine						0.1 (skin)
Nitromethane						
2-Nitronaphthalene						

Table 4 (cont.) NIOSH Recommended Exposure Limits (REL).

Chemical Name	NIOSH REL TWA (ppm)	NIOSH REL TWA (mg/m^3)	NIOSH Ceiling REL (ppm)	NIOSH Ceiling REL (mg/m^3)	NIOSH REL STEL (ppm)	NIOSH REL STEL (mg/m^3)
1-Nitropropane	25	90				
2-Nitropropane						
N-Nitrosodimethylamine						
m-Nitrotoluene	2 (skin)	11 (skin)				
o-Nitrotoluene	2 (skin)	11 (skin)				
p-Nitrotoluene	2 (skin)	11 (skin)				
Nitrous oxide	25	46				
Nonane	200	1050				
1-Nonanethiol			0.5 (15min)	3.3 (15min)		
Octachloronaphthalene		0.1 (skin)				0.3 (skin)
1-Octadecanethiol			0.5 (15min)	5.9 (15min)		
Octane	75 (15min)	350 (15min)	385 (15min)	1800 (15min)		
1-Octanethiol			0.5 (15min)	3.0 (15min)		
Oil mist (mineral)		5				10
Osmium tetroxide	0.0002	0.002			0.0006	0.006
Oxalic acid		1				2
Oxygen difluoride			0.05	0.1		

Ozone			0.1	0.2		
Paraffin wax fume		2				
Paraquat (Paraquat dichloride)		0.1 (resp/skin)				
Parathion		0.05 (skin)				
Particulates not otherwise regulated						
Pentaborane	0.005	0.01			0.015	0.03
Pentachloroethane	(handle with caution in the workplace)					
Pentachloronaphthalene		0.5 (skin)				
Pentachlorophenol		0.5 (skin)				
Pentaerythritol		10 (total) 5 (resp)				
n-Pentane	120 (15min)	350 (15min)	610 (15min)	1800 (15min)		
1-Pentanethiol			0.5 (15min)	2.1 (15min)		
2-Pentanone	150	530				
Perchloromethyl mercaptan	0.1	0.8				
Perchloryl fluoride	3	14			6	28
Perlite		10 (total) 5 (resp)				

Table 4 (cont.) NIOSH Recommended Exposure Limits (REL).

Chemical Name	NIOSH REL TWA (ppm)	NIOSH REL TWA (mg/m³)	NIOSH Ceiling REL (ppm)	NIOSH Ceiling REL (mg/m³)	NIOSH REL STEL (ppm)	NIOSH REL STEL (mg/m³)
Petroleum distillates (naphtha)		350 (15min)		1800 (15min)		
Phenol	5 (15min)	19 (15min)	15.6 (15min)	60 (15min)		
Phenothiazine		5 (skin)				
p-Phenylene diamine		0.1 (skin)				
Phenyl ether-biphenyl mixture (vapor)	1	7				
Phenyl ether (vapor)	1	7				
Phenyl glycidyl ether			1 (15min)	6 (15min)		
Phenylhydrazine			0.14 (2hour/ skin)	0.6 (2hour/ skin)		
N-Phenyl-beta-naphthylamine						
Phenylphosphine			0.05	0.25		
Phorate		0.05 (skin)				0.2 (skin)
Phosdrin®	0.01 (skin)	0.1 (skin)			0.03 (skin)	0.3 (skin)

Phosgene	0.1 (15min)	0.4 (15min)	0.2 (15min)	0.8 (15min)		
Phosphine	0.3	0.4			1	1
Phosphoric acid		1				3
Phosphorus oxychloride	0.1	0.6			0.5	3
Phosphorus pentachloride		1				
Phosphorus pentasulfide		1				3
Phosphorus trichloride	0.2	1.5			0.5	3
Phosphorus (yellow)		0.1				
Phthalic anhydride	1	6				
m-Phthalodinitrile		5				
Picloram						
Picric acid		0.1 (skin)				0.3 (skin)
Pindone		0.1				
Piperazine dihydrochloride		5				
Plaster of Paris		10 (total) 5 (resp)				
Platinum		1				
Platinum (soluble salts, as Pt)		0.002				
Portland cement		10 (total) 5 (resp)				
Potassium cyanide (as CN)			4.7 (10min)	5 (10min)		
Potassium hydroxide				2		

Table 4 (cont.) NIOSH Recommended Exposure Limits (REL).

Chemical Name	NIOSH REL TWA (ppm)	NIOSH REL TWA (mg/m³)	NIOSH Ceiling REL (ppm)	NIOSH Ceiling REL (mg/m³)	NIOSH REL STEL (ppm)	NIOSH REL STEL (mg/m³)
Propane	1000	1800				
Propane sultone						
1-Propanethiol			0.5 (15min)	1.6 (15min)		
Propargyl alcohol	1 (skin)	2 (skin)				
beta-Propiolactone						
Propionic acid	10	30			15	45
Propionitrile	6	14				
Propoxur		0.5				
n-Propyl acetate	200	840			250	1050
n-Propyl alcohol	200 (skin)	500 (skin)			250 (skin)	625 (skin)
Propylene dichloride						
Propylene glycol dinitrate	0.05 (skin)	0.3 (skin)				
Propylene glycol mono-methyl ether	100	360			150	540
Propylene imine	2 (skin)	5 (skin)				
Propylene oxide						
n-Propyl nitrate	25	105			40	170
Pyrethrum		5				

Pyridine	5	15				
Quinone	0.1	0.4				
Resorcinol	10	45			20	90
Rhodium (metal fume and insoluble compounds, as Rh)		0.1				
Rhodium (soluble compounds, as Rh)		0.001				
Ronnel		10				
Rosin core solder, pyrolysis products (as formaldehyde)		0.1				
Rotenone		5				
Rouge						
Selenium		0.2				
Selenium hexafluoride	0.05					
Silica, amorphous		6				
Silica, crystalline (as respirable dust)		0.05				
Silicon		10 (total) 5 (resp)				
Silicon carbide		10 (total 5 (resp)				
Silicon tetrahydride	5	7				
Silver (metal dust and soluble compounds, as Ag)		0.01				

Table 4 (cont.) NIOSH Recommended Exposure Limits (REL).

Chemical Name	NIOSH REL TWA (ppm)	NIOSH REL TWA (mg/m^3)	NIOSH Ceiling REL (ppm)	NIOSH Ceiling REL (mg/m^3)	NIOSH REL STEL (ppm)	NIOSH REL STEL (mg/m^3)
Soapstone (containing less than 1% quartz)		6 (total) 3 (resp)				
Sodium aluminum fluoride (as F)		2.5				
Sodium azide			0.1 (as HN3 on skin)	0.3 (as NaN3 on skin)		
Sodium bisulfite		5				
Sodium cyanide (as CN)			4.7 (10min)	5 (10min)		
Sodium fluoride (as F)		2.5				
Sodium fluoroacetate		0.05 (skin)				0.15 (skin)
Sodium hydroxide				2		
Sodium metabisulfite		5				
Starch		10 (total) 5 (resp)				
Stibine	0.1	0.5				
Stoddard solvent		350 (15min)		1800 (15min)		
Strychnine		0.15				
Styrene	50	215			100	425

Subtilisins						0.00006 (60min)
Succinonitrile	6	20				
Sucrose		10 (total 5 (resp)				
Sulfur dioxide	2	5			5	13
Sulfur hexafluoride	1000	6000				
Sulfuric acid		1				
Sulfur monochloride			1	6		
Sulfur pentafluoride			0.01	0.1		
Sulfur tetrafluoride			0.1	0.4		
Sulfuryl fluoride	5	20			10	40
Sulprofos		1				
2, 4, 5-T		10				
Talc (containing no asbestos and less than 1% quartz)		2 (resp)				
Tantalum (metal and oxide dust, as Ta)		5				10
TEDP		0.2 (skin)				
Tellurium		0.1				
Tellurium hexafluoride	0.02	0.2				
Temephos		10 (total) 5 (resp)				

Table 4 (cont.) NIOSH Recommended Exposure Limits (REL).

Chemical Name	NIOSH REL TWA (ppm)	NIOSH REL TWA (mg/m³)	NIOSH Ceiling REL (ppm)	NIOSH Ceiling REL (mg/m³)	NIOSH REL STEL (ppm)	NIOSH REL STEL (mg/m³)
TEPP		0.05 (skin)				
m-Terphenyl			0.5	5		
o-Terphenyl			0.5	5		
p-Terphenyl			0.5	5		
2, 3, 7, 8-Tetrachloro-dibenzo-p-dioxin						
1, 1, 1, 2-Tetrachloro-2, 2-difluoroethane	500	4170				
1, 1, 2, 2-Tetrachloro-1, 2-difluoroethane	500	4170				
1, 1, 1, 2-Tetrachloroethane	(handle with caution in the workplace)					
1, 1, 2, 2-Tetrachloroethane	1 (skin)	7 (skin)				
Tetrachloroethylene	(minimize occupational exposure)					

Tetrachloronaphthalene		2 (skin)				
Tetraethyl lead (as Pb)		0.075 (skin)				
Tetrahydrofuran	200	590			250	735
Tetramethyl lead (as Pb)		0.075 (skin)				
Tetramethyl succinonitrile	0.5 (skin)	3 (skin)				
Tetranitromethane	1	8				
Tetrasodium pyrophosphate		5				
Tetryl		1.5 (skin)				
Thallium (soluble compounds, as Tl)		0.1 (skin)				
4, 4′-Thiobis(6-tert-butyl-m-cresol)		10 (total) 5 (resp)				
Thioglycolic acid	1 (skin)	4 (skin)				
Thionyl chloride			1	5		
Thiram		5				
Tin		2				
Tin (organic compounds, as Sn)		0.1 (skin)				
Tin(II) oxide (as Sn)		2				
Tin(IV) oxide (as Sn)		2				
Titanium dioxide						

Table 4 (cont.) NIOSH Recommended Exposure Limits (REL).

Chemical Name	NIOSH REL TWA (ppm)	NIOSH REL TWA (mg/m³)	NIOSH Ceiling REL (ppm)	NIOSH Ceiling REL (mg/m³)	NIOSH REL STEL (ppm)	NIOSH REL STEL (mg/m³)
o-Tolidine				0.02 (60min/skin)		
Toluene	100	375			150	560
Toluenediamine						
Toluene-2, 4-diisocyanate						
m-Toluidine						
o-Toluidine						
p-Toluidine						
Tributyl phosphate	0.2	2.5				
Trichloroacetic acid	1	7				
1, 2, 4-Trichlorobenzene			5	40		
1, 1, 2-Trichloroethane	10 (skin)	45 (skin)				
Trichloroethylene						
Trichloronaphthalene		5 (skin)				
1, 2, 3-Trichloropropane	10 (skin)	60 (skin)				
1, 1, 2-Trichloro-1, 2, 2-trifluoroethane	1000	7600			1250	9500
Triethylamine						

Trifluorobromomethane	1000	6100				
Trimellitic anhydride	0.005	0.04				
Trimethylamine	10	24			15	36
1, 2, 3-Trimethylbenzene	25	125				
1, 2, 4-Trimethylbenzene	25	125				
1, 3, 5-Trimethylbenzene	25	125				
Trimethyl phosphite	2	10				
2, 4, 6-Trinitrotoluene		0.5 (skin)				
Triorthocresyl phosphate		0.1 (skin)				
Triphenylamine		5				
Triphenyl phosphate		3				
Tungsten		5				10
Tungsten carbide (cemented)						
Tungsten (soluble compounds, as W)		1				3
Turpentine	100	560				
1-Undecanethiol			0.5 (15min)	3.9 (15min)		
Uranium (insoluble compounds, as U)		0.2				0.6
Uranium (soluble compounds, as U)		0.05				

Table 4 (cont.) NIOSH Recommended Exposure Limits (REL).

Chemical Name	NIOSH REL TWA (ppm)	NIOSH REL TWA (mg/m³)	NIOSH Ceiling REL (ppm)	NIOSH Ceiling REL (mg/m³)	NIOSH REL STEL (ppm)	NIOSH REL STEL (mg/m³)
n-Valeraldehyde	50	175				
Vanadium dust				0.05 mg V/ m³ (15min)		
Vanadium fume				0.05 mg V/ m³ (15min)		
Vegetable oil mist		10 (total) 5 (resp)				
Vinyl acetate			4 (15min)	15 (15min)		
Vinyl bromide						
Vinyl chloride						
Vinyl cyclohexene dioxide	10 (skin)	60 (skin)				
Vinyl fluoride	1		5			
Vinylidene chloride						
Vinylidene fluoride	1		5			
Vinyl toluene	100	480				
VM & P Naphtha		350 (15min)		1800 (15min)		
Warfarin		0.1				
Welding fumes						
Wood dust		1				

m-Xylene	100	435			150	655
o-Xylene	100	435			150	655
p-Xylene	100	435			150	655
m-Xylene-alpha, alpha′-diamine				0.1 (skin)		
Xylidine	2 (skin)	10 (skin)				
Yttrium		1				
Zinc chloride fume		1				2
Zinc oxide		5 (dust) 5 (fume)		15 (dust)		10 (fume)
Zinc stearate		10 (total) 5 (resp)				
Zirconium compounds (as Zr)		5				10

Table 5 NIOSH Immediately Dangerous to Life and Health (IDLH) Values.

Chemical Name	Original IDLH	Revised IDLH
Acetaldehyde	10,000 ppm	2,000 ppm
Acetic acid	1,000 ppm	50 ppm
Acetic anhydride	1,000 ppm	200 ppm
Acetone	20,000 ppm	2,500 ppm [LEL]
Acetonitrile	4,000 ppm	500 ppm
Acetylene tetrabromide	10 ppm	8 ppm
Acrolein	5 ppm	2 ppm
Acrylamide	Unknown	60 mg/m^3
Acrylonitrile	500 ppm	85 ppm
Aldrin	100 mg/m^3	25 mg/m^3
Allyl alcohol	150 ppm	20 ppm
Allyl chloride	300 ppm	250 ppm
Allyl glycidyl ether	270 ppm	50 ppm
2 Aminopyridine	5 ppm	5 ppm [Unch]
Ammonia	500 ppm	300 ppm
Ammonium sulfamate	5,000 mg/m^3	1,500 mg/m^3
n-Amyl acetate	4,000 ppm	1,000 ppm
sec-Amyl acetate	9,000 ppm	1,000 ppm
Aniline	100 ppm	100 ppm [Unch]
o-Anisidine	50 mg/m^3	50 mg/m^3 [Unch]
p-Anisidine	50 mg/m^3	50 mg/m^3 [Unch]
Antimony compounds (as Sb)	80 mg Sb/m^3	50 mg Sb/m^3
ANTU	100 mg/m^3	100 mg/m^3 [Unch]
Arsenic (inorganic compounds, as As)	100 mg As/m^3	5 mg As/m^3
Arsine	6 ppm	3 ppm
Azinphosmethyl	20 mg/m^3	10 mg/m^3
Barium (soluble compounds, as Ba)	1,100 mg Ba/m^3	50 mg Ba/m^3
Benzene	3,000 ppm	500 ppm
Benzoyl peroxide	7,000 mg/m^3	1,500 mg/m^3

Table 5 (cont.) NIOSH Immediately Dangerous to Life and Health (IDLH) Values.

Chemical Name	Original IDLH	Revised IDLH
Benzyl chloride	10 ppm	10 ppm [Unch]
Beryllium compounds (as Be)	10 mg Be/m^3	4 mg Be/m^3
Boron oxide	N.E.	2,000 mg/m^3
Boron trifluoride	100 ppm	25 ppm
Bromine	10 ppm	3 ppm
Bromoform	Unknown	850 ppm
1,3-Butadiene	20,000 ppm [LEL]	2,000 ppm [LEL]
2-Butanone	3,000 ppm	3,000 ppm [Unch]
2-Butoxyethanol	700 ppm	700 ppm [Unch]
n-Butyl acetate	10,000 ppm	1,700 ppm [LEL]
sec-Butyl acetate	10,000 ppm	1,700 ppm [LEL]
tert-Butyl acetate	10,000 ppm	1,500 ppm [LEL]
n-Butyl alcohol	8,000 ppm	1,400 ppm [LEL]
sec-Butyl alcohol	10,000 ppm	2,000 ppm
tert-Butyl alcohol	8,000 ppm	1,600 ppm
n-Butylamine	2,000 ppm	300 ppm
tert-Butyl chromate	30 mg/m^3 (as CrO3)	15 mg Cr(VI)/m^3
n-Butyl glycidyl ether	3,500 ppm	250 ppm
n-Butyl mercaptan	2,500 ppm	500 ppm
p-tert-Butyltoluene	1,000 ppm	100 ppm
Cadmium dust (as Cd)	50 mg Cd/m^3	9 mg Cd/m^3
Cadmium fume (as Cd)	9 mg Cd/m^3	9 mg Cd/m^3 [Unc h]
Calcium arsenate (as As)	100 mg As/m^3	5 mg As/m^3
Calcium oxide	Unknown	25 mg/m^3
Camphor (synthetic)	200 mg/m^3	200 mg/m^3 [Unch]
Carbaryl	600 mg/m^3	100 mg/m^3
Carbon black	N.E.	1,750 mg/m^3
Carbon dioxide	50,000 ppm	40,000 ppm
Carbon disulfide	500 ppm	500 ppm [Unch]

Table 5 (cont.) NIOSH Immediately Dangerous to Life and Health (IDLH) Values.

Chemical Name	Original IDLH	Revised IDLH
Carbon monoxide	1,500 ppm	1,200 ppm
Carbon tetrachloride	300 ppm	200 ppm
Chlordane	500 mg/m^3	100 mg/m^3
Chlorinated camphene	200 mg/m^3	200 mg/m^3 [Unch]
Chlorinated diphenyl oxide	Unknown	5 mg/m^3
Chlorine	30 ppm	10 ppm
Chlorine dioxide	10 ppm	5 ppm
Chlorine trifluoride	20 ppm	20 ppm [Unch]
Chloroacetaldehyde	100 ppm	45 ppm
alpha-Chloroacetophenone	100 mg/m^3	15 mg/m^3
Chlorobenzene	2,400 ppm	1,000 ppm
o-Chlorobenzylidene malononitrile	2 mg/m^3	2 mg/m^3 [Unch]
Chlorobromomethane	5,000 ppm	2,000 ppm
Chlorodiphenyl (42% chlorine)	10 mg/m^3	5 mg/m^3
Chlorodiphenyl (54% chlorine)	5 mg/m^3	5 mg/m^3 [Unch]
Chloroform	1,000 ppm	500 ppm
1-Chloro-1-nitropropane	2,000 ppm	100 ppm
Chloropicrin	4 ppm	2 ppm
beta-Chloroprene	400 ppm	300 ppm
Chromic acid and chromates	30 mg/m^3 (as CrO3)	15 mg Cr(VI)/m^3
Chromium (II) compounds [as Cr(II)]	N.E.	250 mg Cr(II)/m^3
Chromium (III) compounds [as Cr(III)]	N.E.	25 mg Cr(III)/m^3
Chromium metal (as Cr)	N.E.	250 mg Cr/m^3
Coal tar pitch volatiles	700 mg/m^3	80 mg/m^3
Cobalt metal, dust and fume (as Co)	20 mg Co/m^3	20 mg Co/m^3 [Unch]

Table 5 (cont.) NIOSH Immediately Dangerous to Life and Health (IDLH) Values.

Chemical Name	Original IDLH	Revised IDLH
Copper (dusts and mists, as Cu)	N.E.	100 mg Cu/m^3
Copper fume (as Cu)	N.E.	100 mg Cu/m^3
Cotton dust (raw)	N.E.	100 mg/m^3
Crag (r) herbicide	5,000 mg/m^3	500 mg/m^3
Cresol (o, m, p isomers)	250 ppm	250 ppm [Unch]
Crotonaldehyde	400 ppm	50 ppm
Cumene	8,000 ppm	900 ppm [LEL]
Cyanides (as CN)	50 mg/m^3 (as CN)	25 mg/m^3 (as CN)
Cyclohexane	10,000 ppm	1,300 ppm [LEL]
Cyclohexanol	3,500 ppm	400 ppm
Cyclohexanone	5,000 ppm	700 ppm
Cyclohexene	10,000 ppm	2,000 ppm
Cyclopentadiene	2,000 ppm	750 ppm
2,4-D	500 mg/m^3	100 mg/m^3
DDT	N.E.	500 mg/m^3
Decaborane	100 mg/m^3	15 mg/m^3
Demeton	20 mg/m^3	10 mg/m^3
Diacetone alcohol	2,100 ppm	1,800 ppm [LEL]
Diazomethane	2 ppm	2 ppm [Unch]
Diborane	40 ppm	15 ppm
Dibutyl phosphate	125 ppm	30 ppm
Dibutyl phthalate	9,300 mg/m^3	4,000 mg/m^3
o-Dichlorobenzene	1,000 ppm	200 ppm
p-Dichlorobenzene	1,000 ppm	150 ppm
Dichlorodifluoromethane	50,000 ppm	15,000 ppm
1,3-Dichloro 5, 5-dimethylhydantoin	Unknown	5 mg/m^3
1,1-Dichloroethane	4,000 ppm	3,000 ppm
1,2-Dichloroethylene	4,000 ppm	1,000 ppm
Dichloroethyl ether	250 ppm	100 ppm

Table 5 (cont.) NIOSH Immediately Dangerous to Life and Health (IDLH) Values.

Chemical Name	Original IDLH	Revised IDLH
Dichloromonofluoromethane	50,000 ppm	5,000 ppm
1,1-Dichloro 1-nitroethane	150 ppm	25 ppm
Dichlorotetrafluoroethane	50,000 ppm	15,000 ppm
Dichlorvos	200 mg/m^3	100 mg/m^3
Dieldrin	450 mg/m^3	50 mg/m^3
Diethylamine	2,000 ppm	200 ppm
2-Diethylaminoethanol	500 ppm	100 ppm
Difluorodibromomethane	2,500 ppm	2,000 ppm
Diglycidyl ether	25 ppm	10 ppm
Diisobutyl ketone	2,000 ppm	500 ppm
Diisopropylamine	1,000 ppm	200 ppm
Dimethyl acetamide	400 ppm	300 ppm
Dimethylamine	2,000 ppm	500 ppm
N,N-Dimethylaniline	100 ppm	100 ppm [Unch]
Dimethyl 1,2-dibromo 2, 2-dichlorethyl phosphate	1,800 mg/m^3	200 mg/m^3
Dimethylformamide	3,500 ppm	500 ppm
1,1-Dimethylhydrazine	50 ppm	15 ppm
Dimethylphthalate	9,300 mg/m^3	2,000 mg/m^3
Dimethyl sulfate	10 ppm	7 ppm
Dinitrobenzene (o, m, p isomers)	200 mg/m^3	50 mg/m^3
Dinitroocresol	5 mg/m^3	5 mg/m^3 [Unch]
Dinitrotoluene	200 mg/m^3	50 mg/m^3
Di sec-octyl phthalate	Unknown	5,000 mg/m^3
Dioxane	2,000 ppm	500 ppm
Diphenyl	300 mg/m^3	100 mg/m^3
Dipropylene glycol methyl ether	Unknown	600 ppm
Endrin	2,000 mg/m^3	2 mg/m^3
Epichlorohydrin	250 ppm	75 ppm
EPN	50 mg/m^3	5 mg/m^3

Table 5 (cont.) NIOSH Immediately Dangerous to Life and Health (IDLH) Values.

Chemical Name	Original IDLH	Revised IDLH
Ethanolamine	1,000 ppm	30 ppm
2-Ethoxyethanol	6,000 ppm	500 ppm
2-Ethoxyethyl acetate	2,500 ppm	500 ppm
Ethyl acetate	10,000 ppm	2,000 ppm [LEL]
Ethyl acrylate	2,000 ppm	300 ppm
Ethyl alcohol	15,000 ppm	3,300 ppm [LEL]
Ethylamine	4,000 ppm	600 ppm
Ethyl benzene	2,000 ppm	800 ppm [LEL]
Ethyl bromide	3,500 ppm	2,000 ppm
Ethyl butyl ketone	3,000 ppm	1,000 ppm
Ethyl chloride	20,000 ppm	3,800 ppm [LEL]
Ethylene chlorohydrin	10 ppm	7 ppm
Ethylenediamine	2,000 ppm	1,000 ppm
Ethylene dibromide	400 ppm	100 ppm
Ethylene dichloride	1,000 ppm	50 ppm
Ethylene glycol dinitrate	500 mg/m^3	75 mg/m^3
Ethyleneimine	100 ppm	100 ppm [Unch]
Ethylene oxide	800 ppm	800 ppm [Unch]
Ethyl ether	19,000 ppm [LEL]	1,900 ppm [LEL]
Ethyl formate	8,000 ppm	1,500 ppm
Ethyl mercaptan	2,500 ppm	500 ppm
N-Ethylmorpholine	2,000 ppm	100 ppm
Ethyl silicate	1,000 ppm	700 ppm
Ferbam	N.E.	800 mg/m^3
Ferrovanadium dust	N.E.	500 mg/m^3
Fluorides (as F)	500 mg F/m^3	250 mg F/m^3
Fluorine	25 ppm	25 ppm [Unch]
Fluorotrichloromethane	10,000 ppm	2,000 ppm
Formaldehyde	30 ppm	20 ppm
Formic acid	30 ppm	30 ppm [Unch]
Furfural	250 ppm	100 ppm

Table 5 (cont.) NIOSH Immediately Dangerous to Life and Health (IDLH) Values.

Chemical Name	Original IDLH	Revised IDLH
Furfuryl alcohol	250 ppm	75 ppm
Glycidol	500 ppm	150 ppm
Graphite (natural)	N.E.	1,250 mg/m^3
Hafnium compounds (as Hf)	Unknown	50 mg Hf/m^3
Heptachlor	700 mg/m^3	35 mg/m^3
n-Heptane	5,000 ppm	750 ppm
Hexachloroethane	300 ppm	300 ppm [Unch]
Hexachloronaphthalene	2 mg/m^3	2 mg/m^3 [Unch]
n-Hexane	5,000 ppm	1,100 ppm [LEL]
2-Hexanone	5,000 ppm	1,600 ppm
Hexone	3,000 ppm	500 ppm
sec Hexyl acetate	4,000 ppm	500 ppm
Hydrazine	80 ppm	50 ppm
Hydrogen bromide	50 ppm	30 ppm
Hydrogen chloride	100 ppm	50 ppm
Hydrogen cyanide	50 ppm	50 ppm [Unch]
Hydrogen fluoride (as F)	30 ppm	30 ppm [Unch]
Hydrogen peroxide	75 ppm	75 ppm [Unch]
Hydrogen selenide (as Se)	2 ppm	1 ppm
Hydrogen sulfide	300 ppm	100 ppm
Hydroquinone	Unknown	50 mg/m^3
Iodine	10 ppm	2 ppm
Iron oxide dust and fume (as Fe)	N.E.	2,500 mg Fe/m^3
Isoamyl acetate	3,000 ppm	1,000 ppm
Isoamyl alcohol (primary)	10,000 ppm	500 ppm
Isoamyl alcohol (secondary)	10,000 ppm	500 ppm
Isobutyl acetate	7,500 ppm	1,300 ppm [LEL]
Isobutyl alcohol	8,000 ppm	1,600 ppm
Isophorone	800 ppm	200 ppm
Isopropyl acetate	16,000 ppm	1,800 ppm

Table 5 (cont.) NIOSH Immediately Dangerous to Life and Health (IDLH) Values.

Chemical Name	Original IDLH	Revised IDLH
Isopropyl alcohol	12,000 ppm	2,000 ppm [LEL]
Isopropylamine	4,000 ppm	750 ppm
Isopropyl ether	10,000 ppm	1,400 ppm [LEL]
Isopropyl glycidyl ether	1,000 ppm	400 ppm
Ketene	Unknown	5 ppm
Lead compounds (as Pb)	700 mg Pb/m^3	100 mg Pb/m^3
Lindane	1,000 mg/m^3	50 mg/m^3
Lithium hydride	55 mg/m^3	0.5 mg/m^3
L.P.G.	19,000 ppm [LEL]	2,000 ppm [LEL]
Magnesium oxide fume	N.E.	750 mg/m^3
Malathion	5,000 mg/m^3	250 mg/m^3
Maleic anhydride	Unknown	10 mg/m^3
Manganese compounds (as Mn)	N.E.	500 mg Mn/m^3
Mercury compounds [except (organo) alkyls, as Hg]	28 mg Hg/m^3	10 mg Hg/m^3
Mercury (organo) alkyl compounds (as Hg)	10 mg Hg/m^3	2 mg Hg/m^3
Mesityl oxide	5,000 ppm	1,400 ppm [LEL]
Methoxychlor	N.E.	5,000 mg/m^3
Methyl acetate	10,000 ppm	3,100 ppm [LEL]
Methyl acetylene	15,000 ppm [LEL]	1,700 ppm [LEL]
Methyl acetylenepropadiene mixture	15,000 ppm	3,400 ppm [LEL]
Methyl acrylate	1,000 ppm	250 ppm
Methylal	15,000 ppm [LEL]	2,200 ppm [LEL]
Methyl alcohol	25,000 ppm	6,000 ppm
Methylamine	100 ppm	100 ppm [Unch]
Methyl (namyl) ketone	4,000 ppm	800 ppm
Methyl bromide	2,000 ppm	250 ppm
Methyl Cellosolve (r)	2,000 ppm	200 ppm
Methyl Cellosolve (r) acetate	4,000 ppm	200 ppm

Table 5 (cont.) NIOSH Immediately Dangerous to Life and Health (IDLH) Values.

Chemical Name	Original IDLH	Revised IDLH
Methyl chloride	10,000 ppm	2,000 ppm
Methyl chloroform	1,000 ppm	700 ppm
Methylcyclohexane	10,000 ppm	1,200 ppm [LEL]
Methylcyclohexanol	10,000 ppm	500 ppm
o-Methylcyclohexanone	2,500 ppm	600 ppm
Methylene bisphenyl isocyanate	100 mg/m^3	75 mg/m^3
Methylene chloride	5,000 ppm	2,300 ppm
Methyl formate	5,000 ppm	4,500 ppm
5-Methyl 3-heptanone	3,000 ppm	100 ppm
Methyl hydrazine	50 ppm	20 ppm
Methyl iodide	800 ppm	100 ppm
Methyl isobutyl carbinol	2,000 ppm	400 ppm
Methyl isocyanate	20 ppm	3 ppm
Methyl mercaptan	400 ppm	150 ppm
Methyl methacrylate	4,000 ppm	1,000 ppm
Methyl styrene	5,000 ppm	700 ppm
Mica	N.E.	1,500 mg/m^3
Molybdenum (insoluble compounds, as Mo)	N.E.	5,000 mg Mo/m^3
Molybdenum (soluble compounds, as Mo)	N.E.	1,000 mg Mo/m^3
Monomethyl aniline	100 ppm	100 ppm [Unch]
Morpholine	8,000 ppm	1,400 ppm [LEL]
Naphtha (coal tar)	10,000 ppm [LEL]	1,000 ppm [LEL]
Naphthalene	500 ppm	250 ppm
Nickel carbonyl (as Ni)	7 ppm	2 ppm
Nickel metal and other compounds (as Ni)	N.E.	10 mg Ni/m^3
Nicotine	35 mg/m^3	5 mg/m^3
Nitric acid	100 ppm	25 ppm
Nitric oxide	100 ppm	100 ppm [Unch]

Table 5 (cont.) NIOSH Immediately Dangerous to Life and Health (IDLH) Values.

Chemical Name	Original IDLH	Revised IDLH
p-Nitroaniline	300 mg/m^3	300 mg/m^3 [Unch]
Nitrobenzene	200 ppm	200 ppm [Unch]
p-Nitrochlorobenzene	1,000 mg/m^3	100 mg/m^3
Nitroethane	1,000 ppm	1,000 ppm [Unch]
Nitrogen dioxide	50 ppm	20 ppm
Nitrogen trifluoride	2,000 ppm	1,000 ppm
Nitroglycerine	500 mg/m^3	75 mg/m^3
Nitromethane	1,000 ppm	750 ppm
1-Nitropropane	2,300 ppm	1,000 ppm
2-Nitropropane	2,300 ppm	100 ppm
Nitrotoluene (oisomer)	200 ppm	200 ppm [Unch]
Nitrotoluene (misomer)	200 ppm	200 ppm [Unch]
Nitrotoluene (p isomers)	200 ppm	200 ppm [Unch]
Octachloronaphthalene	Unknown	Unknown [Unch]
Octane	5,000 ppm	1,000 ppm [LEL]
Oil mist (mineral)	N.E.	2,500 mg/m^3
Osmium tetroxide (as Os)	1 mg Os/m^3	1 mg Os/m^3 [Unch]
Oxalic acid	500 mg/m^3	500 mg/m^3 [Unch]
Oxygen difluoride	0.5 ppm	0.5 ppm [Unch]
Ozone	10 ppm	5 ppm
Paraquat	1.5 mg/m^3	1 mg/m^3
Parathion	20 mg/m^3	10 mg/m^3
Pentaborane	3 ppm	1 ppm
Pentachloronaphthalene	Unknown	Unknown [Unch]
Pentachlorophenol	150 mg/m^3	2.5 mg/m^3
n-Pentane	15,000 ppm [LEL]	1,500 ppm [LEL]
2-Pentanone	5,000 ppm	1,500 ppm
Perchloromethyl mercaptan	10 ppm	10 ppm [Unch]
Perchloryl fluoride	385 ppm	100 ppm

Table 5 (cont.) NIOSH Immediately Dangerous to Life and Health (IDLH) Values.

Chemical Name	Original IDLH	Revised IDLH
Petroleum distillates (naphtha)	10,000 ppm	1,100 ppm [LEL]
Phenol	250 ppm	250 ppm [Unch]
p-Phenylene diamine	Unknown	25 mg/m^3
Phenyl ether (vapor)	N.E.	100 ppm
Phenyl etherbiphenyl mixture (vapor)	N.E.	10 ppm
Phenyl glycidyl ether	Unknown	100 ppm
Phenylhydrazine	295 ppm	15 ppm
Phosdrin	4 ppm	4 ppm [Unch]
Phosgene	2 ppm	2 ppm [Unch]
Phosphine	200 ppm	50 ppm
Phosphoric acid	10,000 mg/m^3	1,000 mg/m^3
Phosphorus (yellow)	N.E.	5 mg/m^3
Phosphorus pentachloride	200 mg/m^3	70 mg/m^3
Phosphorus pentasulfide	750 mg/m^3	250 mg/m^3
Phosphorus trichloride	50 ppm	25 ppm
Phthalic anhydride	10,000 mg/m^3	60 mg/m^3
Picric acid	100 mg/m^3	75 mg/m^3
Pindone	200 mg/m^3	100 mg/m^3
Platinum (soluble salts, as Pt)	N.E.	4 mg Pt/m^3
Portland cement	N.E.	5,000 mg/m^3
Propane	20,000 ppm [LEL]	2,100 ppm [LEL]
n-Propyl acetate	8,000 ppm	1,700 ppm
n-Propyl alcohol	4,000 ppm	800 ppm
Propylene dichloride	2,000 ppm	400 ppm
Propylene imine	500 ppm	100 ppm
Propylene oxide	2,000 ppm	400 ppm
n-Propyl nitrate	2,000 ppm	500 ppm
Pyrethrum	5,000 mg/m^3	5,000 mg/m^3 [Unch]
Pyridine	3,600 ppm	1,000 ppm

Table 5 (cont.) NIOSH Immediately Dangerous to Life and Health (IDLH) Values.

Chemical Name	Original IDLH	Revised IDLH
Quinone	300 mg/m^3	100 mg/m^3
Rhodium (metal fume and insoluble compounds, as Rh)	N.E.	100 mg Rh/m^3
Rhodium (soluble compounds, as Rh)	N.E.	2 mg Rh/m^3
Ronnel	5,000 mg/m^3	300 mg/m^3
Rotenone	Unknown	2,500 mg/m^3
Selenium compounds (as Se)	Unknown	1 mg Se/m^3
Selenium hexafluoride	5 ppm	2 ppm
Silica, amorphous	N.E.	3,000 mg/m^3
Silica, crystalline (respirable dust)	N.E.	
cristobalite/tridymite:		25 mg/m^3
quartz/tripoli:		50 mg/m^3
Silver (metal dust and soluble compounds, as Ag)	N.E.	10 mg Ag/m^3
Soapstone	N.E.	3,000 mg/m^3
Sodium fluoroacetate	5 mg/m^3	2.5 mg/m^3
Sodium hydroxide	250 mg/m^3	10 mg/m^3
Stibine	40 ppm	5 ppm
Stoddard solvent	29,500 mg/m^3	20,000 mg/m^3
Strychnine	3 mg/m^3	3 mg/m^3 [Unch]
Styrene	5,000 ppm	700 ppm
Sulfur dioxide	100 ppm	100 ppm [Unch]
Sulfuric acid	80 mg/m^3	15 mg/m^3
Sulfur monochloride	10 ppm	5 ppm
Sulfur pentafluoride	1 ppm	1 ppm [Unch]
Sulfuryl fluoride	1,000 ppm	200 ppm
2,4,5-T	Unknown	250 mg/m^3
Talc	N.E.	1,000 mg/m^3
Tantalum (metal and oxide dust, as Ta)	N.E.	2,500 mg Ta/m^3

Table 5 (cont.) NIOSH Immediately Dangerous to Life and Health (IDLH) Values.

Chemical Name	Original IDLH	Revised IDLH
TEDP	35 mg/m³	10 mg/m³
Tellurium compounds (as Te)	N.E.	25 mg Te/m³
Tellurium hexafluoride	1 ppm	1 ppm [Unch]
TEPP	10 mg/m³	5 mg/m³
Terphenyl (o, m, p isomers)	Unknown	500 mg/m³
1,1,1,2-Tetrachloro 2, 2-difluoroethane	15,000 ppm	2,000 ppm
1,1,2,2-Tetrachloro 1, 2-difluoroethane	15,000 ppm	2,000 ppm
1,1,2,2-Tetrachloroethane	150 ppm	100 ppm
Tetrachloroethylene	500 ppm	150 ppm
Tetrachloronaphthalene	Unknown	Unknown [Unch]
Tetraethyl lead (as Pb)	40 mg Pb/m³	40 mg Pb/m³ [Unch]
Tetrahydrofuran	20,000 ppm [LEL]	2,000 ppm [LEL]
Tetramethyl lead (as Pb)	40 mg Pb/m³	40 mg Pb/m³ [Unch]
Tetramethyl succinonitrile	5 ppm	5 ppm [Unch]
Tetranitromethane	5 ppm	4 ppm
Tetryl	N.E.	750 mg/m³
Thallium (soluble compounds, as Tl)	20 mg Tl/m³	15 mg Tl/m³
Thiram	1,500 mg/m³	100 mg/m³
Tin (inorganic compounds, as Sn)	400 mg Sn/m³	100 mg Sn/m³
Tin (organic compounds, as Sn)	Unknown	25 mg Sn/m³
Titanium dioxide	N.E.	5,000 mg/m³
Toluene	2,000 ppm	500 ppm
Toluene 2,4-diisocyanate	10 ppm	2.5 ppm
o-Toluidine	100 ppm	50 ppm
Tributyl phosphate	125 ppm	30 ppm
1,1,2-Trichloroethane	500 ppm	100 ppm

Table 5 (cont.) NIOSH Immediately Dangerous to Life and Health (IDLH) Values.

Chemical Name	Original IDLH	Revised IDLH
Trichloroethylene	1,000 ppm	1,000 ppm [Unch]
Trichloronaphthalene	Unknown	Unknown [Unch]
1,2,3-Trichloropropane	1,000 ppm	100 ppm
1,1,2-Trichloro 1,2,2-trifluoroethane	4,500 ppm	2,000 ppm
Triethylamine	1,000 ppm	200 ppm
Trifluorobromomethane	50,000 ppm	40,000 ppm
2,4,6-Trinitrotoluene	1,000 mg/m^3	500 mg/m^3
Triorthocresyl phosphate	40 mg/m^3	40 mg/m^3 [Unch]
Triphenyl phosphate	N.E.	1,000 mg/m^3
Turpentine	1,500 ppm	800 ppm
Uranium (insoluble compounds, as U)	30 mg U/m^3	10 mg U/m^3
Uranium (soluble compounds, as U)	20 mg U/m^3	10 mg U/m^3
Vanadium dust	70 mg/m^3 (as V2O5)	35 mg V/m^3
Vanadium fume	70 mg/m^3 (as V2O5)	35 mg V/m^3
Vinyl toluene	5,000 ppm	400 ppm
Warfarin	350 mg/m^3	100 mg/m^3
Xylene (oisomer)	1,000 ppm	900 ppm
Xylene (misomer)	1,000 ppm	900 ppm
Xylene (p isomers)	1,000 ppm	900 ppm
Xylidine	150 ppm	50 ppm
Yttrium compounds (as Y)	N.E.	500 mg Y/m^3
Zinc chloride fume	4,800 mg/m^3	50 mg/m^3
Zinc oxide	2,500 mg/m^3	500 mg/m^3
Zirconium compounds (as Zr)	500 mg Zr/m^3	25 mg Zr/m^3

Table 6A ACGIH IDLH.

Chemical Name	IDLH (ppm)	IDLH (mg/m^3)
Acetaldehyde	10000	
Acetic acid	1000	
Acetic anhydride	1000	
Acetone	20000	
Acetonitrile	4000	
Acetylene tetrabromide	10	
Acrolein	5	
Acrylamide	unknown	
Acrylonitrile	500	
Aldrin		100
Allyl alcohol	150	
Allyl chloride	300	
Allyl glycidyl ether	270	
2-Aminopyridine	5	
Ammonia	500	
Ammonium sulfamate		5000
n-Amyl acetate	4000	
sec-Amyl acetate	9000	
Aniline	100	
o-Anisidine		50
p-Anisidine		50
Antimony compounds (as Sb)		80 Sb/m^3
ANTU		100
Arsenic (inorganic compounds, as As)		100 As/m^3
Arsine	6	
Azinphosmethyl		20
Barium (soluble compounds, as Ba)		1100 mg Ba/m^3
Benzene	3000	
Benzoyl peroxide		7000
Benzyl chloride	10	

Table 6A (cont.) ACGIH IDLH.

Chemical Name	IDLH (ppm)	IDLH (mg/m³)
Beryllium compounds (as Be)		10 mg Be/m³
Boron oxide	no evidence	
Boron trifluoride	100	
Bromine	10	
Bromoform	unknown	
1,3-Butadiene	20000 (LEL)	
2-Butanone	3000	
2-Butoxyethanol	700	
n-Butyl acetate	10000	
sec-Butyl acetate	10000	
tert-Butyl acetate	10000	
n-Butyl alcohol	8000	
sec-Butyl alcohol	10000	
tert-Butyl alcohol	8000	
n-Butylamine	2000	
tert-Butyl chromate		30 mg CrO3/m³
n-Butyl glycidyl ether	3500	
n-Butyl mercaptan	2500	
p-tert-Butyltoluene	1000	
Cadmium dust (as Cd)		50 mg Cd/m³
Cadmium fume (as Cd)		
Calcium arsenate (as As)		100 mg As/m³
Calcium oxide	unknown	
Camphor (synthetic)		200
Carbaryl		600
Carbon black	no evidence	
Carbon dioxide	50000	
Carbon disulfide	500	
Carbon monoxide	1500	
Carbon tetrachloride	300	
Chlordane		500

Table 6A (cont.) ACGIH IDLH.

Chemical Name	IDLH (ppm)	IDLH (mg/m^3)
Chlorinated camphene		200
Chlorinated diphenyl oxide	unknown	
Chlorine	30	
Chlorine dioxide	unknown	
Chlorine trifluoride	20	
Chloroacetaldehyde	100	
alpha-Chloroacetophenone		100
Chlorobenzene	2400	
o-Chlorobenzylidene malononitrile		2
Chlorobromomethane	5000	
Chlorodiphenyl (42% chlorine)		10
Chlorodiphenyl (54% chlorine)		
Chloroform	1000	
1-Chloro-1-nitropropane	2000	
Chloropicrin	4	
beta-Chloroprene	400	
Chromic acid and chromates		30 (as CrO3)
Chromium (II) compounds [as Cr(II)]	no evidence	
Chromium (III) compounds [as Cr(III)]	no evidence	
Chromium metal (as Cr)	no evidence	
Coal tar pitch volatiles		700
Cobalt metal, dust and fume (as Co)		20 mg Co/m^3
Copper (dusts and mists, as Cu)	no evidence	
Copper fume (as Cu)	no evidence	
Cotton dust (raw)	no evidence	
Crag (r) herbicide		5000
Cresol (o, m, p isomers)	250	
Crotonaldehyde	400	
Cumene	8000	
Cyanides (as CN)		50

Table 6A (cont.) ACGIH IDLH.

Chemical Name	IDLH (ppm)	IDLH (mg/m³)
Cyclohexane	10000	
Cyclohexanol	3500	
Cyclohexanone	5000	
Cyclohexene	10000	
Cyclopentadiene	2000	
2,4-D		500
DDT	no evidence	
Decaborane		100
Demeton		20
Diacetone alcohol	2100	
Diazomethane	2	
Diborane	40	
Dibutyl phosphate	125	
Dibutyl phthalate		9300
o-Dichlorobenzene	1000	
p-Dichlorobenzene	1000	
Dichlorodifluoromethane	50000	
1,3-Dichloro 5,5-dimethylhydantoin	unknown	
1,1-Dichloroethane	4000	
1,2-Dichloroethylene	4000	
Dichloroethyl ether	250	
Dichloromonofluoromethane	50000	
1,1-Dichloro 1-nitroethane	150	
Dichlorotetrafluoroethane	50000	
Dichlorvos		200
Dieldrin		450
Diethylamine	2000	
2-Diethylaminoethanol	500	
Difluorodibromomethane	2500	
Diglycidyl ether	25	
Diisobutyl ketone	2000	

Table 6A (cont.) ACGIH IDLH.

Chemical Name	**IDLH (ppm)**	**IDLH (mg/m^3)**
Diisopropylamine	1000	
Dimethyl acetamide	400	
Dimethylamine	2000	
N,N-Dimethylaniline	100	
Dimethyl 1,2-dibromo 2, 2-dichlorethyl phosphate		1800
Dimethylformamide	3500	
1,1-Dimethylhydrazine	50	
Dimethylphthalate		9300
Dimethyl sulfate	10	
Dinitrobenzene (o, m, p isomers)		200
Dinitroocresol		5
Dinitrotoluene		200
Di sec-octyl phthalate	unknown	
Dioxane	2000	
Diphenyl		300
Dipropylene glycol methyl ether	unknown	
Endrin		2000
Epichlorohydrin	250	
EPN		50
Ethanolamine	1000	
2-Ethoxyethanol	6000	
2-Ethoxyethyl acetate	2500	
Ethyl acetate	10000	
Ethyl acrylate	2000	
Ethyl alcohol	15000	
Ethylamine	4000	
Ethyl benzene	2000	
Ethyl bromide	3500	
Ethyl butyl ketone	3000	
Ethyl chloride	20000	
Ethylene chlorohydrin	10	

Table 6A (cont.) ACGIH IDLH.

Chemical Name	**IDLH (ppm)**	**IDLH (mg/m³)**
Ethylenediamine	2000	
Ethylene dibromide	400	
Ethylene dichloride	1000	
Ethylene glycol dinitrate		500
Ethyleneimine	100	
Ethylene oxide	800	
Ethyl ether	19000 (LEL)	
Ethyl formate	8000	
Ethyl mercaptan	2500	
N-Ethylmorpholine	2000	
Ethyl silicate	1000	
Ferbam	no evidence	
Ferrovanadium dust	no evidence	
Fluorides (as F)		500 mg F/m³
Fluorine	25	
Fluorotrichloromethane	10000	
Formaldehyde	30	
Formic acid	30	
Furfural	250	
Furfuryl alcohol	250	
Glycidol	500	
Graphite (natural)	no evidence	
Hafnium compounds (as Hf)	unknown	
Heptachlor		700
n-Heptane	5000	
Hexachloroethane	300	
Hexachloronaphthalene		2
n-Hexane	5000	
2-Hexanone	5000	
Hexone	3000	
sec Hexyl acetate	4000	
Hydrazine	80	

Table 6A (cont.) ACGIH IDLH.

Chemical Name	**IDLH (ppm)**	**IDLH (mg/m³)**
Hydrogen bromide	50	
Hydrogen chloride	100	
Hydrogen cyanide	50	
Hydrogen fluoride (as F)	30	
Hydrogen peroxide	75	
Hydrogen selenide (as Se)	2	
Hydrogen sulfide	300	
Hydroquinone	unknown	
Iodine	10	
Iron oxide dust and fume (as Fe)	no evidence	
Isoamyl acetate	3000	
Isoamyl alcohol (primary)	10000 (LEL)	
Isoamyl alcohol (secondary)	10000 (LEL)	
Isobutyl acetate	7500	
Isobutyl alcohol	8000	
Isophorone	800	
Isopropyl acetate	16000	
Isopropyl alcohol	12000	
Isopropylamine	4000	
Isopropyl ether	10000	
Isopropyl glycidyl ether	1000	
Ketene	unknown	
Lead compounds (as Pb)		700 mg Pb/m³
Lindane	1000	
Lithium hydride		55
L.P.G.	19000 (LEL)	
Magnesium oxide fume	no evidence	
Malathion		5000
Maleic anhydride	unknown	
Manganese compounds (as Mn)	no evidence	
Mercury compounds [except (organo) alkyls, as Hg]		28 mg Hg/m³

Table 6A (cont.) ACGIH IDLH.

Chemical Name	IDLH (ppm)	IDLH (mg/m^3)
Mercury (organo) alkyl compounds(as Hg)		10 mg Hg/m^3
Mesityl oxide	5000	
Methoxychlor	no evidence	
Methyl acetate	10000	
Methyl acetylene	15000 (LEL)	
Methyl acetylenepropadiene mixture	15000	
Methyl acrylate	1000	
Methylal	15000 (LEL)	
Methyl alcohol	25000	
Methylamine	100	
Methyl (namyl) ketone	4000	
Methyl bromide	2000	
Methyl Cellosolve (r)	2000	
Methyl Cellosolve (r) acetate	4000	
Methyl chloride	10000	
Methyl chloroform	1000	
Methylcyclohexane	10000	
Methylcyclohexanol	10000	
o-Methylcyclohexanone	2500	
Methylene bisphenyl isocyanate		100
Methylene chloride	5000	
Methyl formate	5000	
5-Methyl 3-heptanone		3000
Methyl hydrazine	50	
Methyl iodide	800	
Methyl isobutyl carbinol	2000	
Methyl isocyanate	20	
Methyl mercaptan	400	
Methyl methacrylate	4000	
Methyl styrene	5000	

Table 6A (cont.) ACGIH IDLH.

Chemical Name	IDLH (ppm)	IDLH (mg/m^3)
Mica	no evidence	
Molybdenum (insoluble compounds, as Mo)	no evidence	
Molybdenum (soluble compounds, as Mo)	no evidence	
Monomethyl aniline	100	
Morpholine	8000	
Naphtha (coal tar)	10000 (LEL)	
Naphthalene	500	
Nickel carbonyl (as Ni)	7	
Nickel metal and other compounds (as Ni)	no evidence	
Nicotine		35
Nitric acid	100	
Nitric oxide	100	
p-Nitroaniline		300
Nitrobenzene	200	
p-Nitrochlorobenzene		1000
Nitroethane	1000	
Nitrogen dioxide	50	
Nitrogen trifluoride	2000	
Nitroglycerine		500
Nitromethane	1000	
1-Nitropropane	2300	
2-Nitropropane	2300	
Nitrotoluene (oisomer)	200	
Nitrotoluene (misomer)	200	
Nitrotoluene (p isomers)	200	
Octachloronaphthalene	unknown	
Octane	5000	
Oil mist (mineral)	no evidence	
Osmium tetroxide (as Os)		1 mg Os/m^3

Table 6A (cont.) ACGIH IDLH.

Chemical Name	**IDLH (ppm)**	**IDLH (mg/m³)**
Oxalic acid		500
Oxygen difluoride	0.5	
Ozone	10	
Paraquat		1.5
Parathion		20
Pentaborane	3	
Pentachloronaphthalene	unknown	
Pentachlorophenol		150
n-Pentane	15000 (LEL)	
2-Pentanone	5000	
Perchloromethyl mercaptan	10	
Perchloryl fluoride	385	
Petroleum distillates (naphtha)	10000	
Phenol	250	
p-Phenylene diamine	unknown	
Phenyl ether (vapor)	no evidence	
Phenyl etherbiphenyl mixture (vapor)	no evidence	
Phenyl glycidyl ether	unknown	
Phenylhydrazine	295	
Phosdrin	4	
Phosgene	2	
Phosphine	200	
Phosphoric acid		10000
Phosphorus (yellow)	no evidence	
Phosphorus pentachloride		200
Phosphorus pentasulfide		750
Phosphorus trichloride	50	
Phthalic anhydride		10000
Picric acid		100
Pindone		200
Platinum (soluble salts, as Pt)	no evidence	

Table 6A (cont.) ACGIH IDLH.

Chemical Name	IDLH (ppm)	IDLH (mg/m^3)
Portland cement	no evidence	
Propane	20000 (LEL)	
n-Propyl acetate	8000	
n-Propyl alcohol	4000	
Propylene dichloride	2000	
Propylene imine	500	
Propylene oxide	2000	
n-Propyl nitrate	2000	
Pyrethrum		5000
Pyridine	3600	
Quinone		300
Rhodium (metal fume and insoluble compounds, as Rh)	no evidence	
Rhodium (soluble compounds, as Rh)	no evidence	
Ronnel		5000
Rotenone	unknown	
Selenium compounds (as Se)	unknown	
Selenium hexafluoride	5	
Silica, amorphous	no evidence	
Silica, crystalline (respirable dust) cristobalite/tridymite	no evidence	
Silica, crystalline (respirable dust) quartz/tripoli:	no evidence	
Silver (metal dust as Ag)	no evidence	
Silver (soluble compounds as Ag)	no evidence	
Soapstone	no evidence	
Sodium fluoroacetate		5
Sodium hydroxide		250
Stibine	40	
Stoddard solvent		29500
Strychnine		3
Styrene	5000	

Table 6A (cont.) ACGIH IDLH.

Chemical Name	IDLH (ppm)	IDLH (mg/m^3)
Sulfur dioxide	100	
Sulfuric acid		80
Sulfur monochloride	10	
Sulfur pentafluoride	1	
Sulfuryl fluoride	1000	
2,4,5-T	unknown	
Talc	no evidence	
Tantalum (metal and oxide dust, as Ta)	no evidence	
TEDP		35
Tellurium compounds (as Te)	no evidence	
Tellurium hexafluoride	1	
TEPP		10
Terphenyl (o, m, p isomers)	unknown	
1,1,1,2-Tetrachloro 2, 2-difluoroethane	15000	
1,1,2,2-Tetrachloro 1, 2-difluoroethane	15000	
1,1,2,2-Tetrachloroethane	150	
Tetrachloroethylene	500	
Tetrachloronaphthalene		2
Tetraethyl lead (as Pb)		40 mg Pb/m^3
Tetrahydrofuran	20000 (LEL)	
Tetramethyl lead (as Pb)		40 mg Pb/m^3
Tetramethyl succinonitrile	5	
Tetranitromethane	5	
Tetryl	no evidence	
Thallium (soluble compounds, as Tl)		20 mg TI/m^3
Thiram		1500
Tin (inorganic compounds, as Sn)		400 mg Sn/m^3
Tin (organic compounds, as Sn)	unknown	
Titanium dioxide	no evidence	

Table 6A (cont.) ACGIH IDLH.

Chemical Name	IDLH (ppm)	IDLH (mg/m³)
Toluene	2000	
Toluene 2,4-diisocyanate	10	
o-Toluidine	100	
Tributyl phosphate	125	
1,1,2-Trichloroethane	500	
Trichloroethylene	1000	
Trichloronaphthalene	unknown	
1,2,3-Trichloropropane	1000	
1,1,2-Trichloro 1,2,2-trifluoroethane	4500	
Triethylamine	1000	
Trifluorobromomethane	50000	
2,4,6-Trinitrotoluene	no evidence	
Triorthocresyl phosphate		40
Triphenyl phosphate	no evidence	
Turpentine	1500	
Uranium (insoluble compounds, as U)		30 mg U/m³
Uranium (soluble compounds, as U)		20 mg U/m³
Vanadium dust		70
Vanadium fume		70
Vinyl toluene		5000
Warfarin		350
Xylene (oisomer)		1000
Xylene (misomer)		1000
Xylene (p isomers)		1000
Xylidine	150	
Yttrium compounds (as Y)	no evidence	
Zinc chloride fume		4800
Zinc oxide	no evidence	
Zirconium compounds (as Zr)		500 mg Zr/m³

http://www.cdc.gov/niosh/idlh/intridl4.html

Table 6C ATSDR Minimal Risk Levels (MRLs) December 2006.

Chemical Name	Route	Duration	MRL	Factors	Endpoint	Draft/ Final	Cover Date
Acenaphthene	Oral	Int.	0.6 mg/kg/day	300	Hepatic	Final	Aug-95
Acetone	Inh.	Acute	26 ppm	9	Neurol.	Final	May-94
		Int.	13 ppm	100	Neurol.		
		Chr.	13 ppm	100	Neurol.		
	Oral	Int.	2 mg/kg/day	100	Hemato.		
Acrolein	Inh.	Acute	0.003 ppm	100	Resp.	Draft	5-Sep
		Int.	0.00004 ppm	300	Resp.		
	Oral	Int.	0.008 mg/kg/day	100	Gastro.		
Acrylonitrile	Inh.	Acute	0.1 ppm	10	Neurol.	Final	Dec-90
	Oral	Acute	0.1 mg/kg/day	100	Develop.		
		Int.	0.01 mg/kg/day	1000	Repro.		
		Chr.	0.04 mg/kg/day	100	Hemato.		
Aldrin	Oral	Acute	0.002 mg/kg/day	1000	Develop.	Final	2-Sep
		Chr.	0.00003 mg/kg/day	1000	Hepatic		
Aluminum	Oral	Int.	1.0 mg/kg/day	30	Neurol.	Draft	6-Sep
		Chr.	1.0 mg/kg/day	90	Neurol.		

Table 6C (cont.) ATSDR Minimal Risk Levels (MRLs) December 2006.

Chemical Name	Route	Duration	MRL	Factors	Endpoint	Draft/ Final	Cover Date
Americium	Rad.	Acute	4 mSv/yr	3	Develop.	Final	4-Oct
		Chr.	1 mSv/yr	3	Other		
Ammonia	Inh.	Acute	1.7 ppm	30	Resp.	Final	4-Oct
		Chr.	0.1 ppm	30	Resp.		
Anthracene	Oral	Int.	10 mg/kg/day	100	Hepatic	Final	Aug-95
Arsenic	Oral	Acute*	0.005 mg/kg/day	10	Gastro.	Draft	5-Sep
*provisional		Chr.	0.0003 mg/kg/day	3	Dermal		
Atrazine	Oral	Acute	0.01 mg/kg/day	100	Body Wt.	Final	3-Sep
		Int.	0.003 mg/kg/day	300	Repro.		
Barium, soluble salts	Oral	Int.	0.7 mg/kg/day	100	Renal	Draft	5-Sep
Benzene	Inh.	Acute	0.009 ppm	300	Immuno.	Draft	5-Sep
		Int.	0.006 ppm	300	Immuno.		
		Chr.	0.003 ppm	10	Immuno.		
Beryllium	Oral	Chr.	0.002 mg/kg/day	300	Gastro.	Final	2-Sep
Bis(chloromethyl) ether	Inh.	Int.	0.0003 ppm	100	Resp.	Final	Dec-89
Bis(2 chloroethyl) ether	Inh.	Int.	0.02 ppm	1000	Body Wt.	Final	Dec-89

Boron	Oral	Int.	0.01 mg/kg/day	1000	Develop.	Final	Jul-92
Bromodichloro methane	Oral	Acute	0.04 mg/kg/day	1000	Hepatic	Final	Dec-89
		Chr.	0.02 mg/kg/day	1000	Renal		
Bromoform	Oral	Acute	0.7 mg/kg/day	100	Hepatic	Final	5-Sep
		Int.	0.2 mg/kg/day	300	Hepatic		
		Chr.	0.02 mg/kg/day	3000	Hepatic		
Bromomethane	Inh.	Acute	0.05 ppm	100	Neurol.	Final	Sep-92
		Int.	0.05 ppm	100	Neurol.		
		Chr.	0.005 ppm	100	Neurol.		
	Oral	Int.	0.003 mg/kg/day	100	Gastro.		
Cadmium	Oral	Chr.	0.0002 mg/kg/day	10	Renal	Final	Jul-99
Carbon disulfide	Inh.	Chr.	0.3 ppm	30	Neurol.	Final	Aug-96
	Oral	Acute	0.01 mg/kg/day	300	Hepatic		
Carbon tetrachloride	Inh.	Int.	0.03 ppm	30	Hepatic	Final	5-Sep
		Chr.	0.03 ppm	30	Hepatic		
	Oral	Acute	0.02 mg/kg/day	300	Hepatic		
		Int.	0.007 mg/kg/day	100	Hepatic		

Table 6C (cont.) ATSDR Minimal Risk Levels (MRLs) December 2006.

Chemical Name	Route	Duration	MRL	Factors	Endpoint	Draft/ Final	Cover Date
Cesium	Rad.	Acute	4 mSv	3	Develop.	Final	4-Oct
		Chr.	1 mSv/yr	3	Other		
Chlordane	Inh.	Int.	0.0002 mg/m^3	100	Hepatic	Final	May-94
		Chr.	0.00002 mg/m^3	1000	Hepatic		
	Oral	Acute	0.001 mg/kg/day	1000	Develop.		
		Int.	0.0006 mg/kg/day	100	Hepatic		
		Chr.	0.0006 mg/kg/day	100	Hepatic		
Chlordecone	Oral	Acute	0.01 mg/kg/day	100	Neurol.	Final	Aug-95
		Int.	0.0005 mg/kg/day	100	Renal		
		Chr.	0.0005 mg/kg/day	100	Renal		
Chlorfenvinphos	Oral	Acute	0.002 mg/kg/day	1000	Neurol.	Final	Sep-97
		Int.	0.002 mg/kg/day	1000	Immuno.		
		Chr.	0.0007 mg/kg/day	1000	Neurol.		
Chlorine dioxide	Inh.	Int.	0.001 ppm	300	Resp.	Final	4-Oct
Chlorite	Oral	Int.	0.1 mg/kg/day	30	Neurol.	Final	4-Oct
Chlorobenzene	Oral	Int.	0.4 mg/kg/day	100	Hepatic	Final	Dec-90
Chlorodibromomethane	Oral	Acute	0.1 mg/kg/day	300	Hepatic	Final	5-Sep
		Chr.	0.09 mg/kg/day	300	Hepatic		

Chloroethane	Inh.	Acute	15 ppm	100	Develop.	Final	Dec-98
Chloroform	Inh.	Acute	0.1 ppm	30	Hepatic	Final	Sep-97
		Int.	0.05 ppm	300	Hepatic		
		Chr.	0.02 ppm	100	Hepatic		
	Oral	Acute	0.3 mg/kg/day	100	Hepatic		
		Int.	0.1 mg/kg/day	100	Hepatic		
		Chr.	0.01 mg/kg/day	1000	Hepatic		
Chloromethane	Inh.	Acute	0.5 ppm	100	Neurol.	Final	Dec-98
		Int.	0.2 ppm	300	Hepatic		
		Chr.	0.05 ppm	1000	Neurol.		
Chlorpyrifos	Oral	Acute	0.003 mg/kg/day	10	Neurol.	Final	Sep-97
		Int.	0.003 mg/kg/day	10	Neurol.		
		Chr.	0.001 mg/kg/day	100	Neurol.		
Chromium(vi), aerosol mists	Inh.	Int.	0.000005 mg/m^3	100	Resp.	Final	Sep-00
Chromium(vi), particulates	Inh.	Int.	0.001 mg/m^3	30	Resp.	Final	Sep-00
Cobalt	Inh.	Chr.	0.0001 mg/m^3	10	Resp.	Final	4-Oct
	Oral	Int.	0.01 mg/kg/day	100	Hemato.		
	Rad.	Acute	4 mSv	3	Develop.		
		Chr.	1 mSv/yr	3	Other		

Table 6C (cont.) ATSDR Minimal Risk Levels (MRLs) December 2006.

Chemical Name	Route	Duration	MRL	Factors	Endpoint	Draft/ Final	Cover Date
Copper	Oral	Acute	0.01 mg/kg/day	3	Gastro.	Final	4-Oct
		Int.	0.01 mg/kg/day	3	Gastro.		
Cresols	Oral	Int.	0.1 mg/kg/day	100	Resp.	Draft	6-Sep
Cyanide, sodium	Oral	Int.	0.05 mg/kg/day	100	Repro.	Final	6-Jul
Cyclotetramethylene tetranitramine (hmx)	Oral	Acute	0.1 mg/kg/day	1000	Neurol.	Final	Sep-97
		Int.	0.05 mg/kg/day	1000	Hepatic		
Cyclotrimethylene-trinitramine (rdx)	Oral	Acute	0.06 mg/kg/day	100	Neurol.	Final	Jun-95
Cyhalothrin	Oral	Acute	0.01 mg/kg/day	100	Gastro.	Final	3-Sep
		Int.	0.01 mg/kg/day	100	Gastro.		
Cypermethrin	Oral	Acute	0.02 mg/kg/day	100	Neurol.	Final	3-Sep
Ddt, p,p′-	Oral	Acute	0.0005 mg/kg/day	1000	Develop.	Final	2-Sep
		Int.	0.0005 mg/kg/day	100	Hepatic		
Di(2-ethylhexyl) phthalate	Oral	Int.	0.1 mg/kg/day	100	Repro.	Final	2-Sep
		Chr.	0.06 mg/kg/day	100	Repro.		
Di-n-butyl phthalate	Oral	Acute	0.5 mg/kg/day	100	Develop.	Final	1-Sep

Di-n-octyl phthalate	Oral	Acute	3 mg/kg/day	300	Hepatic	Final	Sep-97
		Int.	0.4 mg/kg/day	100	Hepatic		
Diazinon	Inh.	Int.	0.01 mg/m^3	30	Neurol.	Draft	6-Sep
	Oral	Acute	0.006 mg/kg/day	100	Neurol.		
		Int.	0.002 mg/kg/day	100	Neurol.		
		Chr.	0.0007 mg/kg/day	100	Neurol.		
Dichlorobenzene, 1,2-	Oral	Acute	0.7 mg/kg/day	100	Hepatic	Final	6-Jul
		Int.	0.6 mg/kg/day	100	Hepatic		
		Chr.	0.3 mg/kg/day	100	Renal		
Dichlorobenzene, 1,3-	Oral	Acute	0.4 mg/kg/day	100	Hepatic	Final	6-Jul
		Int.	0.02 mg/kg/day	100	Endocr.		
Dichlorobenzene, 1,4-	Inh.	Acute	2 ppm	10	Ocular	Final	6-Jul
		Int.	0.2 ppm	100	Hepatic		
		Chr.	0.01 ppm	30	Resp.		
	Oral	Int.	0.07 mg/kg/day	100	Hepatic		
		Chr.	0.07 mg/kg/day	100	Hepatic		
Dichlorvos	Inh.	Acute	0.002 ppm	100	Neurol.	Final	Sep-97
		Int.	0.0003 ppm	100	Neurol.		
		Chr.	0.00006 ppm	100	Neurol.		
	Oral	Acute	0.004 mg/kg/day	1000	Neurol.		
		Int.	0.003 mg/kg/day	10	Neurol.		
		Chr.	0.0005 mg/kg/day	100	Neurol.		

Table 6C (cont.) ATSDR Minimal Risk Levels (MRLs) December 2006.

Chemical Name	Route	Duration	MRL	Factors	Endpoint	Draft/ Final	Cover Date
Dieldrin	Oral	Int.	0.0001 mg/kg/day	100	Neurol.	Final	2-Sep
		Chr.	0.00005 mg/kg/ day	100	Hepatic		
Diethyl phthalate	Oral	Acute	7 mg/kg/day	300	Repro.	Final	Jun-95
		Int.	6 mg/kg/day	300	Hepatic		
Diisopropyl methylphosphonate	Oral	Int.	0.8 mg/kg/day	100	Hemato.	Final	Oct-98
(DIMP)		Chr.	0.6 mg/kg/day	100	Hemato.		
Disulfoton	Inh.	Acute	0.006 mg/m^3	30	Neurol.	Final	Aug-95
		Int.	0.0002 mg/m^3	30	Neurol.		
	Oral	Acute	0.001 mg/kg/day	100	Neurol.		
		Int.	0.00009 mg/kg/ day	100	Develop.		
		Chr.	0.00006 mg/kg/ day	1000	Neurol.		
Endosulfan	Oral	Int.	0.005 mg/kg/day	100	Immuno.	Final	Sep-00
		Chr.	0.002 mg/kg/day	100	Hepatic		
Endrin	Oral	Int.	0.002 mg/kg/day	100	Neurol.	Final	Aug-96
		Chr.	0.0003 mg/kg/day	100	Neurol.		

Ethion	Oral	Acute	0.002 mg/kg/day	30	Neurol.	Final	Sep-00
		Int.	0.002 mg/kg/day	30	Neurol.		
		Chr.	0.0004 mg/kg/day	150	Neurol.		
Ethylbenzene	Inh.	Int.	1.0 ppm	100	Develop.	Final	Jul-99
Ethylene glycol	Inh.	Acute	0.5 ppm	100	Renal	Final	Sep-97
	Oral	Acute	2.0 mg/kg/day	100	Develop.		
		Chr.	2.0 mg/kg/day	100	Renal		
Ethylene oxide	Inh.	Int.	0.09 ppm	100	Renal	Final	Dec-90
Fluoranthene	Oral	Int.	0.4 mg/kg/day	300	Hepatic	Final	Aug-95
Fluorene	Oral	Int.	0.4 mg/kg/day	300	Hepatic	Final	Aug-95
Fluoride, sodium	Oral	Chr.	0.05 mg/kg/day	3	Musculo.	Final	3-Sep
Fluorine	Inh.	Acute	0.01 ppm	10	Resp.	Final	3-Sep
Formaldehyde	Inh.	Acute	0.04 ppm	9	Resp.	Final	Jul-99
		Int.	0.03 ppm	30	Resp.		
		Chr.	0.008 ppm	30	Resp.		
	Oral	Int.	0.3 mg/kg/day	100	Gastro.		
		Chr.	0.2 mg/kg/day	100	Gastro.		
Fuel oil no.2	Inh.	Acute	0.02 mg/m^3	1000	Neurol.	Final	Jun-95
Guthion	Inh.	Acute	0.02 mg/m^3	30	Neurol.	Draft	6-Sep
		Int.	0.01 mg/m^3	30	Neurol.		

Table 6C (cont.) ATSDR Minimal Risk Levels (MRLs) December 2006.

Chemical Name	Route	Duration	MRL	Factors	Endpoint	Draft/ Final	Cover Date
		Chr.	0.01 mg/m^3	30	Neurol.		
	Oral	Acute	0.01 mg/kg/day	100	Neurol.		
		Int.	0.003 mg/kg/day	100	Neurol.		
		Chr.	0.003 mg/kg/day	100	Neurol.		
Heptachlor	Oral	Int.	0.0001 mg/kg/day	300	Immuno.	Draft	5-Sep
Hexachlorobenzene	Oral	Acute	0.008 mg/kg/day	300	Develop.	Final	2-Sep
		Int.	0.0001 mg/kg/day	90	Repro.		
		Chr.	0.00005 mg/kg/ day	300	Develop.		
Hexachlorobutadiene	Oral	Int.	0.0002 mg/kg/day	1000	Renal	Final	May-94
Hexachlorocyclohexane, alpha-	Oral	Chr.	0.008 mg/kg/day	100	Hepatic	Final	5-Sep
Hexachlorocyclohexane, beta-	Oral	Acute	0.05 mg/kg/day	100	Neurol.	Final	5-Sep
		Int.	0.0006 mg/kg/day	300	Hepatic		
Hexachlorocyclohexane, gamma-	Oral	Acute	0.003 mg/kg/day	300	Develop.	Final	5-Sep
		Int.	0.00001 mg/kg/ day	1000	Immuno.		

Hexachlorocyclopentadiene	Inh.	Int.	0.01 ppm	30	Resp.	Final	Jul-99
		Chr.	0.0002 ppm	90	Resp.		
	Oral	Int.	0.1 mg/kg/day	100	Renal		
Hexachloroethane	Inh.	Acute	6 ppm	30	Neurol.	Final	Sep-97
		Int.	6 ppm	30	Neurol.		
	Oral	Acute	1 mg/kg/day	100	Hepatic		
		Int.	0.01 mg/kg/day	100	Hepatic		
Hexamethylene diisocyanate	Inh.	Int.	0.00003 ppm	30	Resp.	Final	Oct-98
		Chr.	0.00001 ppm	90	Resp.		
Hexane, n-	Inh.	Chr.	0.6 ppm	100	Neurol.	Final	Jul-99
Hydrazine	Inh.	Int.	0.004 ppm	300	Hepatic	Final	Sep-97
Hydrogen fluoride	Inh.	Acute	0.02 ppm	30	Resp.	Final	3-Sep
Hydrogen sulfide	Inh.	Acute	0.07 ppm	27	Resp.	Final	6-Jul
		Int.	0.02 ppm	30	Resp.		
Iodide	Oral	Acute	0.01 mg/kg/day	1	Endocr.	Final	4-Oct
		Chr.	0.01 mg/kg/day	1	Endocr.		
Ionizing radiation, n.o.s.	Rad.	Acute	4 mSv	3	Neurol.	Final	Sep-99
		Chr.	1 mSv/yr	3	Other		
Isophorone	Oral	Int.	3 mg/kg/day	100	Other	Final	Dec-89

Table 6C (cont.) ATSDR Minimal Risk Levels (MRLs) December 2006.

Chemical Name	Route	Duration	MRL	Factors	Endpoint	Draft/ Final	Cover Date
		Chr.	0.2 mg/kg/day	1000	Hepatic		
Jp-4	Inh.	Int.	9 mg/m^3	300	Hepatic	Final	Jun-95
Jp-5/jp-8	Inh.	Int.	3 mg/m^3	300	Hepatic	Final	Oct-98
Jp-7	Inh.	Chr.	0.3 mg/m^3	300	Hepatic	Final	Jun-95
Kerosene	Inh.	Int.	0.01 mg/m^3	1000	Hepatic	Final	Jun-95
Malathion	Inh.	Acute	0.2 mg/m^3	100	Neurol.	Final	3-Sep
		Int.	0.02 mg/m^3	1000	Resp.		
	Oral	Int.	0.02 mg/kg/day	10	Neurol.		
		Chr.	0.02 mg/kg/day	100	Neurol.		
Manganese	Inh.	Chr.	0.00004 mg/m^3	500	Neurol.	Final	Sep-00
Mercuric chloride	Oral	Acute	0.007 mg/kg/day	100	Renal	Final	Mar-99
		Int.	0.002 mg/kg/day	100	Renal		
Mercury	Inh.	Chr.	0.0002 mg/m^3	30	Neurol.	Final	Mar-99
Methoxychlor	Oral	Int.	0.005 mg/kg/day	1000	Repro.	Draft	Sep-00
Methyl parathion	Oral	Int.	0.0007 mg/kg/day	300	Neurol.	Final	1-Sep
		Chr.	0.0003 mg/kg/day	100	Hemato.		
Methyl-t-butyl ether	Inh.	Acute	2 ppm	100	Neurol.	Final	Aug-96
		Int.	0.7 ppm	100	Neurol.		

		Chr.	0.7 ppm	100	Renal		
	Oral	Acute	0.4 mg/kg/day	100	Neurol.		
		Int.	0.3 mg/kg/day	300	Hepatic		
Methylene chloride	Inh.	Acute	0.6 ppm	100	Neurol.	Final	Sep-00
		Int.	0.3 ppm	90	Hepatic		
		Chr.	0.3 ppm	30	Hepatic		
	Oral	Acute	0.2 mg/kg/day	100	Neurol.		
		Chr.	0.06 mg/kg/day	100	Hepatic		
Methylmercury	Oral	Chr.	0.0003 mg/kg/day	4	Develop.	Final	Mar-99
Mirex	Oral	Chr.	0.0008 mg/kg/day	100	Hepatic	Final	Aug-95
N-nitrosodi-n-propylamine	Oral	Acute	0.095 mg/kg/day	100	Hepatic	Final	Dec-89
Naphthalene	Inh.	Chr.	0.0007 ppm	300	Resp.	Final	5-Sep
	Oral	Acute	0.6 mg/kg/day	90	Neurol.		
		Int.	0.6 mg/kg/day	90	Neurol.		
Nickel	Inh.	Int.	0.0002 mg/m^3	30	Resp.	Final	5-Sep
		Chr.	0.00009 mg/m^3	30	Resp.		
Pentachlorophenol	Oral	Acute	0.005 mg/kg/day	1000	Develop.	Final	1-Sep
		Int.	0.001 mg/kg/day	1000	Repro.		
		Chr.	0.001 mg/kg/day	1000	Endocr.		
Perchlorates	Oral	Chr.	0.0007 mg/kg/day	10	Endocr.	Draft	5-Sep
Permethrin	Oral	Acute	0.3 mg/kg/day	100	Neurol.	Final	3-Sep

Chemical Name	Route	Duration	MRL	Factors	Endpoint	Draft/ Final	Cover Date
		Int.	0.2 mg/kg/day	100	Neurol.		
Phenol	Inh.	Acute	0.02 ppm	30	Resp.	Draft	6-Sep
	Oral	Acute	0.6 mg/kg/day	100	Body Wt.		
Phosphorus, white	Inh.	Acute	0.02 mg/m^3	30	Resp.	Final	Sep-97
	Oral	Int.	0.0002 mg/kg/day	100	Repro.		
Polybrominated biphenyls (pbbs)	Oral	Acute	0.01 mg/kg/day	100	Endocr.	Final	4-Oct
Polybrominated diphenyl ethers (pbdes), lower brominated	Inh	Int.	0.006 mg/m^3	90	Endocr.	Final	4-Oct
	Oral	Acute	0.03 mg/kg/day	30	Endocr.		
		Int.	0.007 mg/kg/day	300	Hepatic		
Pbde, decabrominated	Oral	Int.	10 mg/kg/day	100	Develop.	Final	4-Oct
Polychlorinated biphenyls (pcbs) (aroclor 1254)	Oral	Int.	0.03 ug/kg/day	300	Neurol.	Final	Nov-00
		Chr.	0.02 ug/kg/day	300	Immuno.		
Propylene glycol dinitrate	Inh.	Acute	0.003 ppm	10	Neurol.	Final	Jun-95
		Int.	0.00004 ppm	1000	Hemato.		
		Chr.	0.00004 ppm	1000	Hemato.		
Propylene glycol	Inh.	Int.	0.009 ppm	1000	Resp.	Final	Sep-97

Refractory ceramic fibers	Inh.	Chr.	0.03 fibers/cc	30	Resp.	Final	4-Oct
Selenium	Oral	Chr.	0.005 mg/kg/day	3	Dermal	Final	3-Sep
Strontium	Oral	Int.	2 mg/kg/day	30	Musculo.	Final	4-Oct
Styrene	Inh.	Chr.	0.06 ppm	100	Neurol.	Final	Sep-92
	Oral	Int.	0.2 mg/kg/day	1000	Hepatic		
Sulfur dioxide	Inh.	Acute	0.01 ppm	9	Resp.	Final	Dec-98
Sulfur mustard	Inh.	Acute	0.0007 mg/m^3	30	Ocular	Final	3-Sep
		Int.	0.00002 mg/m^3	30	Ocular		
	Oral	Acute	0.5 µg/kg/day	1000	Develop.		
		Int.	0.07 µg/kg/day	300	Gastro.		
Tetrachloroethylene	Inh.	Acute	0.2 ppm	10	Neurol.	Final	Sep-97
		Chr.	0.04 ppm	100	Neurol.		
	Oral	Acute	0.05 mg/kg/day	100	Develop.		
Tin, inorganic	Oral	Int.	0.3 mg/kg/day	100	Hemato.	Final	5-Sep
Tin, dibutyl-	Oral	Int.	0.005 mg/kg/day	1000	Immunol.	Final	39330
Tin, tributyl-	Oral	Int.	0.0003 mg/kg/day	100	Immunol.	Final	5-Sep
		Chr.	0.0003 mg/kg/day	100	Immunol.		
Titanium tetrachloride	Inh.	Int.	0.01 mg/m^3	90	Resp.	Final	Sep-97
		Chr.	0.0001 mg/m^3	90	Resp.		

Table 6C (cont.) ATSDR Minimal Risk Levels (MRLs) December 2006.

Chemical Name	Route	Duration	MRL	Factors	Endpoint	Draft/ Final	Cover Date
Toluene	Inh.	Acute	1 ppm	10	Neurol.	Final	Sep-00
		Chr.	0.08 ppm	100	Neurol.		
	Oral	Acute	0.8 mg/kg/day	300	Neurol.		
		Int.	0.02 mg/kg/day	300	Neurol.		
Toxaphene	Oral	Acute	0.005 mg/kg/day	1000	Hepatic	Final	Aug-96
		Int.	0.001 mg/kg/day	300	Hepatic		
Trichloroethylene	Inh.	Acute	2 ppm	30	Neurol.	Final	Sep-97
		Int.	0.1 ppm	300	Neurol.		
	Oral	Acute	0.2 mg/kg/day	300	Develop.		
Uranium, highly soluble salts	Inh.	Int.	0.0004 mg/m^3	90	Renal	Final	Sep-99
		Chr.	0.0003 mg/m^3	30	Renal		
	Oral	Int.	0.002 mg/kg/day	30	Renal		
Uranium, insoluble compounds	Inh.	Int.	0.008 mg/m^3	30	Renal	Final	Sep-99
Vanadium	Inh.	Acute	0.0002 mg/m^3	100	Resp.	Final	Jul-92
	Oral	Int.	0.003 mg/kg/day	100	Renal		
Vinyl acetate	Inh.	Int.	0.01 ppm	100	Resp.	Final	Jul-92
Vinyl chloride	Inh.	Acute	0.5 ppm	30	Develop.	Final	6-Jul

		Int.	0.03 ppm	30	Hepatic		
	Oral	Chr.	0.003 mg/kg/day	30	Hepatic		
Xylenes, mixed	Inh.	Acute	2 ppm	30	Neurol.	Draft	5-Sep
		Int.	0.6 ppm	90	Neurol.		
		Chr.	0.05 ppm	300	Neurol.		
	Oral	Acute	1 mg/kg/day	100	Neurol.		
		Int.	1 mg/kg/day	300	Neurol.		
		Chr.	0.6 mg/kg/day	300	Neurol.		
Zinc	Oral	Int.	0.3 mg/kg/day	3	Hemato.	Final	5-Sep
		Chr.	0.3 mg/kg/day	3	Hemato.		
1-methylnaphthalene	Oral	Chr.	0.07 mg/kg/day	1000	Resp.	Final	5-Sep
2-methylnaphthalene	Oral	Chr.	0.04 mg/kg/day	100	Resp.	Final	5-Sep
1,1-dichloroethene	Inh.	Int.	0.02 ppm	100	Hepatic	Final	May-94
	Oral	Chr.	0.009 mg/kg/day	1000	Hepatic		
1,1-dimethylhydrazine	Inh.	Int.	0.0002 ppm	300	Hepatic	Final	Sep-97
1,1,1-trichloroethane	Inh.	Acute	2 ppm	100	Neurol.	Final	6-Jul
		Int.	0.7 ppm	100	Neurol.		
	Oral	Int.	20 mg/kg/day	100	Body Wt.		
1,1,2-trichloroethane	Oral	Acute	0.3 mg/kg/day	100	Neurol.	Final	Dec-89
		Int.	0.04 mg/kg/day	100	Hepatic		

Table 6C (cont.) ATSDR Minimal Risk Levels (MRLs) December 2006.

Chemical Name	Route	Duration	MRL	Factors	Endpoint	Draft/ Final	Cover Date
1,1,2,2-tetrachloroethane	Oral	Int.	0.5 mg/kg/day	100	Hepatic	Draft	6-Sep
1,2-dibromo-3-chloropropane	Inh.	Int.	0.0002 ppm	100	Repro.	Final	Sep-92
	Oral	Int.	0.002 mg/kg/day	1000	Repro.		
1,2-dichloroethane	Inh.	Chr.	0.6 ppm	90	Hepatic	Final	1-Sep
	Oral	Int.	0.2 mg/kg/day	300	Renal		
1,2-dichloroethene, cis-	Oral	Acute	1 mg/kg/day	100	Hemato.	Final	Aug-96
		Int.	0.3 mg/kg/day	100	Hemato.		
1,2-dichloroethene, trans-	Inh.	Acute	0.2 ppm	1000	Hepatic	Final	Aug-96
		Int.	0.2 ppm	1000	Hepatic		
	Oral	Int.	0.2 mg/kg/day	100	Hepatic		
1,2-dichloropropane	Inh.	Acute	0.05 ppm	1000	Resp.	Final	Dec-89
		Int.	0.007 ppm	1000	Resp.		
	Oral	Acute	0.1 mg/kg/day	1000	Neurol.		
		Int.	0.07 mg/kg/day	1000	Hemato.		
		Chr.	0.09 mg/kg/day	1000	Hepatic		
1,2-dimethylhydrazine	Oral	Int.	0.0008 mg/kg/day	1000	Hepatic	Final	Sep-97

1,2,3-trichloropropane	Inh.	Acute	0.0003 ppm	100	Resp.	Final	Sep-92
	Oral	Int.	0.06 mg/kg/day	100	Hepatic		
1,3-dichloropropene	Inh.	Int.	0.008 ppm	30	Resp.	Draft	6-Sep
		Chr.	0.007 ppm	30	Resp.		
	Oral	Int.	0.04 mg/kg/day	100	Gastro.		
		Chr.	0.03 mg/kg/day	100	Gastro.		
1,3-dinitrobenzene	Oral	Acute	0.008 mg/kg/day	100	Repro.	Final	Jun-95
		Int.	0.0005 mg/kg/day	1000	Hemato.		
1,4-dioxane	Inh.	Acute	2 ppm	30	Ocular	Final	6-Jul
		Int.	1 ppm	30	Hepatic		
		Chr.	1 ppm	30	Hepatic		
	Oral	Acute	4 mg/kg/day	100	Resp.		
		Int.	0.6 mg/kg/day	100	Hepatic		
		Chr.	0.1 mg/kg/day	100	Hepatic		
2-butoxyethanol	Inh.	Acute	6 ppm	9	Hemato.	Final	Oct-98
		Int.	3 ppm	9	Hemato.		
		Chr.	0.2 ppm	3	Hemato.		
	Oral	Acute	0.4 mg/kg/day	90	Hemato.		
		Int.	0.07 mg/kg/day	1000	Hepatic		

Table 6C (cont.) ATSDR Minimal Risk Levels (MRLs) December 2006.

Chemical Name	Route	Duration	MRL	Factors	Endpoint	Draft/ Final	Cover Date
2,3-dichloropropene	Inh.	Acute	0.002 ppm	90	Resp.	Draft	6-Sep
2,3,4,7,8-pentachloro-dibenzofuran	Oral	Acute	0.001 ug/kg/day	3000	Immuno.	Final	May-94
		Int.	0.00003 ug/kg/ day	3000	Hepatic		
2,3,7,8-tetrachlorodibenzo-p-dioxin	Oral	Acute	0.0002 ug/kg/day	21	Immuno.	Final	Dec-98
		Int.	0.00002 ug/kg/ day	30	Lymphor.		
		Chr.	0.000001 ug/kg/ day	90	Develop.		
2,4-dichlorophenol	Oral	Int.	0.003 mg/kg/day	100	Immuno.	Final	Jul-99
2,4-dinitrophenol	Oral	Acute	0.01 mg/kg/day	100	Body Wt.	Final	Aug-95
2,4-dinitrotoluene	Oral	Acute	0.05 mg/kg/day	100	Neurol.	Final	Dec-98
		Chr.	0.002 mg/kg/day	100	Hemato.		
2,4,6-trinitrotoluene	Oral	Int.	0.0005 mg/kg/day	1000	Hepatic	Final	Jun-95
2,6-dinitrotoluene	Oral	Int.	0.004 mg/kg/day	1000	Hemato.	Final	Dec-98
4-chlorophenol	Oral	Acute	0.01 mg/kg/day	100	Hepatic	Final	Jul-99

4,4′-methylenebis(2-chloroaniline)	Oral	Chr.	0.003 mg/kg/day	3000	Hepatic	Final	May-94
4,4′-methylenedianiline	Oral	Acute	0.2 mg/kg/day	300	Hepatic	Final	Oct-98
		Int.	0.08 mg/kg/day	100	Hepatic		
4,6-dinitro-o-cresol	Oral	Acute	0.004 mg/kg/day	100	Neurol.	Final	Aug-95
		Int.	0.004 mg/kg/day	100	Neurol.		

Table 10A. ACGIH TLV.

Chemical Name	**TLV (ppm)**	**TLV (mg/m³)**
Acetic Acid	10	
Acetic Anhydride	5	
Acetone	750	
Acetonitrile	40	
Acetylene tetrabromide	1	
Acrolein	0.1	
Aldrin		0.25
Allyl alcohol	2	
Allyl chloride	1	
Allyl glycidyl ether	5	
Allyl propyl disulfide	2	
alpha-Alumina (td)		10
Aluminum, metal (td)		10
2-Aminopyridine	0.5	
Ammonia	25	
Ammonium sulfamate (td)		10
n-Amyl acetate	100	
sec-Amyl acetate	125	
Aniline and homologs	2	
Anisidine o-, p-isomers	0.1	
Antimony & compounds		0.5
ANTU		0.3
Arsine	0.05	
Azinphos-methyl		0.2
Barium, sol. comp.		0.2
Barium sulfate (td)		10
Benomyl (td)		10
Benzoyl peroxide		5
Benzyl chloride	1	
Bismuth telluride undoped (td)		10
Boron oxide (td)		10

Table 10A (cont.) ACGIH TLV.

Chemical Name	**TLV (ppm)**	**TLV (mg/m³)**
Bromine	0.1	
Bromoform	0.5	
1,3 Butadiene	2	
2-Butanone (MEK)	200	
2-Butoxyethanol	25	
n-Butyl-acetate	150	
sec-Butyl acetate	200	
tert-Butly acetate	200	
sec-Butyl alcohol	100	
tert-Butyl alcohol	100	
n-Butyl glycidyl ether	25	
Butyl mercaptan	0.5	
p-tert-Butyl-1-toluene	1	
Calcium carbonate (td)		10
Calcium hydroxide (td)		5
Calcium oxide		2
Calcium silicate (td)		10
Calcium sulfate (td)		10
Camphor, synthetic		12
Carbaryl		5
Carbon black		3.5
Carbon dioxide	5000	
Carbon Disulfide	10	
Carbon monoxide	25	
Cellulose (td)		10
Chlorodane		0.5
Chlorinated camphene		0.5
Chlorinated diphenyl oxide		0.5
Chlorine	0.5	
Chlorine dioxide	0.1	
a-chloroacetophenone	0.05	

Table 10A (cont.) ACGIH TLV.

Chemical Name	**TLV (ppm)**	**TLV (mg/m³)**
Chlorobenzene	10	
Chlorobromomethane	200	
Chlorodiphenyl (42% Cl)		1
Chlorobiphenyl (54% Cl)		0.5
1-Chloro-1-nitro- propane	2	
Chloropicrin	0.1	
beta-Chloroprene	10	
2-Chloro-6-trichloro- methyl) pyridine (td)		10
Chromium (III) compounds		0.5
Clopidol (td)		10
Coal tar pitch volatiles		0.2
Copper fume		0.2
Copper dusts and mists		1
Cotton dust		0.2
Cresol, all isomers	5	
Crotonaldehyde	2	
Cumene	50	
Cyclohexane	300	
Cyclohexanol	50	
Cyclohexanone	25	
Cyclohexene	300	
Cyclopentadiene	75	
2,4-D		10
Decaborane	0.05	
Demeton	0.01	
Diacetone alcohol	50	
Diazomethane	0.2	
Diborane	0.1	
Dibutyl phosphate	1	
Dibutyl phthalate		5

Table 10A (cont.) ACGIH TLV.

Chemical Name	**TLV (ppm)**	**TLV (mg/m³)**
o-Dichlorobenzene	25	
p-Dichlorobenzene	10	
1,1Dichloroethane	100	
1,2-Dichloroethylene	200	
Dichloroethyl ether	5	
Dichloromono- fluomethane	10	
1,1 Dichloro-1- nitroethane	2	
Dichlorotetra- fluoroethane	1000	
Dichlorvos (DDVP)	0.1	
Dicyclopentadienyl iron (td)		10
Dieldrin		0.25
Diethylamine	5	
2-Diethylaminoethanol	2	
Difluorodibromomethane	100	
Diglycidyl ether (DGE)	0.1	
Diisobutyl ketone	25	
Diisopropylamine	5	
Dimethyl acetamide	10	
Dimethylamine	5	
Dimethylaniline (N,N-Dimethylaniline)	5	
Dimethyl-1,2-dibromo- 2,2-dichloroethyl phosphate		3
Dimethylformamide	10	
1,1-Dimethylhydrazine	0.01	
Dimethylphthalate		5
Dimethyl sulfate	0.1	
Dinitrobenzene (all isomers)	0.15	
Dinitro-o-cresol		0.2
Dinitrotoluene		0.15
Dioxane (Diethylene dioxide)	25	
Diphenyl (Biphenyl)	0.2	

Table 10A (cont.) ACGIH TLV.

Chemical Name	**TLV (ppm)**	**TLV (mg/m³)**
Dipropylene glycol methyl ether	100	
Di-sec octyl phthalate (Di-(2-ethylhexyl) phthalate)		5
Emery (td)		10
Endosulfan		0.1
Endrin		0.1
Epichlorohydrin	2	
EPN		0.1
Ethanolamine	3	
2-Ethoxyethanol (Cellosolve)	5	
2-Ethoxyethyl acetate (cellosolve acetate)	5	
Ethyl acetate	400	
Ethyl acrylate	5	
Ethyl alcohol (Ethanol)	1000	
Ethylamine	5	
Ethyl amyl ketone (5- Methyl-3-heptanone)	25	
Ethyl benzene	100	
Ethyl bromide	5	
Ethyl butyl ketone (3- Heptanone)	50	
Ethyl chloride	100	
Ethyl ether	400	
Ethyl formate	100	
Ethyl mercaptan	0.5	
Ethyl silicate	10	
Ferbam (td)		10
Ferrovanadium dust		1
Fluorides (as F)		2.5
Fluorine	1	
Formic acid	5	
Furfural	2	

Table 10A (cont.) ACGIH TLV.

Chemical Name	**TLV (ppm)**	**TLV (mg/m³)**
Furfuryl alcohol	10	
Grain dust (oat, wheat, barley)		4
Glycerin (td)		10
Glycidol	25	
Graphite, synthetic (td)		2
Gypsum (td)		10
Hafnium		0.5
Heptachlor		0.05
Heptane (n-Heptane)	400	
Hexachloroethane	1	
Hexachloronaph- thalene		0.2
n-Hexane	50	
2-Hexanone (Methyl n-butyl ketone)	5	
Hexone (Methyl isobutyl ketone)	50	
sec-Hexyl acetate	50	
Hydrazine	0.01	
Hydrogen peroxide	1	
Hydrogen selenide (as Se)	0.05	
Hydroquinone		2
Iron oxide fume		5
Isoamyl acetate	100	
Isoamyl alcohol (primary and secondary)	100	
Isobutyl acetate	150	
Isobutyl alcohol	50	
Isopropyl acetate	250	
Isopropyl alcohol	400	
Isopropylamine	5	
Isopropyl ether	250	
Isopropyl glycidyl ether (IGE)	50	
Kaolin (td)		2
Ketene	0.5	

Table 10A (cont.) ACGIH TLV.

Chemical Name	TLV (ppm)	TLV (mg/m^3)
Limestone (td)		10
Lindane		0.5
Lithium hydride		0.025
L.P.G. (Liquefied petroleum gas)	1000	
Magnesite (td)		10
Magnesium oxide fume (total particulate)		10
Malathion (td)		10
Maleic anhydride	0.25	
Manganese compounds (as Mn)		0.2
Marble (td)		10
Mesityl oxide	15	
Methoxychlor (td)		10
2-Methoxyethanol (Methyl cellosolve)	5	
2-Methoxyethyl acetate (Methyl cellosolve acetate)	5	
Methyl acetate	200	
Methyl acetylene (propyne)	1000	
Methyl acetylene- propadiene mixture (MAPP)	1000	
Methyl acrylate	10	
Methylal (Dimetoxy- methane)	1000	
Methyl alcohol	200	
Methylamine	5	
Methyl n-amyl ketone	50	
Methyl bromide	5	
Methyl chloroform (1,1,1- Trichloroethane)	350	
Methylcyclohexane	400	
Methylcyclohexanol	50	
o-Methylcyclohexanone	50	
Methylene bisphenyl isocyanate (MDI)	0.005	
Methyl formate	100	

Table 10A (cont.) ACGIH TLV.

Chemical Name	TLV (ppm)	TLV (mg/m^3)
Methyl hydrazine (Mono- methyl hydrazine)	0.01	
Methyl iodide	2	
Methyl isoamyl ketone	50	
Methyl isobutyl carbinol	25	
Methyl isocyanate	0.02	
Methyl mercaptan	0.5	
Methyl methacrylate	100	
alpha-Methyl styrene	50	
Molybdenum (as Mo) Soluble compounds		5
Molybdenum (as Mo) Insoluble compounds (td)		10
Morpholine	20	
Naphthalene	10	
Nickel carbonyl (as Ni)	0.05	
Nickel, metal and insoluble compounds (as Ni)		1
Nickel, soluble compounds (as Ni)		0.1
Nicotine		0.5
Nitric acid	2	
Nitric oxide	25	
p-Nitroaniline		3
Nitrobenzene	1	
p-Nitrochlorobenzene	0.1	
Nitroethane	100	
Nitrogen dioxide	3	
Nitrogen trifluoride	10	
Nitroglycerin	0.05	
Nitromethane	20	
1-Nitropropane	25	
2-Nitropropane	10	
Nitrotoluene (all isomers)	2	

Table 10A (cont.) ACGIH TLV.

Chemical Name	**TLV (ppm)**	**TLV (mg/m^3)**
Octachloronaphthalene		0.1
Octane	300	
Oil mist, mineral		5
Osmium tetroxide (as Os)	0.0002	
Oxalic acid		1
Paraquat (rd)		0.2
Parathion		0.1
Particulates not other- wise regulated (PNOR) (td)		10
Particulates not other- wise regulated (PNOR) (rf)		3
Pentaborane	0.005	0.5
Pentachloronaphthalene		0.5
Pentachlorophenol		0.5
Pentaerythritol (td)		10
Pentane	600	
2-Pentanone (Methyl propyl ketone)	200	
Perchloromethyl mercaptan	0.1	
Perchloryl fluoride	3	
Perlite (td)		10
Petroleum distillates (Naphtha) (Rubber Solvent)	300	
Phenol	5	
p-Phenylene diamine		0.1
Phenyl ether, vapor	1	
Phenyl glycidyl ether (PGE)	0.1	
Phenylhydrazine	0.1	
Phosdrin (mevinphos)	0.01	
Phosgene (Carbonyl chloride)	0.1	
Phosphine	0.3	
Phosphoric acid		1
Phosphorus (yellow)	0.02	

Table 10A (cont.) ACGIH TLV.

Chemical Name	**TLV (ppm)**	**TLV (mg/m³)**
Phosphorus pentachloride	0.1	
Phosphorus pentasulfide		1
Phosphorus trichloride	0.2	
Phthalic anhydride	1	
Picloram (td)		10
Picric acid		0.1
Pindone (2-Pivalyl-1, 3-indandione)		0.1
Plaster of Paris (td)		10
Platinum (as Pt) Soluble salts		0.002
Portland cement (td)		10
n-Propyl acetate	200	
n-Propyl alcohol	200	
Propylene dichloride	75	
Propylene imine	2	
Propylene oxide	20	
n-Propyl nitrate	25	
Pyrethrum		5
Pyridine	5	
Quinone	0.1	
Rhodium, metal fume and insoluble compounds		1
Rhodium, soluble compounds		0.01
Ronnel		10
Rotenone		5
Rouge (td)		10
Selenium compounds (as Se)		0.2
Selenium hexafluoride (as Se)	0.05	
Silicon (td)		10
Silicon carbide (td)		10
Silver, metal and soluble compounds (as Ag)		0.01
Sodium fluoro- acetate		0.05

Table 10A (cont.) ACGIH TLV.

Chemical Name	TLV (ppm)	TLV (mg/m³)
Starch (td)		10
Stibine	0.1	
Stoddard solvent	100	
Strychnine		0.15
Sucrose (td)		10
Sulfur dioxide	2	
Sulfur hexafluoride	1000	
Sulfuric acid		1
Sulfuryl fluoride	5	
2,4,5-T (2,4,5-trichloro- phenoxyacetic acid)		10
Tantalum, metal and oxide dust		5
TEDP (Sulfotep)		0.2
Tellurium and com- pounds (as Te)		0.1
Tellurium hexafluoride (as Te)	0.02	
Temephos (td)		10
TEPP (Tetraethyl pyrophosphate)	0.004	
1,1,1,2-Tetrachloro-2, 2-difluoroethane	500	
1,1,2,2-Tetrachloro-1, 2-difluoroethane	500	
1,1,2,2-Tetrachlo- roethane	1	
Tetrachloronaph- thalene		2
Tetraethyl lead (as Pb)		0.1
Tetrahydrofuran	200	
Tetramethyl lead (as Pb)		0.15
Tetramethyl succinonitrile	0.5	
Tetranitromethane	0.005	
Tetryl (2,4,6-Trinitro- phenylmethylni- tramine)		1.5
Thallium, soluble compounds (as Tl)		0.1
4,4′-Thiobis (6-tert, Butyl-m-cresol) (td)		10
Thiram		1

Table 10A (cont.) ACGIH TLV.

Chemical Name	**TLV (ppm)**	**TLV (mg/m³)**
Tin, inorganic compounds (except oxides) (as Sn)		2
Tin, organic com- pounds (as Sn)		0.1
Titanium dioxide (td)		10
Toluene-2,4- diisocyanate (TDI)	0.005	
o-Toluidine	2	
Tributyl phosphate	0.2	
1,1,2-Trichloroethane	10	
Trichloronaphthalene		5
1,2,3-Trichloropropane	10	
1,1,2-Trichloro-1,2,2- trifluoroethane	1000	
Triethylamine	1	
Trifluorobromo- methane	1000	
2,4,6-Trinitrotoluene (TNT)		0.5
Triorthocresyl phosphate		0.1
Triphenyl phosphate		3
Turpentine	100	0.2
Uranium (as U) Soluble compounds		0.2
Uranium (as U) Insoluble compounds		0.2
Vanadium (rd) (as V2O5)		0.05
Vanadium (fume) (as V2O5)		0.05
Vegetable oil mist (td)		10
Vinyl toluene	50	
Warfarin		0.1
Xylenes (o-, m-, p-isomers)	100	
Xylidine	0.5	
Yttrium		1
Zinc chloride fume		1
Zinc oxide fume		5
Zinc oxide (td)		10
Zirconium compounds (as Zr)		5

**Reference: http://www.labsafety.com/refinfo/ezfacts/ezf232.htm

Appendix B

Forms

NEW EMPLOYEE SAFETY ORIENTATION TRAINING CHECKLIST

(To be completed by employee & supervisor; return to payroll)

Employee Name ______________________ Job Title ______________

Supervisor's Name ______________________ Department ______________

ALL EMPLOYEES WILL BE TRAINED ON THE FOLLOWING TOPICS:

The Injury & Illness Prevention Program (IIPP)

- ☐ Discussed "Report of Unsafe Condition or Hazard" Form
- ☐ Employee has received "Code of Safe Practices" and forwarded signed acknowledgement to payroll.
- ☐ Informed of the duties and responsibilities of safety officiers, safety committees, management and employees.
- ☐ Material Safety Data Sheets (MSDS) 800-451-8346
- ☐ Chemical safety / Personal Protective Equipment (P.P.E.)
- ☐ Location of safety manual (Injury & illness prevention plan)
- ☐ Reporting of work-related injuries (3-step process)

Fire Safety, Emergency & Disaster Preparedness

- ☐ Designated evacuation assembly points
- ☐ Emergency action plans
- ☐ Emergency escape routes
- ☐ List of emergency phone numbers
- ☐ Types of fires
- ☐ Types of fire extinguishers

Locations

- ☐ Location of fire alarm
- ☐ Location of safety postings
- ☐ Locations and use of fire extinguishers
- ☐ Location of Automatic External Defibrillator (A.E.D.)
- ☐ Location of natural gas shut-off

Certifications (If applicable)

- ☐ Fire extinguisher
- ☐ CPR
- ☐ Automatic External Defibrillator (A.E.D.)
- ☐ First aid

Ergonomics Program

- ☐ Overview of RMIs (Repetitive Motion Injuries)
- ☐ Proper lifting
- ☐ Safe work practices
- ☐ Workstation evaluation

Uniforms / Attire

- ☐ Discuss appropriate attire
- ☐ Discuss appropriate footwear

Other Required Training

- ☐ Hazard communication
- ☐ Bloodborne pathogen
- ☐ Ladder
- ☐ Hand cart / dolly
- ☐ List: ______________
- ☐ List: ______________
- ☐ List: ______________
- ☐ List: ______________

Record of Safety Orientation Training

Signature of Trainer ______________________ Date ______________

Signature of Employee ______________________ Date ______________

Forward signed copy to payroll to be placed in employee's personnel file 12/06

DEPARTMENTAL NEW EMPLOYEE SAFETY ORIENTATION

Employee Name:____________________________________

Job Title: __________________________ SSN: ____________

Topics 1-4 are contained in the departmental Emergency Operations Plan. Reviewing this plan during the orientation will more than meet the requirements of these first few topics.

☐ 1. **Reporting Emergencies**

Tell and show the new employee(s) the police, medical, and fire emergency reporting number(s) for their work area.

General Campus	*Police - Medical - Fire*	*5566*

The emergency number should be posted on all telephones.

Your dept., div., unit, worksite, etc. *Name*		
Location	*Emergency Service(s)*	*Phone#*

☐ 2. **Emergency Evacuation**

Walk new employees through the appropriate emergency evacuation route for their work area. Also point out the secondary emergency evacuation route to be used if the primary route is blocked. Show them where to assemble after evacuation. Discuss special evacuation needs and plans with disabled employees. (Campus building evacuation floor plans are available from Campus Safety.)

☐ 3. **Local Fire Alarm Signaling System**

Show new employees where fire alarm pull stations are and instruct them in their use. Let them know that activating the pull station sounds an alarm in the building to alert other occupants to evacuate. Describe what the alarm in your building sounds like (a bell, chimes, a slow whoop).

- Tell your new employees that they must leave the building immediately upon hearing the alarm, closing doors behind them.
- When employees discover a fire, they should first pull the nearest fire alarm pull station and then exit the alarmed area. If possible, employees should follow up with a telephone call from a safe location to provide more details.
- On Campus: The activation of a fire alarm pull station also sends a signal to Campus Safety and Saratoga Springs Fire Department showing the location of the emergency.

☐ 4. **Portable Fire Extinguishers**
Show the employee(s) where portable fire extinguishers are located. Tell them to use a portable fire extinguisher only if:

- they have been trained to use them,
- the fire alarm has been sounded first,
- the fire is small (wastebasket size), and
- they have a clear evacuation route.

☐ 5. **Department Reporting Procedures**
Tell your new employee(s) to immediately report accidents, incidents, near misses, motor vehicle accidents, and any unsafe conditions or acts to (*usually their supervisor*):

Name:	Phone:
Location:	Room:

☐ a. **Reporting Accidents and Incidents**
Explain that after they immediately report on-the-job accidents, they have to fill out a College accident incident report form.

Explain the form and tell them where the forms are located. All accidents must be reported on this form regardless of the extent of personal injury.

Reporting all accidents and incidents helps the College and the employing department initiate effective safety programs and accident prevention measures.

☐ b. **Reporting Motor Vehicle Accidents**
All automobile accidents in College-owned vehicles must also be reported to the Department of Campus Safety (X5566) immediately, whether or not there appears to be personal injury or property damage.

☐ c. **Reporting Unsafe Conditions and Acts**
Along with immediately reporting unsafe conditions and acts to their supervisors or the person noted above, employees may report safety problems to Campus Safety.

Explain that employees should take responsibility for correcting unsafe conditions when feasible, e.g., wiping up small, nontoxic spills and removing tripping hazard.

☐ 6. **Workers' Compensation and Industrial Insurance**
Tell employees that work-related injuries or illnesses resulting in medical expenses or time loss are covered by New York State's Workers' Compensation. To establish a Workers' Compensation claim, employees must fill out the appropriate paperwork. Contact Human Resources for additional information. Explain, also, that prompt reporting of accidents to you, the supervisor, will make the claims process easier and may allow you to find them modified work during their recovery.

☐ 7. **First Aid**
Tell new employees where first aid kits are located. Explain what actions employees should take if they or others are injured. If safety showers or eye wash stations are located in your department, show new employees where they are and instruct them in their use.

☐ 8. **Hazard Communication (Chemical Safety)**
(Worker Right-to-Know. HazCom)
(Refer to the College HazCom Plan located in Facilities Services and Campus Safety)

☐ a. **General (all employees)**

- Tell new employees where hazardous materials are used or stored in their work area.
- Explain the labeling system for these materials.
- Show employees where material safety data sheets (MSDS) are located or explain how they can obtain an MSDS.
- If new employees will be working with hazardous materials, tell them they will receive training in the safe handling of these materials or conduct the training at this time, if appropriate.

  Hazard Communication training is conducted by supervisors or a designated departmental trainer.

- Inform new employees that hazardous materials emergencies, such as spills or releases too big for them to clean up, are to be reported to:

	Who	Phone
Small Spills		
Large Spills or releases		

Report large spills or releases to Campus Safety- at X5566.

- Explain the hazardous materials waste disposal procedures that apply in your area

☐ **b. Specific Worksites**

Office Staff
For staff whose only chemical exposures are in an office environment

- Discuss hazard information and protection measures for products they will work with.
- Explain an MSDS and tell employees where they are located or how to obtain them.

Laboratory Staff
The laboratory supervisor or staff must provide additional training, specific to the chemicals in the laboratory.

Non-Laboratory Hazardous Chemicals
Employees who work with chemicals in non-laboratory environments must receive detailed hazard communication training from their supervisor or designated departmental HazCom trainer. (Employees who fall into this category include maintenance, custodial/housekeeping, food service and printing and copy/duplicating employees.)

☐ **9. Worksite Warning Signs and Labels**
Explain to all new employees the meaning of warning signs, tags, and labels used in their work area.

☐ **10. Personal Protective Equipment (PPE)**

Check the personal protective equipment needed for this job.

	Gloves		Hard Hats
	Safety Glasses, Goggles, Face Shields		Hearing Protectors
	Personal Protective Clothing		Fall Protection
	Orange Safety Vest		Safety Shoes
	Respirator		

Explain precisely the use, care, cleaning, and storage of any personal protective equipment the new employee will be required to use on the job. Stress the need for strict adherence to department, division, unit, and/or lab policy on the use of PPE.

☐ **11. Employee Safety and Health Training**

Use the following list to indicate the safety and health training classes the new employee will be required to take for their job. Recommended classes could also be marked but priority must be given to arranging the required health and safety training classes.

☐ **12. Safety and Health Committee(s) and/or Safety Meetings**

Tell new employees about the Organizational and University-wide Health and Safety. Committees and about the departmental health and safety committee and safety meetings, if applicable. Tell them who their safety committee representatives are and how to contact them.

☐ **13. Safety Bulletin Board**

Point out the departmental safety bulletin board and tell them what items can be found on the board.

- *Other safety notices, newsletters, safety and health committee minutes, etc. should be posted here also.*

☐ **14. Departmental/Worksite Safety Practices and Rules**

Conduct an on-the-job review of the practices necessary to perform the initial job assignments in a safe manner. Employees should understand that supervisors will provide job safety instruction and inspection on a continuing basis. Review safety rules for your department (e.g., non-smoking areas, working alone, safe use of chemicals, biohazards, radioactive materials, etc).

☐ **15. Tour Department/Facility Reviewing Worksite Hazards**

Encourage your employees to ask questions and to develop a sense of safety consciousness.

Index

Also of Interest

Check out these forthcoming related titles coming soon from Scrivener Publishing

Coal Ash Management Practices in the Power Industry, by Nicholas P. Cheremisinoff and Anton Davletshin, October 2010, ISBN 9780470625972.
The most comprehensive and up-to-date coverage of the safety practices, safe removal, and other environmental issues associated with coal ash management.

Ethics in Engineering, by James Speight and Russell Foote, December 2010, ISBN 9780470626023.
Covers the most thought-provoking ethical questions in engineering.

Zero-Waste Engineering, by Rafiqul Islam, February 2011, ISBN 9780470626047.
In this controvercial new volume, the author explores the question of zero-waste engineering and how it can be done, efficiently and profitably.

The Greening of Petroleum Operations, by Rafiqul Islam, June 2010, ISBN 9780470625903.
A breakthrough treatment of one of the most difficult and sought-after subjects of the modern era: sustainable energy.

Energy Storage: A New Approach, by Ralph Zito, July 2010, ISBN 9780470625910.
Exploring the potential of reversible concentrations cells, the author of this groundbreaking volume reveals new technologies to solve the global crisis of energy storage.

Formulas and Calculations for Drilling Engineers, by Robello Samuel, September 2010, ISBN 9780470625996.
The only book every drilling engineer must have, with all of the formulas and calculations that the engineer uses in the field.

Fundamentals of LNG Plant Design, by Saeid Mokhatab, David Messersmith, Walter Sonne, and Kamal Shah, August 2011.
The only book of its kind, detailing LNG plant design, as the world turns more and more to LNG for its energy needs.

Flow Assurance, by Boyun Guo and Rafiqul Islam, September 2011, ISBN 9780470626085.
Comprehensive and state-of-the-art guide to flow assurance in the petroleum industry.